NEW FORMS OF URBAN

T0253482

A deft handling of ongoing discussions about statistical methodology for measuring the populations and demographic characteristics of cities, New Forms of Urbanization *tackles the issues of comparability, areal units, methods of aggregation, and measures of urban and rural in standards for the presentation of urban statistics.* New Forms of Urbanization *does so ably by bringing in perspectives from all over the world. This book should help address the global disparities in data, paving the way to greater efforts to streamline and coordinate approaches.*
David R. Rain, US Census Bureau, Washington DC, USA

This book should be of major interest to scholars, students and government officials interested in urban planning, demography, geography, and other social sciences concerned with the dramatic ongoing process of urbanization sweeping the globe.
Professor Richard Bilsborrow, University of North Carolina, USA

The International Union for the Scientific Study of Population (IUSSP), originally set up in 1928, was reconstituted in its present form in 1947. Its principal objectives are to:

- encourage research into demographic issues, population problems, and population and development relationships;
- stimulate interest in population questions among governments, international and national organizations, the scientific community and the general public;
- foster exchanges between population specialists worldwide and in different disciplines;
- disseminate scientific knowledge on population as widely as possible.

The principal ways through which the IUSSP currently achieves its aims are:

- a General Conference held every four years, the most recent being that in Salvador, Brazil, in 2001;
- large regional population conferences, organized as the opportunity arises;
- scientific activities, taken forward under the auspices of its Council;
- publication of conference proceedings and scientific reports.

Further details about IUSSP and its activities are available on its www.iussp.org website.

New Forms of Urbanization
Beyond the Urban-Rural Dichotomy

Edited by

TONY CHAMPION
*School of Geography, Politics and Sociology,
University of Newcastle upon Tyne, United Kingdom*

and

GRAEME HUGO
*Department of Geographical and Environmental Studies,
University of Adelaide, Australia*

for

International Union for the Scientific Study of Population

Routledge
Taylor & Francis Group
LONDON AND NEW YORK

First published 2004 by Ashgate Publishing

2 Park Square, Milton Park, Abingdon, Oxon OX14 4RN
711 Third Avenue, New York, NY 10017, USA

Routledge is an imprint of the Taylor & Francis Group, an informa business

First issued in paperback 2016

British Library Cataloguing in Publication Data
New forms of urbanization : beyond the urban-rural
    dichotomy
    1. Urbanization 2. Cities and towns - Growth 3. Rural-urban migration
    I. Champion, A. G (Anthony Gerard) II. Hugo, Graeme
    III. International Union for the Scientific Study of Population
    307.2'6

Library of Congress Cataloging-in-Publication Data
New forms of urbanization : beyond the urban-rural dichotomy / edited by Tony Champion
    and Graeme Hugo.
        p. em.
    Revised papers commissioned for and discussed at meetings of the Working Group on
Urbanization of the IUSSP (International Union for the Scientific Study of Population)
held in Bellagio, Italy.
    Includes bibliographical references and index.
    ISBN 978-0-7546-3588-8
    1.Urbanization—Cross-cultural studies—Congresses. 2.Rural-urban migration—Cross
cultural studies—Congresses. 3. Cities and towns—Growth—Cross-cultural studies—
Congresses. 4. Human settlements—Statistics—Congresses. 5. Urbanization—Statistics—
Congresses. I. Champion, A. G. (Anthony Gerard) II. Hugo, Graeme.

HT151.N478 2003
307.76—dc22
                                                                        20003356294
ISBN 978-0-7546-3588-8 (hbk)
ISBN 978-1-138-25483-1 (pbk)

Transferred to Digital Printing in 2014

Contents

PART I: INTRODUCTION

PART II: REGIONAL PERSPECTIVES ON SETTLEMENT CHANGE

PART III: CASE STUDIES

PART IV: CONCEPTUALIZING SETTLEMENT SYSTEMS

PART V: MOVING FROM THE CONCEPTUAL TO THE OPERATIONAL

PART VI: THE WAY FORWARD

List of Figures

List of Plates

List of Tables

List of Contributors

Philippe Bocquier, Institute for Research and Development, Paris, France, and French Institute for Research in Africa (Institut français de recherche en Afrique, IFRA), Nairobi, Kenya.

Larry S. Bourne, Department of Geography and Program in Planning, and Centre for Urban and Community Studies, University of Toronto, Canada.

David L. Brown, Department of Development Sociology, and Polson Institute for Global Development, Cornell University, Ithaca, USA.

Tony Champion, School of Geography, Politics and Sociology, University of Newcastle upon Tyne, England.

Mike Coombes, Centre for Urban and Regional Development Studies (CURDS), University of Newcastle upon Tyne, England.

John B. Cromartie, Economic Research Service, US Department of Agriculture, Washington, DC, USA.

José Marcos Pinto da Cunha, Population Studies Center (NEPO), University of Campinas (UNICAMP), São Paulo, Brazil.

Véronique Dupont, Institute of Research for Development, Paris, France.

James D. Fitzsimmons, Geographic Studies and Information Resources, International Programs Center, US Census Bureau, Washington DC, USA.

William H. Frey, Population Studies Center, University of Michigan, Ann Arbor, Michigan, and Brookings Institution, Washington DC, USA.

Gustavo Garza, Center for Demographic and Urban Development Studies, El Colegio de México, Mexico City, Mexico.

Keith Halfacree, Department of Geography, University of Wales, Swansea, Wales.

Graeme Hugo, Department of Geographical and Environmental Studies and National Key Center in Social Applications of Geographical Information Systems, University of Adelaide, Adelaide, Australia.

Gavin W. Jones, Demography and Sociology Program, Research School of Social Sciences, Australian National University, Canberra, Australia.

Alfredo E. Lattes, Centro de Estudios de Población (CENEP), Buenos Aires, Argentina.

Denise Pumain, UMR Géographie-Cités, CNRS, and Université Paris 1, Paris, France.

Michael R. Ratcliffe, Population Distribution Branch, Population Division, US Census Bureau, Washington DC, USA.

Jorge Rodríguez, División de Población, Centro Latinoamericano y Caribeño de Demografía (CELADE), Santiago, Chile.

Jim Simmons, Department of Geography, University of Toronto, Canada.

Miguel Villa, División de Población, Centro Latinoamericano y Caribeño de Demografía (CELADE), Santiago, Chile.

John R. Weeks, Department of Geography and International Population Center, San Diego State University, San Diego CA, USA.

Yu Zhu, Institute of Geography, Fujian Normal University, Fuzhou, Fujian Province, People's Republic of China, and Demography and Sociology Program, Australian National University, Canberra, Australia.

Hania Zlotnik, United Nations Population Division, New York, USA.

Preface and Acknowledgements

When observers take stock towards the end of the twenty-first century, they are very likely to be part of a world that, to all intents and purposes, is fully urbanized, at least in the sense that the vast majority of people will be leading lives that are essentially 'urban' in style as opposed to those that are conventionally seen as 'rural'. These observers will also probably be able to confirm that the greatest increase in the proportion of the world's population residing in 'urban areas' occurred during the period that we are now living in. This is certainly the picture suggested by the main source of worldwide intelligence on this major transformation, namely the estimates produced by the United Nations Population Division and published biennially in its *World Urbanization Prospects* reports. According to its most recent version, the *2001 Revision* (UN, 2003, p.5),[1] the year 2007 looks set to be when the planet's number of urban residents overtakes the number of rural dwellers. Moreover, according to its data, the rise in the urban share of world population will be at its greatest magnitude ever during the coming 20 years, averaging 0.44 percentage points a year before slowing somewhat after 2025. Alongside this is the seemingly inexorable growth in the number of large cities, with the number of 'megacities' with 10 million or more people expected to reach 21 by 2015 (compared to 5 in 1975 and 14 in 1995) and with the total number of 'millionaire cities' projected to be 554 by then (up from 195 in 1975 and 348 in 1995). Not for nothing, therefore, did the 1996 Habitat report (UN, 1996) focus on *An Urbanizing World* and look forward to 'the urban century'.

At the same time, while these types of statistics are cited frequently and – in most cases – are quoted in terms that suggest that they are accepted as hard fact, there is a substantial and growing body of opinion that questions their accuracy and indeed challenges their relevance and meaningfulness. The 1996 Habitat report itself urges caution in using the data published in *World Urbanization Prospects*: 'The proportion of the world's population currently living in urban centers is best considered not as a precise percentage (i.e. 45.2 per cent in 1995) but as being between 40 and 55 per cent, depending on the criteria used to define what is "an urban center"' (UN, 1996, p.14). It goes on (p.15) to illustrate how much the size of some of the world's largest cities can vary according to the definition used; for instance, for Tokyo, which is considered by the UN to be the world's largest city with 25 million people in its 'agglomeration' in 1990 but which can also be represented in terms of its administrative city (8.2 million in that year), the Tokyo prefecture (11.9 million), the Greater Tokyo Metropolitan Area (31.6 million) and the National Capital Region (39.2 million). The UN Population Division itself is keenly aware of these sorts of problems, hence the amount of effort that it invests in

---

[1] For details of references, see the bibliography at the end of the book.

trying to provide population data not just for administrative cities but also for agglomerations defined on consistent builtup-area criteria. Yet even this latter approach is increasingly being criticized for failing to incorporate the entirety of cities, as these settlements – whether naturally or as a result of building controls – have sprouted noncontiguous development. Indeed, the often-reported slowdown in the overall growth rate of the world's largest cities and associated reports of deconcentration towards intermediate-sized and smaller cities may be largely a statistical artefact arising from this failure. Certainly, there is a widespread feeling that the conventional methods of monitoring urbanization and city growth are not portraying the full picture of the changes now affecting individual urban centres and their wider settlement systems, and indeed that they may no longer be capturing the most important dimensions of these changes. Not least is this the case among the contributions to the present volume.

It was this sense of unease that prompted the actions that have led to the preparation of this volume. This work began in 1999 when the Council of the International Union for the Scientific Study of Population (IUSSP) set up a Working Group on Urbanization, as recommended by an Exploratory Mission charged with sounding out the need for a review of current ways of monitoring urbanization and analyzing urban population change. The latter had reported in no uncertain terms that it was time for a fundamental rethink, noting particularly that there seemed to have been very little change in approach over the previous quarter of a century. This lack of progress was in spite of the fact that the IUSSP's previous initiative in this area – its Committee on Urbanization and Population Redistribution chaired by Sidney Goldstein in 1971-74 – had made a range of recommendations for improving on contemporary practice. If anything, there had been some deterioration since then, in the sense that there was now an even lower level of reporting of population characteristics on any form of settlement-system basis. Fewer countries were routinely furnishing the UN with data on fertility, mortality and nuptiality broken down by urban and rural areas, let alone providing such information for size groups of cities as recommended by the Goldstein Committee. There had also been a diminution in the coverage of countries supplying census-based data on the demographic, social, educational and economic characteristics of their urban and rural populations. Perhaps even more surprisingly given the escalating level of concern about the rapid growth of large cities, there appeared to be only very fragmentary intelligence about the dynamics of urban growth in terms of the direct components of population change, let alone about the processes underlying them. In these observations, the IUSSP's Mission was very much in step with the thinking that lay behind the decision by the US National Academy of Sciences's Committee on Population to set up a Panel on Urban Population Dynamics, which also began its activities in 1999 (National Research Council, 2003).

In this context, the Working Group on Urbanization found itself faced with a potential agenda of colossal proportions and, given its limited remit of organizing 'one major activity', had to decide on its main priority. In the end, its core members – comprising Tony Champion (chair), Graeme Hugo and Alfredo Lattes – decided that most fundamental to improving our understanding of contemporary patterns of human settlement change was the better conceptualization and definition of settlement patterns. It was seen as particularly important to try and move beyond the simple urban-rural dichotomy which was still the only basis being used by the UN to publish subnational data on population characteristics, apart from its population counts for cities. Indeed, it was felt that one reason for the declining number of countries now supplying the UN with census data on this basis, as just mentioned, was the increasing number for which this simple distinction was no longer deemed by national statistical offices to be an important dimension of their settlement systems.

In choosing to focus on issues of definition and measurement, it was recognized that many important research questions identified by the IUSSP's Exploratory Mission would necessarily be sidelined. For instance, is urban mortality still declining, and what life-expectancy differentials exist between different types of urban area? What relationship exists between urbanisation and fertility, and what mechanisms lead to falling fertility in urban areas? How is migration affecting the size and composition of urban populations, and what is the relative importance of internal and international migration? While issues such as these are mentioned in the contributions to this book, their treatment here is subordinated to the central question of the geographical frameworks within which these questions are most appropriately studied. In restricting its focus in this way, the Working Group was comforted by the knowledge that the NAS Panel on Urban Population Dynamics was addressing the much wider agenda and indeed had decided to undertake primary research that would go beyond the urban-rural dichotomy and examine variations in demographics across size groups of cities. This latter work, however, had to take as given the 'city' definitions used by the countries under study, so the Working Group's review of the adequacy of current approaches to defining cities would help in the interpretation of the Panel's findings. Moreover, the choice of this focus was not only approved by IUSSP Council, but was subsequently endorsed by the Rockefeller Foundation in selecting the Working Group's bid in its competition for use of its Conference and Study Center at Bellagio, Italy.

This book, reflecting the central concerns of the Working Group, therefore starts from the premise that the traditional approach to studying urbanization trends, based on the urban-rural dichotomy, has lost much of its relevance owing to the major changes that have affected settlement patterns in recent decades. Notwithstanding this, its main thesis is that place still matters in the internal demography of countries and that it remains very important to take a settlement-based approach in studying the processes and patterns of demographic, economic and social development. Hence the need for this search for, and assessment of, alternative ways of defining and classifying human settlement that can form the basis for identifying and analysing between-place differences in population trends and related characteristics.

As such, the book aims to answer the following sorts of questions:

- Why is it important to study demographic, economic and social development through the lens of the settlement system?
- What are the contemporary dynamics of the settlement system, as far as can be seen from both existing statistics and other observations?
- How far are these dynamics captured in traditional rural-urban classification schemes?
- What steps have already been taken by national statistical agencies, other organizations and academics to develop representations of the settlement system that go beyond the simple rural-urban dichotomy?

In addition, recognizing that the scientific environment has moved on considerably since the IUSSP's previous attempt at examining these sorts of issues 30 years ago, the book addresses the question of how far new technologies of data collection, storage, manipulation and analysis can assist in the operationalization of these alternative representations and suggest further possible approaches.

The organization of the book, fashioned at an early stage of the Working Group's preparations for its bid to the Rockefeller Foundation, revolves around the set of commissioned papers that form its main contents – papers that were precirculated to the participants at the Bellagio meeting and then revised in the light of the week of discussions there. Part I sets the scene, beginning with an introduction by the editors that builds on this Preface by giving more detail about the purpose of the book and the themes running through it. Two further chapters document the lessons from past reviews and provide an overview of our present state of knowledge on urbanization and city growth across the world, based on the latest update from the United Nations Population Division. Part II presents regional perspectives on settlement change. Its four chapters look at parts of the world arranged by historical experience of urbanization, starting with the USA because of its long-established position at the leading edge of settlement-system change in the more developed world and continuing with continent-wide reviews of Latin America, Asia and Africa. Part III is composed of four case studies that go into more detail about the nature of new forms of urbanization in less developed countries. These deal respectively with the transformation of the urban hierarchy in Mexico, developments in urban form and population distribution in Delhi's metropolitan area, urbanization patterns and processes in metropolitan Brazil, and the role of *in situ* rural-urban transformation in China's recent urbanization. The eight chapters in Parts II and III thus have the twin aims of describing the main developments in urbanization and settlement-system change in their areas and of assessing how adequately existing statistical frameworks capture the principal dimensions of these evolving patterns.

The remainder of the book adopts a more reflective mode in an attempt to discern how best to plug the gap between existing statistical frameworks and what actually appears to be happening on the ground as identified by the earlier chapters. The purpose of Part IV is to consider the wider issues that need to be addressed in developing new ways of conceptualizing settlement systems. Two chapters approach this task from the urban end of the spectrum, one taking an 'evolutionary' approach to settlement systems and the other arguing for an 'urban system' perspective. A further two chapters examine the challenge from the rural end, one discussing the extent and nature of continuing distinctions between 'the urban' and 'the rural' from a North American perspective, the other thinking through the meaning of 'rurality' from a mainly European viewpoint. Then, Part V attempts to provide a bridge between the conceptual and the operational aspects of the challenge. The first of its three chapters argues for the need to recognise multiple dimensions of settlement systems instead of trying to devise a single classification and provides examples of measures used to represent these separate dimensions. The second demonstrates the potentially valuable role of new technology in identifying and classifying settlements, while the third documents the latest experience of reviewing the USA's metropolitan area standards, pointing to the political as well as scientific issues that needed to be overcome in that exercise. The final part of the book comprises a single chapter which, written after the Bellagio meeting, draws together the main findings of the commissioned chapters and considers the way forward in the light of the ideas that dominated the discussions at that meeting.

As such, the book is aimed at two main audiences. Its central thrust is towards the research community, but broadly defined and not just to members of academe. Indeed, one of its primary aims is to alert professionals who are routinely involved in the collection, processing and analysis of population data at subnational levels that, unless their geographical frameworks are able to capture current developments in their countries' settlement systems, their findings are likely to misrepresent what is actually occurring and thereby provide a less than optimal basis for anticipating future trends. A related objective is to spur academics into renewed efforts aimed at improving our understanding of the patterns and especially the processes of urbanization and city change, particularly taking advantage of the data collected in the latest round of population and housing censuses. The hope is that, within their various regions and countries of interest, academic researchers will be encouraged by this book to engage with relevant professionals employed in national statistical offices, non-governmental agencies and possibly even private business, so as to evaluate existing ways of representing settlement patterns and – if felt necessary – design and test alternatives in the light of users' requirements.

Secondly, the book is aimed at all those who wish to obtain insights into the latest trends in urbanization and settlement-system change around the world and into what some of the world's leading researchers consider to be their significance in terms of new forms of urban development and emerging challenges for policy makers. In practice, it is anticipated that this will be particularly attractive to a student audience, especially postgraduates and final-year undergraduates who not only want to be informed about these matters but are also equipped to take a critical position on the issues raised. The disciplinary range targeted in this respect, however, is wider than the population experts mentioned above and includes urban geographers, economists, planners and sociologists, together with others involved in place-based studies, whether urban, rural or regional. Moreover, as is evident from its aims and contents, the book is geared up for a fully international readership. This is not to say that it tries to provide a comprehensive survey of developments in all parts of the world: other recent books have a fuller coverage of different national contexts, notably Geyer's (2002) *International Handbook of Urban Systems*. Rather, its intention is to demonstrate enough of the main features of current and evolving patterns of urbanization to be able to critique existing ways of depicting urban growth and suggest alternative approaches. Even so, although the book is skewed towards the less developed world where the most impressive changes now seem to be taking place, many of the issues that these raise are similar to those encountered in more developed countries now or in the past and need to be discussed in this wider context.

In all that has led up to the completion of this book, the editors have many causes for gratitude. On behalf of the Working Group, they acknowledge the important initial role played by the members of IUSSP's Exploratory Mission that recommended the Group's establishment: Gavin Jones, John Oucho, Ronald Skeldon, Miguel Villa and Hania Zlotnik. They are indebted to the support of the IUSSP secretariat, notably Jane Verrall, Landis Mackellar, Elizabeth Omoluabi and Delphine Lebugle. They are extremely grateful for the major contribution made by the A.W. Mellon Foundation towards the travel and administrative costs of the Working Group, and to Carolyn Makinson for her encouragement of the project. They wish to record their huge appreciation for the support of the Rockefeller Foundation in making available its Bellagio Center and in meeting the travel expenses of participants from developing countries, and especially to Susan Garfield and Amanda Bergbower in New York for facilitating the award and to Gianna Celli and Nadia Gilardoni for helping to make the stay at the Villa Serbelloni so pleasant and rewarding.

Most of all, however, the editors are very grateful to those that participated so enthusiastically and diligently in the academic work on which this book is grounded, including five who do not appear in the list of contributors but performed a vital role as invited discussants at the Bellagio meeting, namely Mike Batty, Ram Bhagat, Dick Bilsborrow, Landis Mackellar, Terry McGee and David Rain. In terms of preparing the book for publication, they would especially like to single out Joanna Rillo for the sterling work she did in producing the camera ready copy and Christine Crothers for redrawing the vast majority of the Figures. Ann Rooke redrew Figures 8.2 and 8.3, and Marilyn Champion prepared the index. Finally, we also pay tribute to the support given by staff at Ashgate, including Carolyn Court, Rosalind Ebdon, Anne Keirby, Melissa Riley-Jones and Valerie Rose and especially to Claire Annals who provided marvellous help in the final stages of revising the camera ready copy. To all these as well as to those whom the book's contributors have acknowledged at the beginning of their chapters, we owe an enormous debt of gratitude, but as ever the responsibility for any last-minute errors and wider shortcomings rests entirely with the editors.

Tony Champion, Newcastle upon Tyne
Graeme Hugo, Adelaide

PART I
INTRODUCTION

Chapter 1

Introduction: Moving Beyond the Urban-Rural Dichotomy

Tony Champion and Graeme Hugo

There is increasing appreciation in the social sciences that context is important in understanding social, economic, cultural, political and demographic processes. An important element in context is the type of place in which people live and work, hence it is important to be able to categorise them according to their situation within the human settlement system. Unfortunately, at present the only such contextual element that is widely captured in standard population data collection systems is a categorisation of areas into either urban or rural. As mentioned in the Preface, this approach was adopted by the United Nations at the outset of its statistical reporting at the end of the 1940s and continues to provide the principal basis for the publication of official statistics on world urbanization trends and projections.

In recent decades, however, the basis of this approach has been increasingly undermined by the massive changes that have been taking place in the size, extent and nature of settlements. In particular, these changes have involved the blurring of the urban-rural distinction. There is no longer any clear dividing line between town and countryside for individual settlements or their inhabitants: indeed, many people reside in one but work in the other. Moreover, in more heavily populated regions, formerly separate cities and towns have merged together into much more extensive urbanized zones. These changes were recognised over 40 years ago in Gottmann's (1961) concept of 'Megalopolis', a phenomenon that today appears to be as common in the less developed as in the more developed regions of the world. As a result, there is now a pressing need to reconceptualise our thinking on human settlement systems and devise new classifications of settlement which capture the diversity of the contemporary and evolving scenes.

Many people, including both academics and governmental bodies, are currently wrestling with the question of the most relevant frameworks to use for monitoring and analysing the changing patterns of settlements and for planning and providing services for their residents. There now exists around the world a wide array of alternative approaches, including extended metropolitan regions, metropolitan areas, functional urban regions, daily urban systems and local labor market areas as well as ways of subdividing these often extensive spatial units into localities and neighborhoods. In addition, in recent years major strides have been made in the information technology for data collection and processing needed for this work. Thus, while the Preface bemoaned that little has changed in the international monitoring of urbanization since the IUSSP's Goldstein Committee on Urbanization and Population Redistribution made its recommendations nearly 30 years ago, the prospects for introducing new ideas seem much more promising today.

It is in this area of the definition and measurement of urbanization that this collection of essays is intended to make its main contribution. The book brings together the ideas of many of the world's leading researchers and practitioners in this field. In preparing their chapters, they were asked to help towards addressing three fundamental questions:

- Given recent trends, are simple urban-rural classifications of population still adequate for capturing the increasing complexity of the human settlement patterns that now influence people's lives and behavior?
- If not, what are the types of human settlement which need to be differentiated in standard classifications of human settlement systems?
- How can these new concepts of human settlement systems be operationalized and measured in standard data collection systems?

In this way, the book's ultimate goal is the development of a new conceptualization of human settlement systems that can be used in standard population data collection systems, especially population censuses, and thereby provide researchers with datasets that permit more meaningful analyses and more accurate projections of sub-national demographic trends than are currently possible.

The remainder of this chapter sets out in more detail the case for a fundamental rethink of our approaches towards defining and classifying human settlement and outlines the key messages of the contributions that follow. First, it provides evidence in support of the statement that spatial context does still matter in the study of population characteristics and demographic behavior within countries. It goes on to present the justification for the traditional approach of differentiating between urban and rural, before demonstrating its shortcomings in handling new forms of urbanization and settlement change. It introduces a number of suggestions that have been put forward for capturing more satisfactorily the principal dimensions of human settlement systems as they are currently evolving. Finally, the chapter explains how the subsequent contributions in the volume help us to better understand the nature of the challenge as well as provide pointers to the way forward.

Space Continues to Matter in Demography

As just mentioned, it has long been recognized that where people live can be a relevant factor in helping to explain demographic behavior. Most obviously, this is exemplified by the great attention given to studying international variations and trends in fertility and mortality, including tracing the 'diffusion' of the demographic transition across the globe (see Chung, 1970, for a study that specifically does this; also, Champion, 2001). At the same time, this statement is just as true intranationally. Indeed, along with measures of economic output, the proportion of people living in urban areas has been found to be associated strongly with countries' progress through the fertility transition, and there has also been much discussion about whether mortality is subject to an 'urban penalty' or not.

At the same time, completion of the passage through the epidemiological and fertility transitions, and indeed through the urban transition too, does not necessarily seem to lead to an erosion of the importance of place in demography. If anything, interest in variations across space has waxed as settlement patterns have become more complex. For instance, as the urban share of a population has risen, increasing evidence has been found of differentiation in demographic behavior by size and type of urban area and also of variations between neighborhoods within individual cities. Similarly, the degree of heterogeneity within areas defined as rural has also increased.

The power of the urban-rural distinction as a demographic discriminator was demonstrated by some of the earliest analyses undertaken by the United Nations, as it began compiling and publishing population data across the world. For instance, in its first major report on 'urban characteristics and trends', published in its Demographic Yearbook for 1952 (UN, 1952, pp.9-19), a clear association was found between level of urbanization and infant mortality rate. Drawing on an analysis of 61 countries with the relevant data, it was shown that five-sixths of those with a majority of their people living in urban areas had infant mortality of under 50 per 1,000 live births, whereas two-thirds of countries with urbanization levels of below 40 per cent recorded rates of 75 or over (UN, 1952, p.18). An even more specific urban-rural comparison was made of fertility, using data on the child-woman ratio. As shown below in Table 1.1, the average number of children under 5 years of age per woman of childbearing age was higher in rural than in urban areas for all the countries studied. This was irrespective of their degree of economic development, although the scale of the differential was generally largest for countries with the greatest dependence on agricultural employment.

According to the latest Demographic Yearbook available at the time of writing (UN, 2001a), urban-rural differentials are still evident on most measures. In terms of fertility, live births per 1000 women aged 15-49 remains markedly lower in urban areas. Only a couple of the countries for which this comparison is possible have an urban fertility rate higher than the rural one, while fully half have an urban rate that is 30 per cent or more below the rural (Table 11 in UN, 2001a). In terms of infant mortality rate (IMR), by contrast, there are almost as many countries where the urban rate is higher than the rural rate as where it is lower. Nevertheless, there are some substantial differences between urban and rural levels. In one in six countries, the urban IMR is 20 per cent or more lower than the rural, while there is a similar proportion where it is 20 per cent or more above the rural level (Table 15 in UN, 2001a). This obviously raises the question of what it might be that distinguishes these groups of countries in terms of the forces at work in their urban and rural areas.

Table 1.1   Children under 5 years of age per 1,000 women 15-49 years old, selected countries

Country and year	Children per 1,000 women		Rural excess	Urban/ rural ratio
	Urban	Rural		
Agricultural countries				
Bulgaria 1934	332	520	188	0.64
El Salvador 1950	493	714	221	0.69
Panama 1940	380	761	381	0.50
Romania 1930	321	597	276	0.54
Turkey 1945	365	599	234	0.61
USSR 1926	411	626	215	0.66
Venezuela 1941	441	671	230	0.66
Semi-industrial countries				
Ceylon 1946	453	558	105	0.81
Finland 1940	195	370	175	0.53
Iceland 1930	440	535	95	0.82
Ireland 1946	366	461	95	0.79
Japan 1947	434	503	69	0.86
Puerto Rico 1940	409	719	310	0.57
South Africa 1936	404	647	243	0.62
Industrial countries				
Australia 1947	353	506	153	0.70
Canada 1951	422	637	215	0.66
Denmark 1948	372	457	85	0.81
Great Britain 1951	332	358	26	0.93
Norway 1930	194	386	192	0.50
Sweden 1945	301	368	67	0.82
USA	383	505	122	0.76

Source: Calculated from UN (1952, p.17, Table F).

Yet perhaps the most impressive feature of these UN tabulations is the very small number of countries that can be examined in this way. For example, in the table on live births per woman aged 15-49, an urban-rural split is provided for only 37 countries, less than one in six of the 228 countries currently recognized by the UN Population Division. Moreover, the majority of these are either Less Developed Countries (18) or former Communist states in Europe (14, including parts of former Yugoslavia), with only five 'western' MDCs being represented (Finland, Greenland, Iceland, Ireland and Japan). Data collection problems may explain a fair proportion of the absences, especially amongst LDCs. Nevertheless, it is also likely to be the case that, for a considerable and no doubt growing number of countries, such a simple dichotomy between urban and rural areas is seen as failing to capture what are now the most significant variations in the settlement system.

Hence the importance of a recent study carried out on behalf of the US National Academy of Sciences Panel on Urban Population Dynamics (National Research Council, 2003). For a substantial sample of LDCs, this study analysed records from Demographic and Health Surveys that were not only coded by urban/rural residence but could also be linked to population size of city. The results for 56 countries with suitable demographic data demonstrated that, while rural areas remained distinctive on most measures, there was also considerable variability according to settlement size. With respect to total fertility rate, for instance, it was found that, while the overall urban rate fell short of the rural one by some 1.4 children per woman, the rate for cities with over 5 million inhabitants was almost two children per woman below that of urban areas with under 100,000 residents. Moreover, there was a regular progression across the five city-size groups recognized.

In terms of selected health measures, the picture was not quite so clearcut. For instance, in terms of young children's weight for height, while the variation across the five city-size groups in a 66-country sample was again greater than the overall difference between urban and rural areas, the progression was not so regular. Though children's weight for height generally increased with city size, in some regions children living in the largest cities were found to be disadvantaged relative to those in intermediate-sized cities of between 100,000 and 5 million people.

These results clearly confirm the importance of looking beyond the simple urban-rural dichotomy when trying to monitor and analyse intranational variations in demographic processes. Taking the results at face value, they raise major questions about the factors influencing fertility behavior and life chances. For instance, what is it about the largest cities that makes them so different from intermediate-sized ones? Again, why is there such a contrast between rural areas and the smallest cities identified by the NAS Panel study?

Of course, the first priority – even when faced with such clear results – is to check that these types of relationships are real rather than due to some other factor and to any inconsistency in the definition of settlements. At least in theory, these patterns could have arisen if the smaller cities have a greater tendency than the larger ones to be located in countries, or regions within them, characterized by higher fertility and mortality rates. Equally, it could possibly be the case that the largest cities are defined on the basis of their more central and wealthier areas, omitting outlying shanty towns and the more mixed urban-rural areas that would normally be included within the boundaries of smaller cities. Just as in the Preface we have already drawn attention to the difficulties that urban-area definition poses for ranking the world's largest cities, so this can cause equal problems for making meaningful comparisons of demographic patterns across city-size groups.

In sum, clearly space continues to matter in demography. There still are important differences between urban and rural areas. In addition, however, the available evidence suggests that differences within the urban population – in this case between size groups of cities – are even greater. Taking this argument further, there may be other ways of defining and classifying urban settlement that might lead to the identification of an even more powerful dimension of between-place variation than city size. Again, as the nature of rural settlement has altered radically during the modernization process, there is a strong case for assessing how sensible it is to continue to treat 'rural' as an undifferentiated entity. It may be too soon to recommend throwing away all conventional practice and starting again from scratch, but as we now turn to look at existing approaches and begin to evaluate their present-day adequacy and relevance, it could perhaps help to try and think the unthinkable.

Challenges to the Simple Dichotomy

Given the empirical evidence in support of going beyond the urban-rural dichotomy, we now outline the main lines of argument that can be used to justify a fundamental review of conventional practice. We begin by providing a reminder of the rationale for differentiating the urban from the rural. Then the case for a rethink is presented. This is argued in terms of three main challenges to this dichotomous approach: firstly, that the distinction between rural and urban areas is becoming increasingly blurred; secondly, that, with a growing variety of users and applications, reliance on a unidimensional classification of settlements is becoming more questionable; and thirdly, that the emergence of new types of urbanization has had implications for settlement systems that cannot be captured through traditional notions of the urban and rural.

The use of the urban-rural dichotomy to differentiate between populations is based on the assumption that there are important contrasts between urban and rural populations. A list of the most commonly quoted contrasts is provided in Table 1.2.

Table 1.2   Some widely accepted traditional stereotypical differences drawn between urban and rural populations

Dimension	Urban	Rural
1. Economy	Dominated by secondary and tertiary activities	Predominantly primary industry and activities supporting it
2. Occupational Structure	Manufacturing, construction, administration and service activities	Agriculture and other primary industry occupations
3. Education Levels and Provision	Higher than national averages	Lower than national averages
4. Accessibility to Services	High	Low
5. Accessibility to Information	High	Low
6. Demography	Low fertility and mortality	High fertility and mortality
7. Politics	Greater representation of liberal and radical elements	Conservative, resistance to change
8. Ethnicity	Varied	More homogeneous
9. Migration Levels	High and generally net inmigration	Low and generally net outmigration

Source: Hugo (1987).

Besides the lower levels of fertility and mortality shown for urban areas in Table 1.2, these are the parts of the nation where agricultural and other primary occupations are less important, where education levels and provision are higher than average, and where accessibility to services and information is also high. Meanwhile, rural areas tend to be more homogeneous in their ethnic composition, more conservative in their political outlook and generally more resistant to change.

It is on this basis that urban-rural distinctions have become a fundamental part of census systems across the world. At one time if not currently, virtually all countries have designated urban areas, treating the remainder of their territory as a rural residual. The rules that they have used for recognizing urban areas are rooted in the types of stereotypical characteristics shown in Table 1.2, usually involving the adoption of one or more of the following criteria:

- A population size threshold.
- Population density.
- Contiguity of builtup area.
- Political status.
- Proportion of the labor force engaged in non-agricultural work.
- Presence of particular services or activities.

Where data are available for small areas like census tracts or enumeration districts, the usual practice is to separate out those that are largely builtup, either directly on the basis of land cover or by reference to population density. These areas are then used as building blocks, which are clustered together – usually on the basis of adjacency or being located within a minimum distance – to form localities. Some threshold, most commonly using population size, is then applied to determine whether the resulting place can be deemed urban. The latter step may also be adopted where there is no suitable small-area data, but in this case it would be applied to an existing administrative unit such as a municipality and may indeed be designated urban or rural solely according to its official political status. Thus the fundamental distinction between urban and rural places is normally in terms of continuously builtup area, population density, and the economic and political functions carried out in those areas.

So, what has been helping to undermine this conventional approach to studying settlement patterns? The biggest threat to the continued relevance of the simple dichotomy is posed by the increased blurring of urban-rural distinctions. This is, of course, not just a recent phenomenon. In 1966, Pahl pointed out that many of the changes occurring in rural Britain were in fact urban in nature. Even before then, as transport and information technology was developing, what were originally urban functions were beginning to locate in rural areas, while mechanization was reducing the labor requirements of the primary and extractive industries that traditionally dominated the countryside. The result is that the population size of a locality has become a less reliable discriminator of whether that place is essentially urban or rural in the way of life of its residents, or at least in certain aspects of their lifestyle. More generally, the meaning of the rural-urban distinction has changed and the rural-urban differences in the variables listed above in Table 1.2 have tended to narrow.

Leading on from this is the second challenge, namely the relevance of a single-dimension classification of settlements that is appropriate for every application. As already seen from Table 1.2, there is a considerable number of facets that in the conventional approach are synthesised to form one binary indicator. It may be that at one time, or at a certain stage of socio-economic development, all these facets mapped on to each other exactly, so that for each one the same clear break existed between what was urban and what was rural. This would perhaps have applied to the situation where town walls marked the boundary of all urban settlement. With the blurring of the urban-rural distinction, however, it is unlikely that these facets are now as conformable as before. Even being quite severely reductionist, one would be hard put to distil the list in Table 1.2 down into fewer than four dimensions – ecological, economic, socio-cultural, institutional.

This issue takes on additional significance when one considers the increasing range of uses that settlement classifications are being put to. Even three decades ago, in a review of the 'broad types of definition and their uses in applied demography', the UN (1973, p.9) acknowledged the existence of at least three different types of 'urban' definition that could be justified in terms of fitness for purpose:

> In common-sense terms, it can be said that definitions of 'urban' localities are of three types, namely administrative, economic or ecological. All three types have their practical uses and it may at times be desirable to make parallel estimates and projections of urban and rural population conforming to different types of definition. Roughly speaking, such data and estimates may be needed for three types of government policy or programme: economic plans, social measures and physical projects. ... In short, economic plans can best be served by population data for economically defined regions associated with major urban centers. ... For the formulation of social policies, it is preferable to define 'urban' populations as those contained within 'urban' administrative areas. ... Physical projects are much more concerned with environmental and traffic-flow management within areas inhabited at 'urban' and other densities. In such contexts it is probably most useful to calculate the population of agglomerations (or 'urbanized areas').

Since the 1970s, users' needs of demographic data have proliferated, not just with a huge increase in evidenced-based policy making and targeting by government but with an even more rapid growth of applications in the private sector.

Thirdly, and reinforcing this idea of different types of areal unit for different purposes, the settlement system has become more complicated: patterns of physical development have evolved and new forms of urbanization have emerged. As acknowledged for some time now, the simple urban-rural scale is inadequate for regions that are predominantly urban, because here the increasing interplay of urban functions has been enlarging the units of urban dominance. According to the UN (1969, p.3), 'As far back as a century ago [1860s], a new trend was identified in western Europe by the "conurbations" resulting from the coalescence of separate towns with their interstitial rural areas'. By the 1960s, it was possible to cite the emergence of even less compact urban forms such as 'metropolitan areas' and 'megalopolis', whose integrity is based primarily on functional criteria that lay stress on the high degree of mutual interdependence of activities. Since then, the identification of new forms of settlement has proceeded apace, perhaps most notably the phenomenon of the 'edge city' but also including terms like 'exurbia', 'peri-metropolitan areas' and 'extended metropolitan regions'.

A key part of these changes is the increasing difficulty of knowing where to position the boundary of any individual settlement – a form of blurring that is somewhat different from, though related to, the one discussed above of deciding whether a settlement, once identified, should be considered urban or not. This is primarily because people are moving about more and dividing their lives between areas conventionally designated urban and rural, to a large extent on a daily basis but also in terms of weekly or seasonal movements. In particular, the extent of commuting between urban and rural areas has increased dramatically (Hugo *et al.*, 1997). The result is the emergence of zones of transition around large urban centers where urban and rural functions are mixed together. Moreover, while this has in the past been seen as primarily a phenomenon of more developed countries, it is fast becoming as true of less developed regions as well.

Finally, the idea that the simple urban-rural dichotomy is inadequate for parts of countries that are now predominantly urban can increasingly be extended to national scale. There are many parts of the world where the level of urbanization is already very high and is now rising only fractionally. According to the latest revision of *World Urbanization Prospects* (UN, 2003), the MDCs now have over 75 per cent of their aggregate population living in urban areas, with some countries much higher than this and with very little scope for further increase – for example, Belgium with a level of 97.5 per cent in 2000. Moreover, these high levels are also been increasingly found elsewhere in the world, with the Latin American region also exceeding 75 per cent in 2000 and with Asia and Africa both expected to pass the 50 per cent mark by 2025. Clearly, there is a growing proportion of countries where this indicator is virtually redundant. How much longer should we delay before bringing forward alternative ways of monitoring settlement-system change across the world, even if this is run alongside the conventional indicator for the next couple of decades?

In sum, even by itself, the high level of urbanization now reached by many parts of the globe serves to undermine the relevance of the simple urban-rural dichotomy. Going beyond this very obvious point, the task of representing the settlement system is being made progressively more difficult by the blurring of urban-rural distinctions, the increasingly multi-dimensional nature of settlement and the emergence of new urban forms. These developments challenge us to look for alternative frameworks for presenting population data and studying the demographic dynamics of settlement in our contemporary world and into the future.

Responding to These Challenges

Given the considerable body of dissatisfaction with conventional approaches that has built up over the years, it is no surprise to find that a fair number of alternatives have been suggested. Most of these have been framed with particular countries in mind, though some have been put forward by supranational bodies of one kind or another. The purpose of this section is to outline the main generic types of alternatives to the simple urban-rural dichotomy, using examples from the existing literature. These comprise:

- the introduction of a third category intermediate between rural and urban;
- treatment of the settlement system as a continuum that can, if necessary, be split into many categories;
- recognition that, rather than a single scale of most rural to most urban, human settlement is multidimensional; and
- various ways of coping with new forms of urbanization, looking beyond the UN's concept of 'urban agglomeration'.

Further examples will be flagged up in the next section, where we highlight key messages from subsequent chapters of this book.

The simplest step, at least at first glance, is to replace the simple dichotomy with a three-fold categorization, i.e. a trichotomy. Two examples, both from supranational bodies, confirm that this idea has been around for some considerable time and is still considered viable. In 1964 the Conference of European Statisticians recommended a three-fold division of settlements based on population size. Those with at least 10,000 inhabitants would be deemed 'urban', those with less than 2,000 would be subsumed within the 'rural' aggregate, and those of 2,000 to 9,999 people would be termed 'semi-urban' (quoted in UN, 1969, p.8). Much more recently, the OECD Group of the Council on Rural Development has actually gone ahead and classified the territories of member countries into three types on the basis of 'degree of rurality' (Dax, 1996). This is based on the proportion of population living in rural communities, with areas with levels of 50 per cent or more being termed 'predominantly rural', 15-50 per cent 'significantly rural' and under 15 per cent 'predominantly urbanized'. While this latter example is obviously referring to a higher spatial level and presupposes that what is a 'rural community' is already defined, it clearly supports the argument that a dichotomous approach is inadequate for some purposes.

The next logical step is to allow more than three categories. Britain and the USA provide numerous examples of this approach to settlement classification, especially in academic studies but also in some that have been designed for governmental purposes. Particularly innovative in its time, not least because it approached its task from the rural end of the spectrum, is Cloke's (1977) 'index of rurality'. This recognized four levels of rurality for those parts of the country which lay outside the larger builtup areas; namely, extreme rural, intermediate rural, intermediate non-rural and extreme non-rural. Similarly, for the USA, Cromartie and Swanson (1996) created a four-level classification of areas beyond the 'metro core'. As shown in Table 1.3, this was based on population size, population density, levels of urbanization, commuting patterns and adjacency.

Table 1.3   Settlement classification for the USA, according to Cromartie and Swanson (1996)

1. Metro Core	Begins with an 'urbanized area' i.e. extent and distribution of the built-up area. If 50 per cent of the spatial unit's population is contained in the urbanized area it is indicated as part of the metro core.
2. Metro Outlying	Areas linked to core by commuting and exhibiting metropolitan character (as measured by population density, percentage urban and recent population growth).
3. Non-Metro Adjacent	Physically adjacent to a metropolitan area with at least 2 per cent of employed labour force commuting to urban core.
4. Non-Metro Non-Adjacent with City	Areas not adjacent to Metro Areas but contain all or part of a city of 10,000 or more residents.
5. Non-Metro Non-Adjacent without City	Access not adjacent to Metro Areas and without a city of 10,000 or more inhabitants.

Source: Cromartie and Swanson (1996, pp. 5-6).

By this stage, one is basically accepting the idea of a full continuum of situations lying between the most rural condition that can be conceived and the most urban. In the words of Lang (1986, p.120), rural and urban 'denote opposite ends of the conceptual continuum with real people and communities falling somewhere between the two hypothetical extremes. Any specific instance in the real world, therefore, can be viewed as demonstrating relative degrees of rurality and its opposite, urbanity, falling somewhere along the continuum between the two extremes'. The simplest, and up till now most commonly used, version of a continuum is, as with the Conference of European Statisticians' recommendation of a trichotomy just mentioned, one based on the single criterion of population size. For instance, as described in more detail in Chapter 2 of this book, after each decennial round of national censuses, the UN's Demographic Yearbook customarily documents the distribution of countries' urban population using 11 size classes ranging from under 200 inhabitants to 500,000 and over. Similarly, its *World Urbanization Prospects* series recognizes five categories that focus on the upper end of the size spectrum ranging from under 0.5 million through to 10 million and over (UN, 2003; see also Chapter 3 of this book).

There is, however, no reason why a continuum-based approach should restrict itself to a single criterion. Indeed, Cromartie and Swanson's approach in Table 1.3 represented a synthesis of two previous classifications of the US settlement system that were more sophisticated attempts at operationalizing the urban-rural continuum. One, by Butler and Beale (1994), had classified US counties into ten categories: four metropolitan and six nonmetropolitan, with the latter being differentiated according to their urban population and degree of adjacency to metro areas. The other, by Ghelfi and Parker (1995), had classified the 3,141 counties and independent cities of the USA firstly into eight categories according to the 'degree of urban influence' and then subdivided each of these to produce a much more detailed 'rural-urban continuum'. In all, this comprises 22 categories of place, ranging from the central cities of metro areas with at least one million people through to those rural counties which neither contain a city nor are adjacent to a metro area. Similarly, Britain's Office of Population Censuses and Surveys devised an 11-fold classification of local authority districts that ranges from Inner London through to 'remoter, mainly rural districts' (see OPCS, 1991, Appendix 5) and has been used in a variety of applications, including testing for the existence of a counterurbanization tendency in population redistribution (Champion, 1992).

In relation to the less developed regions of the world, a particularly interesting – if ultimately frustrating – example of developing a rural-urban continuum is provided by Indonesia. Here the status of each rural village (*desa*) and urban subcommunity (*kelurahan*) is determined on the basis of three criteria: population density, proportion of households engaged in agricultural production and number out of 15 designated 'urban' facilities that are represented in the settlement. As shown in Table 1.4, these areas are graded on each of these three criteria and the results summed to produce an overall score, with a maximum of 30.

Table 1.4   Scores given to Indonesia's *desa* for three criteria used to determine whether they should be classified as urban

Score given	Population density per km^2	Percentage of households engaged in agriculture	Total urban facilities*
1	Less than 500	over 95	0
2	500 - 999	91 - 95	0
3	1,000 - 1,499	86 - 90	1
4	1,500 - 1,999	76 - 85	2
5	2,000 - 2,499	66 - 75	3
6	2,500 - 2,999	56 - 65	4
7	3,000 - 3,499	46 - 55	5
8	3,500 - 3,999	36 - 45	6
9	4,000 - 4,999	26 - 35	7
10	5,000 or more	25 or less	8 or more

* Primary school, junior and senior high school, cinema, hospital, maternity hospital, clinic, road negotiable by motorised four-wheel-drive vehicle, post office or telephone, shopping centre, bank, factory, restaurant, public electricity, party equipment renting service.

Source: Biro Pusat Statistik (1979, p.5).

The official Indonesian practice is that areas scoring 23 or above are classified as urban and those with 17 or less as rural, while those with scores in between are field-checked to see if they should be classified as urban or rural. There is therefore an implicit suggestion here that each area can be given a measure of its urbanness and also that it is possible to identify the cut-off where a place is said to be urban as opposed to rural. Yet ultimately the latter must be a rather arbitrary decision. Maybe in this case there are strong policy-related reasons for needing to distinguish between urban and rural places, such as there being different administrative powers or funding regimes for these two types of area. On the face of it, however, this approach would seem to provide an ideal basis for a more detailed classification of settlement. Even if the full scale of 3 to 30 was deemed too cumbersome to retain, at minimum perhaps the areas with scores of 18 through 22 could be defined as 'intermediate' or 'transitional' between urban and rural, while a classification using more than three categories would appear a distinct possibility.

Thus far, the solutions examined, whether using the simple dichotomy or some form of continuum, have restricted themselves to deriving a single-dimensional classification revolving around the concepts of urban and rural. As mentioned in the previous section, however, it is quite common to synthesize several variables into this single scale, as in the case of Indonesia just described. The usual justification for this approach is that 'urbanness' and 'ruralness' are multifaceted, as depicted in Table 1.2. An alternative interpretation, however, is that the various criteria being used are measuring different things, any one of which may be of more interest to some users than a notionally 'general-purpose' classification. The very fact that three separate criteria are used in the Indonesian case in Table 1.4 signifies that the three do not map on to each other conformably: if they had done so, just one of them would have been sufficient to distinguish the *desa* that are urban. At this point, therefore, it is relevant to consider the possibility of adopting a range of alternative, fit-for-purpose classifications of settlement, indeed perhaps with none of them explicitly based on a scale that should be labelled rural to urban.

This point has been made particularly clearly by Coombes and Raybould (2001). They argue (p.224) that, '... in an increasingly complex pattern of settlement, linked with socio-economic polarization, no single measure can represent all of the distinct aspects of settlement structure that will be of interest to public policy'. They go on to suggest that there are at least three key dimensions to modern human settlement patterns that are quite distinct from each other and which are all important for policy makers to take into account when they are designing programs or allocating resources. These are settlement size (ranging from metropolitan to hamlet), degree of concentration (ranging from dense to sparse) and level of accessibility (ranging from central to remote). In their view, these three dimensions need to be recognised and measured individually: 'It is inappropriate to try to proxy any of the three with either of the others' (Coombes and Raybould, 2001, p.224).

In more detail, the first of Coombes and Raybould's dimensions – settlement size – is the most commonly used means for differentiating urban localities from rural ones, with larger places being associated with a greater heterogeneity, bigger economies of scale, a wider range of facilities and traditionally a very different lifestyle from that experienced in the smallest settlements. This settlement-specific measure is distinguished from 'concentration', which they also refer to as the 'intensity of settlement' and define in terms of the wider geographical context within which a place is situated. As such, it can be represented by an indicator that takes account of regional population density and the degree to which that population is located in a few larger settlements or more evenly spread. This, in its turn, is set apart from 'accessibility', which is seen as a complex variable that should attempt to measure residents' ease of reaching a wide range of essential amenities.

Given the wealth of studies that have analyzed settlement systems in terms of population size and density, it is perhaps worth demonstrating that accessibility, too, can be associated with demographic patterns. An example of the discriminatory power of this dimension is provided in Table 1.5, which shows the variation in selected fertility and mortality measures between settlements in non-metropolitan Australia grouped by degree of accessibility to facilities. The Accessibility/Remoteness Index of Australia (ARIA) is based on the 11,338 population localities identified on the 1:250,000 topographic map series and uses road distances measured between each of these and the nearest of four different levels of service centre (Bamford *et al.*, 1999). A maximum of 3.0 was scored for each level and the scores for each locality were summed to produce a continuous variable from 0 (highest accessibility) to 12 (highest remoteness), but here the ARIA values are aggregated into five categories to facilitate interpretation. The relationship with the demographic indicators is just as evident here for accessibility as was described earlier in the chapter in relation to the city-size dimension. Of course, in such a broad analysis as this, it is likely that accessibility will be correlated with size and density to some extent, but there will also be situations where one of the three dimensions is largely independent of the other two, such as in the case of small settlements located close to a city with many facilities.

Table 1.5   Total Fertility Rate (TFR), Infant Mortality Rate (IMR) and Standardized Mortality Rate (SMR) for males and females aged 15-64, for Australia's localities classified by degree of accessibility/remoteness, 1992-1995

Accessibility/remoteness grouping	TFR	IMR	SMR Males 15-64	SMR Females 15-64
Very accessible	1.79	5.8	96	97
Accessible	2.15	7.1	118	102
Moderately accessible	2.30	6.3	116	106
Remote	2.43	8.0	128	126
Very remote	2.51	13.4	201	258

Source: Glover *et al.* (1999, pp. 135, 140, 144 and 182).

If settlements were routinely scored on a variety of dimensions, it would be possible to gauge the independent effect of each of these dimensions on their demographics. One way of undertaking such an analysis would be to crosstabulate the characteristics of settlements before calculating the demographic variables. It was one of the suggestions made by the UN (1973) that population data should be compiled for a classification of national territory based on a combination of urban-rural and metro-nonmetro definitions. Even the most basic result – four categories comprising metropolitan urbanized, metropolitan rural, nonmetro urbanized and nonmetro rural – was felt to be a very important step that 'may eventually serve most practical purposes best' (UN, 1973, p.12). The value of a four-fold system was endorsed by the IUSSP's Committee in its final report in 1975 (Goldstein and Sly, 1975a, pp. 12-13; see also Chapter 2 of this book). Indeed it was this idea that Ghelfi and Parker (1995) were applying in their 22-fold classification of US counties mentioned above, though their study included the urban-rural dimension only for the nonmetro part of America.

This leads us on, finally, to responding to the challenge of the new forms of urbanization. In one sense, the way forward is easy. The four-fold classification just described provides a model of how to handle a situation in which a new dimension of settlement-system change emerges alongside more traditional expressions. Thus a settlement definition based on functional criteria like employment structure or commuting patterns, designed to cope with the enlargement of the units of urban dominance, can be used in parallel with physically-defined urban areas and indeed with administratively-defined settlements. Population data can either be produced separately for each of these three 'takes' on the settlement system so as to provide the intelligence needed for specific uses such as the 'economic plans', 'physical projects' and 'social policies' mentioned by UN (1973, p.9), or be presented for some crosstabulation of these frameworks like that suggested in the same report. In practice, however, so far there does not seem to have been any international reporting of population statistics along these latter lines. Given the passage of 30 years since this recommendation was made, this book needs to try to discover why this is the case.

Much harder, however, are the tasks of identifying what the most important dimensions of the settlement system are and of establishing the criteria for delineating the territorial units that represent these best. This is especially the case in the context of newly-evolving 'settlement spaces' where the new forms are as yet found in only embryonic state. Hand in hand with this challenge, in this fast-changing world, is the need for an assessment of the utility – even continued validity and relevance – of traditional methods of representing settlement patterns, not least the simple urban-rural dichotomy. Herein lies the primary purpose of this book, namely to document the main types of change in settlement systems around the world and to focus in on particular examples of new forms of what, for want of a better word, we are at this stage calling 'urbanization'. In doing this, the dual aim is to get a better idea of the conceptual and definitional challenges posed by the recent developments and to seek out ways of dealing with these, whether already in operation or still an academic construct.

Outline of the Book

While the previous section has introduced a sample of existing ideas about moving beyond the urban-rural dichotomy, this final part of our introduction highlights the main observations and suggestions to be found in the following chapters of this book. As mentioned in the Preface, the remaining two chapters in Part I provide contextual information about the UN's work on urbanization and its latest results. The chapters in Parts II and III provide broad regional perspectives on recent developments in urbanization and then home in on case studies that illustrate particular types of urban change at national and subnational levels. Those in Parts IV and V step back to consider the wider issues involved in developing new ways of conceptualizing settlement systems and then attempt to provide a bridge between the conceptual and operational aspects of this challenge, including assessing the potential of new GIS and remote sensing technologies. Lastly, in Part VI, the final chapter attempts to set out the way forward.

In more detail, Chapter 2 introduces the UN's work on urbanization, describing its main published outputs in this area, examining the key features of its approach to the definition and measurement of urbanization, and surveying the recommendations made by two reviews of UN practice published in the early 1970s. Champion points out that both the UN's own review and the report of the IUSSP's Goldstein Committee set out some clear pointers to the way forward, including several of the ideas outlined in the previous section of this chapter. This observation serves two purposes. In the first place, it prompts us to ask why, despite this, there seems to have been very little change since then in the approach used in the global monitoring and analysis of urbanization. Secondly, it provides a baseline against which to assess the novelty of suggestions made in later chapters.

Chapter 3 presents an overview of world urbanization, using the UN's most recent set of estimates and projections of world urbanization. Zlotnik uses the *2001 Revision* of *World Urbanization Prospects* to document the truly remarkable increases since 1950 in the number of people living in urban areas. She also draws attention to the rise of the very large city, though emphasising that it looks as if megacities of 10 million or more inhabitants will not be accounting for more than a small share of the world's urban population for the foreseeable future. The growth of small cities (under 500,000 people) and the urbanization of rural settlements are expected to make a far more important contribution to future urban population growth. The sheer scale of these changes underlines the importance of identifying the best tools with which to study them. The UN Population Division is clearly very aware of the limitations of the data that it has at its disposal and would especially like to see improvement in methods of identifying and delimiting individual localities. This applies across the full spectrum of settlement size, not least to the largest cities where – according to some of the later chapters – conventional representations are failing to keep pace with the new realities of urban growth.

The aim of the four chapters in Part II is to add detail to and critique the overall picture presented by the UN data in a set of essentially continental 'bites' starting with the USA and moving on to Latin America, Asia and Sub-Saharan

Africa. Chapter 4 is not only the most focused in spatial coverage but also in theme, with Frey arguing that the two principal dimensions of settlement classification adopted in the USA many decades ago for the official reporting of population statistics are now rapidly fading in their discriminatory power. The distinction between city and suburb is being eroded because the latter is taking on many of the features of the former and becoming more heterogeneous in both economic activity and socio-demographic profile. Secondly, metro-nonmetro differences are receding, as rural territory is becoming more strongly integrated into the national economy and as nonmetro areas are attracting increasing numbers of retirees and exurbanites. Among Frey's suggestions for coping with the new America are a classification of metro areas based on the importance of migration growth, especially immigration, and an extended suburban community typology that allows for the emergence of high-density centers and also differentiates between inner and outer elements. In a parting shot, he takes this further by suggesting abandoning morphology and demographic characteristics in defining metro areas and instead using commuting clusters as the basis for recognizing 'community areas' – one of the ideas considered by a recent review of US metropolitan area standards, which is examined more fully in Chapter 18.

Despite covering three major regions of the less developed world that exhibit substantial differences in urbanization level, Chapters 5, 6 and 7 convey very similar messages. In particular, these concern the high degree of diversity within their regions, the many variations and changes in urban definitions that make comparisons between countries and over time very tricky, and the general inadequacy of current approaches for capturing the latest settlement-system developments. All three are able to demonstrate how our perspectives of the changing settlement systems are affected by the use of inadequate concepts and by the limitations of the current data collection systems. In fact, in a test that applies different threshold sizes of urban place, Lattes, Rodríguez and Villa find that the ranking of urbanization level for a selection of Latin American countries does not vary greatly, but they admit that this is only a partial test as they have had to take the definitions of each settlement as given. In relation to the latter, they argue for the need to go back to basics, beginning at block-level in the city and using a common software like REDATAM. Likewise, Jones ends his review of Asian urbanization by referring to the new opportunities for small-area analysis provided by new technology and by more national censuses moving from a sample-survey to full-count basis. He also presents some glaring examples of inconsistencies in urban definitions, notably contrasting Thailand and the Philippines and then using the case of Cambodia to highlight the limitations of definitions based solely on administrative criteria. Both the Latin American and Asian chapters also point to the emergence of very large urban concentrations and raise issues about how these extended metropolitan regions can best be handled in statistical reporting. In the case of Sub-Saharan Africa, though, Bocquier finds the morphology of cities rather classical and, instead, is much more exercised by the phenomenon of population circulation between rural and urban areas and what this means for attempts at differentiating settlements according to lifestyle. The story in relation to the adequacy of statistics is, however, very similar. Indeed, Africa would appear to be

even less well served than the other two regions – a very unsatisfactory situation
that Bocquier tries to see in a positive light by suggesting that this might actually
lead to the faster uptake of new ways of studying urbanization than may prove
possible in the other two regions. Accurate monitoring into the future would help
to see whether Bocquier is right in suggesting that urbanization levels across much
of Africa will level off at 50-55 per cent rather the 80 per cent plus likely to be
reached by 2030 in Latin America and the MDRs in aggregate.

Next, the four chapters of Part III comprise a set of country-based studies in
which authors examine the specific aspects of recent urban transformations that
they have highlighted in their research. Mexico is taken as an example of a large
country that switched from predominantly rural to majority urban within barely a
lifetime and where the principal interest now is in the changing distribution of
population within the urban system. In Chapter 8, Garza points to the emergence of
a 'polycentric' urban hierarchy in the 1990s, by which he id referring to the
proliferation of large cities at the same time as the growth rate of the four major
centers slows, leading to a filling out and deepening of the Mexican urban system.
Even so, Mexico City remains larger than the eight other million-plus cities
combined and, with accelerating metropolitanization, is evolving into the core of a
much larger urban conglomerate that comprises two of these other cities as well as
many smaller ones. Garza concludes that tools need to be developed in order to
clarify the new kinds of spatial organization, especially measures that could help to
make sense of areas taking on 'megalopolitan' features and also those that would
allow the identification of urban populations regardless of locality size.

Chapter 9 focuses on a single metropolitan area, namely Delhi in India, and
analyses its dynamics from two interrelated perspectives: firstly, the evolving
metropolitan forms arising from 'peri-urbanization' and 'rurbanization', including
the expansion of suburbs, the formation of new residential quarters in surrounding
rural areas and the creation of satellite towns; and secondly, population
redistribution within the metropolitan area. Clearly evident from Dupont's account
are the increasing interweaving of urbanized zones and countryside, the blurring of
the distinction between rural and urban categories, and the development of new,
physically separate residential areas in the 'rural' fringes that are functionally
linked to the metropolitan center. The multinodal, quasi-continuous urban area that
has resulted calls for revision of the limits of the official agglomeration so as to
encompass the nearby new towns. Moreover, many inhabitants of this wider area
appear to be neither exclusively urban nor exclusively rural, underscoring the
inadequacy of the dichotomous classification of human settlements here as
elsewhere and reinforcing the need for recognizing some form of intermediate
category.

Cunha's account of the processes of urbanization and metropolitanization in
Brazil, presented in Chapter 10, combines themes from both the previous case
studies. As in Mexico, some of the country's secondary cities now appear to be
growing more rapidly than the largest ones. Yet there is again uncertainty about
whether the latter are really comparable with the former or suffer from having
boundaries that are less likely to keep pace with the extending settlement.
Secondly, focusing in on the chapter's main study area of São Paulo State, Cunha

highlights the growth of 'new rural areas' which are places, located especially in the officially designated metropolitan areas, that seem to be in a process of transition from rural to urban worlds. He reports the results of his reworking of census data for a more refined settlement classification than the urban-rural dichotomy traditionally used by the Census Office, identifying an intermediate or transitional zone that parallels the suggestion for Delhi.

In the final case study, Chapter 11 examines the phenomenon of *in situ* urbanization identified in China, especially using the example of Fujian province. Zhu defines this as where rural settlements transform themselves into urban or quasi-urban ones without much population movement. As such, it is closely related to the more widely-used term 'reclassification', but whereas the latter implies a change of official status associated with a qualitative shift from one side of the dichotomy to the other, *in situ* urbanization refers to the whole transition process from rural to urban regardless of any administrative change. While Zhu acknowledges that the changes in urban definition introduced by China's 2000 census are a great improvement on previous practice, he is critical of their still being based on the traditional builtup-area-based dichotomy. He argues for a more refined classification of settlement types, preferably based on a multidimensional scoring system. He then puts forward detailed proposals for revising the criteria for the designation of towns and cities. He also recommends that special treatment is given to the way in which members of the 'floating population' are considered in settlement statistics, suggesting that registration details from the *Hukou* system are used to resolve their double residential identities.

Moving into Part IV of the book, the emphasis shifts from assessing existing frameworks and making specific suggestions for dealing with their perceived inadequacies towards thinking through the wider issues that need to be addressed in developing new ways of conceptualizing settlement systems. In Chapter 12, Pumain argues for an evolutionary approach to settlement systems, criticizing existing typologies as too often being based on static reasoning. As settlement becomes more complex, the simple urban-rural dichotomy has to be replaced by a number of categories that are defined in terms of the sizes of settlements, their functional types and especially their evolutionary trajectories within the system. This approach considers that each settlement is a subsystem that cannot be described merely in terms of the current characteristics of its own residents. Also, a settlement acquires specific collective properties during its development, which need to be included as criteria in a classification. While Pumain justifies her case solely by reference to European examples, she maintains that the same basic approach can, and indeed should, be applied to other continents, taking into account different stages and phases of the urbanization process.

Bourne and Simmons, in Chapter 13, start from a similar position, arguing that our conventional ways of approaching the urban question have been overwhelmed by the increasing complexity and diversity of the global urbanization process. They also point to the complex transformations of space and place attributable to economic restructuring, social and demographic change, and the introduction of new technologies. They are as concerned as Pumain to shift the focus from static descriptions of the attributes of individual places to the dynamics of those places set within a larger and more flexible systems framework. Moreover, the latter can be allowed to transcend national boundaries, with the core of the concept resting on the importance of flows among urban nodes of people, goods, capital, profits, information and ideas. As an example of this approach in practice, they describe Canada's Metropolitan Influence Zones methodology which classifies all nonmetro areas outside the urbanized territory in terms of the degree of influence exerted by urban areas.

The next two chapters tackle the task of settlement classification from the rural end of the spectrum, with both arguing vehemently that even in highly urbanized countries, the rural is still important. There remains, however, the question of what exactly it represents. Brown and Cromartie, in Chapter 14, are highly critical of traditional approaches that treat rural as a single undifferentiated entity. Firstly, what is conventionally viewed as rural, or more specifically nonmetropolitan, needs to be subdivided. They describe a new US classification scheme, which they feel represents an important step in recognizing rural diversity. Secondly, they present a scheme for conceptualizing rurality that is mulitdimensional in nature, reflecting economic, institutional and cultural factors as well as the standard ecological criteria of population size, density and accessibility. These steps serve to emphasise that rurality is a variable and not a discrete dichotomy.

In Chapter 15, Halfacree goes even further by rejecting the idea that the rural is merely what is left over after urban areas have been delineated and arguing that it should be seen as an integral part of settlement systems. He points out that, despite the widespread view that the rural is an outdated concept, it simply does not go away but instead has a social and cultural significance today that may be as great as it has ever been. To make sense of this, he draws a distinction between attempts to maintain a largely 'material' understanding of the term, rooted in the presence or absence of a relatively distinct rural 'locality', and efforts to dematerialize the concept through placing it within the realm of imagination. He then draws out the range of meanings attributed to 'rural' across Europe, in particular contrasting the importance of the term in Britain with its much lower significance as a device for the understanding of space in Germany.

The three chapters in Part V move on to discuss some of the more practical challenges and opportunities. In Chapter 16, Coombes points to the complexity of the processes which are re-shaping settlement patterns, particularly in advanced economies. This leads to the creation of urban systems that cannot be captured by the simple urban-rural categories. Coombes illustrates this complexity with the example that place A can be more 'urban' than place B in some respects, but more 'rural' than it in others. He identifies what he sees as the key dimensions of modern settlement patterns that have to be represented when devising statistical indicators of settlement patterns. In terms of ways of demarcating urban from rural areas, he looks first at options for bricks-and-mortar definitions and then at measures that take account of the wider context within which any urban or rural area is set. The latter can be varied according to data availability, not necessarily needing commuting or other interaction data, but he also draws attention to a number of other implementation issues, including the question of how several indicators can be combined in a multidimensional approach to depicting settlement patterns. The latter partly conditions his final recommendation of a two-dimensional classification which builds on the longstanding measure of settlement size, though this relative simplicity also reflects his belief that users prefer transparent and stable methods.

In Chapter 17, Weeks reviews the use of remote sensing and geographic information systems for identifying the underlying properties of urban environments. This builds on the fact that urban places represent built environments that are physically distinguishable from the surrounding natural environment and are thus readily identified through the use of remotely sensed sources such as satellite images. He then goes on to suggest that variability in the built environment is associated with variability in human behaviour in urban places. He maintains that such variability in urban places can be captured through the classification of remotely-sensed images and then analyzed within a geographic information system. This is illustrated with an example from Egypt, in which variables derived from satellite images are combined with census data to improve our understanding of the spatial variability in human behaviour within the urban environment. Weeks goes on to suggest ways in which this type of analysis could be used to measure and understand phenomena such as urban sprawl and the multi-nucleation of metropolitan areas. At the same time, while ending on an upbeat note about the valuable insights that this additional data source can provide, he also recognizes that these are still early days. There remains much scope for the wider testing and refinement of techniques, and there are also resource issues relating to the cost of the data and to the expertise needed to manipulate and interpret it.

In Chapter 18, a case study of the fifth and latest review of metropolitan area standards in the United States provides clear insights into the issues involved and the types of questions as yet not fully resolved. Fitzsimmons and Ratcliffe recount how all aspects of the previous standards were within the scope of the review and demonstrate that none of the previous (1990) provisions remained unchanged. Among the main issues addressed were the geographic units that would be used as the areas' building blocks, the measures to be employed in grouping those units into statistical areas, and the frequency with which the resultant areas would be updated. The biggest change was the introduction of 'micropolitan areas', defined in the same way as metropolitan areas but based on a densely-settled urban core of just 10,000 people or more rather than the 50,000 threshold that was retained for the metros. Another important development was the grouping together of these core-based areas into a 'combined statistical area' where justified by employment interchange between them. Among the issues that they indicate as meriting further attention are the possibility of using units smaller than counties as the geographic building blocks, the challenge of subdividing statistical areas into different types of locale and the question of whether the standards should exhaust national space rather than leave an extensive undifferentiated residual area. Plenty of scope therefore remains for further work, which can be informed by analysis of new data from the 2000 census and other sources such as the American Community Survey. At the same time, their account provides a fascinating insight into the political difficulties encountered in trying to revise a system that had been widely used for government funding decisions for decades – another consideration, in addition to the methodological, technical and resource-related hurdles mentioned in the two previous chapters, that needs to be borne in mind when thinking about the way ahead.

In the final chapter that comprises Part VI of the book, we therefore try to steer a middle course between the great potential that exists for improving the reporting of population data for the settlement systems of the world and what appears to us as likely to be possible within, say, the next 5-10 years. Perhaps the most fundamental conclusions drawn from the previous chapters are that there is indeed a very valid case for classifying the *places* where people live, that conventional ways of doing this leave a lot to be desired and that it should be more widely recognized that an individual settlement can be classified in more than one way. Flexibility should be the new watchword, and this can be helped immeasurably if settlement definition procedures can take advantage of data for smaller geographic building blocks than the administrative districts that are still commonly used. These small areas provide the key to the more accurate delineation of settlements at any one time, as well as facilitating the updating of settlement boundaries and the assembling of comparable historical data for the newly defined areas. In addition, they make it possible to provide data for alternative conceptualizations of settlements, whether this is designed to generate customized classifications for particular applications or merely to test which version seems the best general-purpose representation. With the latter in mind, the chapter sketches out what the main elements of a generic classification for the twenty-first century might look like. Finally, after discussing the potential provided by GIS and remote sensing and rehearsing the resourcing and access issues relating to these, we conclude with a restatement of the case for change and with a call for action. Recommendations are provided for three groups: relevant international agencies, the statistical offices of individual nations and the academic research community. In particular, we look towards the setting-up of country-based case studies in which academics and statistical office staff can work together devising and testing settlement definitions and classifications, whilst also liaising with other users and with the international agencies that currently have such a difficult task in producing a meaningful and reliable picture of urbanization and settlement-system change across the world.

Chapter 2

Lest We Re-invent the Wheel: Lessons from Previous Experience

Tony Champion

This chapter reviews the work of the United Nations (UN) in acting as the world's main source of published international comparative statistics on urbanization. It begins with an account of the types of information currently made available through UN publications, which reveals the prevalence of the simple urban-rural dichotomy for presenting demographic and social data, just as was the case half a century ago when the UN's reporting began. In seeking an explanation for such strong continuity in the face of the rapid pace of urbanization and settlement system change over this period, there follows a description of the guidelines that the UN provides to national statistical offices for the collection and presentation of data on population. This includes assessments of how much these have been revised and updated since the start of this procedure in the 1950s and of whether there has been any uplift over time in the proportion of countries reporting data on the basis of the UN's preferred approach. Concluding that there has been only marginal change in these two respects, the chapter then demonstrates that this cannot be attributed to a lack of awareness of the limitations of this work. It does this by summarizing the results of two sets of reports that appeared over a quarter of a century ago: one from within the UN itself, the other from a Committee set up by the International Union for the Scientific Study of Population (IUSSP) and the direct precursor of this volume.

The underlying aim of the chapter is to provide a marker against which the novelty of the observations and suggestions made in subsequent chapters of this book can be assessed. At the same time, where it is found that previous recommendations have not been widely acted on, it is important to try and discover why this should have been the case. For instance, even as early as the UN's 1952 Demographic Yearbook, it was recognized that the settlement system should be considered as a continuum from small clusters to large agglomerations, yet the main element of the UN's reporting on population distribution trends continues to be based on the urban-rural dichotomy. In similar vein, the IUSSP's Committee advocated a multi-dimensional approach to disaggregating the settlement system, suggesting a four-way classification based on the metropolitan-nonmetropolitan dimension combined with the urban-rural divide. Why, therefore, does there appear to have been such limited progress towards producing international comparative statistics on a more refined basis such as this? Whatever the reason, the mere fact of having to ask this question stresses the need for the practical viability and political acceptability of any new proposals to be clearly thought out and justified.

The UN's Work in the Area of Urbanization

Virtually all the information that is publicly available about urbanization at the global scale, and indeed most of that for the world's major regions and other supranational groupings, is taken directly from UN publications or is produced as a result of further analyses of data originating from the UN. This is hardly surprising, given the huge scale of the twin tasks of assembling data from each individual country and of processing that information to fit as closely as possible to the UN's output formats. The purpose of this section is to describe briefly these standard outputs, so as to provide a feel for the nature and range of the intelligence that they provide about urbanization and urban growth around the world. Along with the following section's discussion of the UN's approach to the practical details of definition and measurement, this will help to reinforce our appreciation of the ambitiousness of the UN's work in this area. Also, however, it will prompt the reader to evaluate the accuracy of the information provided and assess its relevance to understanding current developments in the settlement systems of those parts of the world in which he or she is most interested. We focus on the two main places where the UN publishes its detailed statistics on urbanization and cities: *World Urbanization Prospects* and *Demographic Yearbook*.

World Urbanization Prospects is produced by the UN's Population Division (UNPD) and is updated every two years. As in previous versions, the latest report in this series – the *2001 Revision* (UN, 2003) – presents data on two types of urbanization measure: the urban (and rural) populations for countries and areas, and the populations of urban agglomerations. The data series for both these start in 1950 and look forward into the medium-term future, with projections for countries and areas through to 2030 and for agglomerations to 2015. These population statistics are given for five-year intervals, being the start and middle year of each decade (e.g. 1990, 1995, 2000), but also include urban and rural population totals for the reference year (2001 in this case).

The data is processed in a number of ways, yielding 17 main tables, ten at the country level and seven for agglomerations. The former contain data for urban, rural and total populations and the urban proportion for each of the years, and for rates of change in these for each five-year period. In all, 228 countries or areas are covered, and aggregations are provided to the world as a whole, its 'development groups' (more developed, less developed and – a subset of the latter – least developed), six major areas (Africa, Asia, Europe, Latin America and the Caribbean, Northern America, Oceania), and region (e.g., five for Africa, namely Eastern, Middle, Northern, Southern and Western).

As regards the seven tables for agglomerations, four present data for 525 cities with at least 750,000 residents in 2000, showing for each its population, the rate of change in this population, and its population as a proportion of the country's urban population and total population – all at five-year intervals for 1950-2015. In addition, one table shows the world's 30 largest agglomerations at each of these years and another gives the population of capital cities in 2001. The final table provides a summary breakdown (for 1975-2015) of the urban population of the world, and each major area and region, for five size classes of agglomeration: 10 million or more inhabitants, 5-10 million, 1-5 million, 500,000 to 1 million, and fewer than 500,000. The last of these is a residual figure, but for the four higher classes, details are given of the total number of agglomerations in each class and the total population of each class, both in thousands and as a proportion of the total urban population. An indication of the insights that can be obtained from both agglomeration- and country-level analyses is presented in Chapter 3 of this book.

The *Demographic Yearbook*, produced by the UN's Statistical Division, is also heavily used by urban researchers and policy makers. Unlike *World Urbanization Prospects*, it contains no statistics on future years, but provides more detail on urbanization and urban populations for the latest year for which data is available and, in some cases, a run of previous years. There are two types of tables: those published annually and 'special topic tables'.

Altogether, 13 of the 25 tables that appear regularly in the Yearbooks have an urban component to them. Three in the 'Population' section present population data in a form similar to that of *World Urbanization Prospects*. Each Yearbook's Table 6 gives urban and total populations and urban share by country for each of the last ten years and also provides urban population and share for males and females separately. Its Table 7 presents a breakdown of countries' urban and rural populations by age as well as sex for the latest available year from 1990. Its Table 8 presents the latest available data on the total population of cities of 100,000 or more inhabitants and for national capitals or seats of government. Subsequent tables of the Yearbook differentiate people by urban and rural residence in the same way as Table 7 but for other characteristics besides sex and age. Three relate to natality, one to foetal mortality, two to infant mortality, three to mortality and one to nuptiality.

The special topic tables that are of most relevance in the present context are those on Population Census Statistics. These are published every ten years or so after each round of censuses. The first of these sets of tables appeared in the 1952 Yearbook and was accompanied by a commentary on 'urban trends and characteristics' (UN, 1952, pp.9-19). The most recent set (at the time of writing) is that published after the censuses that took place between 1985 and the early 1990s, which are spread across the Yearbooks for 1993 and 1994. Table 2.1 lists the titles of the tables, indicating the wide range of items covered: population numbers, age/sex composition, ethnic group, language, religion, literacy, educational attainment, school attendance, economic activity, industry, occupation and marital status.

Two particular features are evident from this table. The first is the heavy emphasis on the urban-rural dichotomy. Almost all these census-based tables restrict themselves to the simple two-way distinction between urban and rural areas of residence. There is only one exception here, namely the use of a classification of localities by size to examine the sex ratio in the population as well as, obviously, the distribution of population by size-class. This latter is far more detailed, with 11 classes being distinguished where this is permitted by the nature of the settlement system (at the high end of the range) and the availability of data (for the smallest localities). The lowest class is of localities of under 200 inhabitants, the next 200-499, with the class intervals above 500 people set at 1, 2, 5, 10, 20, 50, 100 and 500 thousand.

Table 2.1   Special topic tables on population census statistics in the UN's Demographic Yearbooks 1993 and 1994

Yearbook/ Table no.	Table title (abbreviated)	N All	N U/R
Demographic and Social Characteristics			
1993/26	Population by single years of age, sex and U/R	77	42
1993/27	Population by national and/or ethnic group, sex and U/R	29	2
1993/28	Population by language, sex and U/R	25	1
1993/29	Population by religion, sex and U/R	29	3
Geographical Characteristics			
1993/30	Population of major civil divisions and U/R	97	49
1993/31	Population in localities by size-class and sex	37	na
Educational Characteristics			
1993/32	Population by literacy, sex, age and U/R	47	10
1993/33	Illiterate and total population 15+ by sex and U/R	47	10
1993/34	Population 15+ by educational attainment, sex and U/R	53	13
1993/35	Population 5-24 by school attendance, sex, age and U/R	49	8
Economic Characteristics			
1994/26	EA population and activity rates by sex, age and U/R	98	24
1994/27	Population not EA by functional category, sex, age and U/R	60	15
1994/28	EA population by industry, sex, age and U/R	79	18
1994/29	EA population by occupation, sex, age and U/R	70	8
1994/30	EA population by status, sex, age and U/R	56	11
1994/31	EA population by status, industry, sex and U/R	51	10
1994/32	EA population by status, occupation, sex and U/R	47	9
1994/33	Female EA population by marital status, age and U/R	36	9

N All Total number of countries included, N U/R Number of countries for which an urban/rural breakdown is provided, EA Economically active, na Not applicable.

Source: UN Demographic Yearbooks for 1993 and 1994.

The other feature revealed by Table 2.1 is the relatively limited number of the world's countries for which any subnational dimension was available on the basis used for these special topic tables. For instance, while statistics on population by single years of age and sex are reported from the censuses of 77 countries, in only 42 of these cases is a breakdown available between their urban and rural components. For most other characteristics, the number of countries where an urban-rural distinction is drawn is much smaller than this. Even where the information provided is just a single figure for total population, as for major civil division, an urban-rural split is reported for no more than 49 of the 97 countries providing statistics at this geographical level.

It is also important to point out that problems of global coverage do not arise merely because of the absence of regular census-taking in some countries. They are also encountered in the annual tables mentioned above. For instance, in the 1999 Yearbook, whereas crude birth rates are reported for 208 countries, separate rates for urban and rural areas are given in only 34 of these cases (UN, 2001a, Table 9). Even for basic variables like sex and age, coverage falls well short of the 228 countries included in *World Urbanization Prospects*. The 1999 Yearbook's Table 6 on total urban population by sex provides data for no more than 124 countries, and Table 7's urban and rural populations by age and sex are for just 103.

These coverage figures, along with the review of the type of information reported on urban populations, allow some conclusions about the current state of intelligence on urbanization and city growth. Most immediately, the limited coverage of subnational data gives an indication of the scale of the challenge faced by the UN Population Division in producing urban population estimates and projections for *World Urbanization Prospects*. Secondly, it is clear that, at least in terms of statistical reporting, the UN gives great emphasis to the simple urban-rural dichotomy, with much less reporting being made on the basis of individual urban agglomerations or localities grouped by size. Out of this arises the question as to why fuller and more detailed information on urban settlements does not appear to be available. Is it because there are many countries for which the urban-rural distinction now appears to have little relevance, or because the problems of identifying and measuring settlements are too great, or some combination of the two? A review of the UN's approach to the definition and measurement of urbanization can help to answer this question.

The UN's Approach to the Definition and Measurement of Urbanization

The UN's approach to the assembly and processing of data on urbanization was first set out some 50 years ago and has subsequently been re-issued every ten years in the guidelines that it publishes in advance of each fresh round of censuses. Two main questions have dominated; namely, how to identify a settlement and how to distinguish urban from rural. In this section, we take a chronological look at the approach taken to each of these questions in turn, and then examine the outcome in terms of the nature of the data that the UN receives from national statistical offices and how much this has changed over the years. The clear theme arising from this review is the remarkably little change in the UN's guidance, together with even some decrease in the apparent ability of countries to supply the recommended data.

UN Guidelines for the Identification of Localities

The challenge of identifying individual settlements, or 'localities' in UN parlance, was evident right from the outset, as is clearly stated in the report on the 1950

round of census results published in the *Demographic Yearbook* for 1952. A flavor
of the problems recognized is given in the following extract (from UN, 1952, p.9):

> The geographic units that are classified into urban and rural categories ... vary widely
> in the degree to which they conform to the demographic unit here referred to as the
> 'agglomeration'. Thus, some countries classify as urban all centres of population which
> have special forms of government and enjoy considerable autonomy in matters of
> taxation, police protection, sanitation, etc. the boundaries of such centres may be drawn
> well outside of what might be considered the limits of the agglomeration *per se*, or they
> may fall well inside limits and therefore exclude the closely settled and clearly urban
> population of the suburban fringe. Only in relatively few countries is the attempt made,
> for census purposes, to count together all the inhabitants of centers of population
> regardless of whether the inhabitants live inside or outside of the official boundary.
> Nevertheless, it is only by delimiting agglomerations as such that comparable data on
> this subject can be obtained.

In the end, the UN was forced to work with three different definitions of
'locality' when it came to the task of presenting data for individual settlements, as
follows:

> a. Agglomerations or clusters of population without regard to official boundaries or
> administrative functions. ...
> b. Localities having fixed boundaries and an administratively recognized 'town' status
> which is usually characterized by some form of local government, operating under a
> charter or under terms of incorporation. ...
> c. Minor civil divisions (often the smallest administrative divisions) which have fixed
> boundaries and which together comprise the entire territory of the given country.
> (UN, 1952, p.25)

Each of these three approaches was recognized to suffer from its own problems in
application. In relation to 'a', while this was the preferred approach, it was found
that there were differences between countries in the minimum size threshold used
for identifying separate settlements. The method adopted in order to achieve
international comparability in this respect was to raise the minimum size to the
level that allowed coverage of all, or at least most, of the countries that reported
details for individual settlements.

With regard to approach 'b', the problem was, while these districts include a
central agglomeration and the surrounding territory that is administered from the
central place, these will quite usually be under- or over-bounded in relation to the
extent of the actual settlement on the ground. More commonly, 'they are separate
cities, wholly urban in character, with "city limits" which may fall inside the edges
of the agglomeration' (UN, 1952, p.9). Less often they include rural areas. Unless
supplementary information is available to allow a more accurate representation of
the real agglomeration (and thus become defined under 'a'), then there is no way of
allowing for the distortions arising from this problem.

Countries for which population data are provided on the basis of a single tier
of undifferentiated territorial units, as in approach 'c', were seen as posing the

biggest problem. This is especially so if the only available data for them is the population count. In some instances, population size could be treated as a surrogate for settlement size, but in most cases this was not possible. Larger settlements commonly comprise many minor civil divisions (and, very probably, an even larger number of census tracts), while at the other end of the settlement scale the civil divisions used in more rural areas may each contain a number of small settlements.

Subsequent rounds of guidance have sought to clarify the definition of locality and steer countries towards the preferred approach. The UN's recommendations for the 1960 censuses (UN, 1958, p.11) define locality as 'a distinct and indivisible population cluster (also designated as agglomeration, inhabited place, populated center, settlement, etc.) of any size, having a name or a locally recognized status and functioning as an integrated social entity. This definition embraces population clusters of all sizes, with or without legal status, including fishing hamlets, mining camps, ranches, farms, market towns, communes, villages, towns, cities and many others.' These recommendations go on to note that localities should not be confused with the smallest administrative divisions of a country, although the two may coincide in some cases. They also suggest that tabulations of the enumerated population by size of locality and of the localities in each size group are essential for the analysis of population distribution. Finally, they note that a preliminary listing and mapping of all identifiable localities is a desirable preparatory step for a census and that it would be useful for each country to present the exact definition of locality used in each census.

The main messages were repeated in the guidance issued before the 1970 census round (UN, 1967, p.51), as follows:

> 232. For census purposes, a locality should be defined as a distinct population cluster (also designated as inhabited place, population center, settlement, etc.) of which the inhabitants live in neighboring buildings and which has a name or a locally recognized status. It thus includes fishing hamlets, mining camps, ranches, farms, market towns, villages, towns, cities and many other population clusters which meet the criteria specified above. Any departure from this definition should be explained in the census report as an aid to the interpretation of the data.

> 233. Localities as defined above should not be confused with the smallest civil divisions of a country. In some cases, the two may coincide. In others, however, even the smallest civil division may contain two or more localities. On the other hand, some large cities or towns may contain two or more civil divisions, which should be considered only segments of a single locality rather than separate localities.

> 234. A large locality of a country (i.e. a city or a town) is often part of an urban agglomeration, which comprises the city or town proper and also the suburban fringe or thickly settled territory lying outside of, but adjacent to, its boundaries but is an additional geographic unit which includes more than one locality.

Jumping forward 30 years, exactly the same wording is used in the most recent set of recommendations (see UN, 1998, p.64, paras 2.49, 2.50, and 2.51). The sole difference between the guidance for the 1970 and the 2000 rounds of censuses is the addition of two sentences at the end of the third and final paragraph, as follows (UN, 1998, p.64): 'In some cases, a single large urban agglomeration may comprise several cities or towns and their suburban fringes. The components of such large agglomerations should be specified in the census results.'

UN Guidelines for Distinguishing Urban from Rural

The challenge of distinguishing urban from rural was also recognized at the outset. In the UN's (1952, p.24) words, 'One of the most difficult problems in presenting internationally comparable demographic data is that of obtaining urban and rural classifications of the population. The designation of areas as urban or rural is so closely bound up with historical, political, cultural and administrative considerations that the process of developing uniform definitions and procedures moves very slowly.' Also acknowledged, even at that very early stage, is the arbitrariness of any line drawn between urban and rural (UN, 1952, p.9):

> One source of difficulty is the fact that the urban-rural classification is usually a dichotomy which divides the population into two parts, one urban, the other rural. Since there is no point in the continuum from small clusters to large agglomerations at which 'rural' ends and 'urban' begins, the line drawn between urban and rural is necessarily an arbitrary one.

One way of getting round this problem was seen to be the tabulation of data for size groups of 'all identifiable agglomerations or clusters of population', as recommended in 1949 by the Fourth Session of the UN's Population Commission (UN, 1952, p.24). Adopting this approach, Table 7 of the 1952 Yearbook presented data on localities or agglomerations classified into ten size groups ranging from 100,000 or more down to under 200 inhabitants. This yearbook also went on to demonstrate the effect of using the 'agglomeration approach' to distinguishing urban from rural. It picked a cut-off point of 2,000 inhabitants and, for a selection of countries, compared the proportion of population living in agglomerations of this size or larger with the proportion defined as urban by those countries' statistical offices. It found percentage point differences ranging from a country where the official urban figure was 25 per cent higher than the agglomeration-based figure to one where it was almost 18 per cent below it (UN, 1952, p.9).

Perhaps because of these results, the UN's subsequent practice has been to run with these two alternative types of measure, as already indicated above in terms of the statistics even now being published in the *Demographic Yearbook* and *World Urbanization Prospects*. Thus, the recommendations for the 1960 census round (UN, 1958, p.11) state:

> Because of the diversity of concepts used in the classification of areas as urban and rural in various countries, it is not practicable to establish uniform definitions of urban and rural populations for international use. It is believed that for purposes of international comparison the classification by size of locality [...] is [...] most nearly adequate and should be used in addition to the urban-rural classification which countries may continue to use for national purposes. The interpretation of any urban-

rural tabulations employed will be facilitated if countries give the definition of urban and of rural areas or populations used for census purposes.

Forty years on, the recommendations issued by the UN for the 2000 round of censuses are very similar. The section devoted to the topic starts with the following statement: 'Because of national differences in the characteristics that distinguish urban from rural areas, the distinction between the urban and the rural population is not yet amenable to a single definition that would be applicable to all countries or, for the most part, even to countries within a region. Where there are no regional recommendations on the matter, countries must establish their own definitions in accordance with their own needs' (UN, 1998, p.64). Then, after noting that the traditional distinctions between urban and rural in terms of standards of living are no longer relevant in many countries, it is suggested that a classification by size of locality can be a useful supplement to the urban-rural dichotomy or that it may even replace the latter when the major concern is with characteristics related only to density along the continuum from the most sparsely settled areas to the most densely builtup localities.

At the same time, the 1998 recommendations contain a fuller discussion of the issues faced in distinguishing urban from rural. In particular, it is recognized that density of settlement may not be a sufficient criterion in countries where large localities are still characterized by a rural way of life. Additional criteria must therefore be used in such cases including, for instance, the proportion of workers not engaged in agriculture, the availability of piped water or electricity, and access to medical care or to schools. It is also noted that in some countries different criteria may have to be used in different parts of the country, if the process of economic transformation that underpins urbanization has not yet expanded to the whole country. However, countries are warned not to use an overly complicated definition of urban, since it may hinder its application to the census and comprehension of it by users of the results.

Lastly, the 1998 recommendations are the first to acknowledge that, 'Images obtained by remote sensing may be of use in the demarcation of boundaries of urban areas when density of habitation is a criterion' (UN, 1998, p.65). However, no further guidance on how density of habitation would be operationalized to distinguish urban from rural areas is provided.

The UN's Approach in Practice

Given this long history of international guidance, it might be expected that a considerable degree of consistency might by now have been achieved in the information provided to the UN by national statistical offices. An analysis of the bases of the data used for the latest edition of *World Urbanization Prospects* indicates that this is far from the case. Indeed, if anything, the situation has deteriorated since the UN prepared its first set of urban and city projections, based on censuses taken between 1955 and 1963 (UN, 1969) – though admittedly the 228 countries covered now is considerably more than the 123 of that first report.

Take, for instance, the 'nomenclature' used to identify the primary units constituting urban areas. As mentioned above, 'locality' is the UN's recommended nomenclature. In the *2001 Revision* (see UN, 2003, p.107, Table 63), just 26 countries use this as their sole basis, with a further 16 using it in conjunction with others, making a total of 18.4 per cent of the 228 counties. This level was even lower than in the UN's 1969 report, when it was used solely or jointly by a total of 35 of the 123 countries, or 23.6 per cent. The most commonly used nomenclature at both times is 'small units specified generically as "cities", "towns", "townships", "villages", etc.', with sole or joint use by 25.4 per cent of countries in the *2001 Revision* and 31.1 per cent in the 1969 report. 'Urban units specified by place name only' is third most popular, rising somewhat from 19.6 per cent in 1969 to 22.8 per cent. Meanwhile, the proportions of countries using 'administrative centers of minor civil divisions' and 'minor civil divisions' remained virtually unchanged, totaling around one-fifth. Finally, and perhaps equally worryingly, one in nine of all countries had supplied no definition at all for the data used for them in the *2001 Revision*, twice the proportion found in the 1969 report.

In relation to the criteria used to determine whether a place is urban or not, there is a somewhat higher allegiance of countries to the UN's recommended approach of population size, but even so such countries are not in the majority. The situation found in the Population Division's work for the *2001 Revision* (UN, 2003, p.107, Table 62) is as follows:

- 109 used administrative criteria, with 89 of these using it as the sole criterion;
- 98 used population size and/or population density, solely in the case of 46 countries, with a variety of cut-off levels on both these measures;
- 27 used economic criteria, notably proportion of labor force employed in non-agricultural activities, sometimes in combination with other criteria, economic or otherwise;
- 24 used criteria relating to the functional nature of urban areas, such as the existence of paved streets, water-supply systems, sewerage systems or electric lighting, often in combination with other criteria;
- 24 did not supply any indication of the basis for their 'urban area' definition;
- 6 countries were deemed by their national statistical offices to be either entirely urban or entirely rural.

Thus, only 43 per cent (98 of 228 countries) made any use of population size, with less than half of these using size and/or density alone. The modal approach was on the basis of administrative criteria, with almost two-fifths of countries (89) using this as their sole criterion.

Comparisons with the situation described in the 1969 report are complicated by the much larger number of countries that did not specify their criteria then (56 of the 123) and by differences in the way the information was presented. The headline figure is that population size was specified as sole criterion by 23 countries, with a further 26 countries using this in conjunction with other criteria, giving a total of 39.8 per cent of the 123 countries. Just one country indicated the use of density, with 10 others using this in conjunction with another criteria (which may have been population size as the numbers using both size and density are not shown). If it is assumed that most of the 56 non-reporting countries did not use population size – and other evidence suggests that many of them used administrative criteria – then we could conclude that the proportion using the UN's recommended approach for identifying urban places has remained more or less the same at around two-fifths. Of course, if it were the case that some of the 56 had used population size then, the picture would be even more discouraging, as that would signify that the proportion using this criterion now is even lower than it was in the 1960s.

Finally, turning to the UN's approach to publishing data on individual cities or urban areas, again there is considerable deviation from the UN's recommendations. The specified preference is for 'urban agglomeration', referring to 'the population contained within the contours of contiguous territory inhabited at urban levels of residential density' (UN, 2003, p.108). This is irrespective of whether that population lives within or outside the administrative boundaries of the main city, which is referred to – rather misleadingly – as the 'city proper'. For only 90 of the 228 countries included in the *2001 Revision* was the data reported on the 'urban agglomeration' basis. This compares with the 109 countries reporting to the UN on the basis of 'city proper'. Thirdly, 13 countries provided data for 'metropolitan areas' (see below), while a further 13 used different definitions for different cities and the remaining three countries gave no details of the definitions used (UN, 2003, p.109, Table 64).

This is a very unsatisfactory basis for reporting city populations, as the UN makes clear. In particular, 'The territory delimited by administrative boundaries ...[does]... not necessarily coincide with the extent of the urbanized territory as delimited by other standards' (UN, 2003, p.108). For instance, this definition may exclude suburban areas, while in some cases two or more adjacent 'cities' may be separately administered, even though forming a single urbanized region. Alternatively, 'cities' may be overbounded in administrative terms, with their boundaries including large tracts of land devoted to agriculture and containing a rural population. Another problem with using administrative boundaries is that these are not continuously being revised to keep up with the growth of the city on the ground, with rather occasional reviews leading to sudden jumps in population size.

At the same time, the UN recognizes that there is more than one way of defining cities in non-administrative terms. Besides the 'urban agglomeration' approach, it also recognizes the concept of 'metropolitan area (or region)' as being superior to 'city proper' for capturing the population dynamics of urban expansion. This concept entails a more extensive definition of the territory of interest than 'urban agglomeration': it 'includes both the contiguous territory inhabited at urban levels of residential density and additional surrounding areas of lower settlement density that are also under the direct influence of the city' (UN, 2003, p.108). A disadvantage with this approach, however, is that it is likely to introduce an upward bias into the population count, as was found by a study carried out as long ago as the 1950s. According to Davis (1959), the world's total population living in cities of 100,000 inhabitants or more in 1955 would have been 353 million if cities were defined in terms of administrative units (i.e. 'city proper'), 440 million if using the UN's preferred basis of 'urban agglomeration' and 519 million if cities had been conceived as urban-dominated regions (i.e. 'metropolitan area').

In sum, despite the effort that the UN has put into its international guidelines over the past half-century, there would appear to have been virtually no progress towards greater comparability. Just as in the 1950s and 1960s, still today the variety of nomenclatures and criteria used to determine what constitutes 'urban places' is bewildering, and there remains widespread reliance on the administrative division of territory to establish their boundaries. This is despite long recognition that neither administrative status nor types of economic activity can be relied upon as a permanent basic criterion to identify urban places and that criteria used in defining the territorial extension of cities are generally not consistent with those adopted to define the urban population. The next section reinforces this final point by reference to two reviews dating from over a quarter of a century ago.

Challenging Conventional Practice: Two Examples from Yesteryear

The need for a review of conventional practice on urban standards comes through very clearly in two sets of reports published over a quarter of a century ago. One of these was by the UN itself as it set out on its task of projecting urban and city populations, while the other set was produced by the IUSSP's Committee on Urbanization and Population Redistribution. The two are, in fact, linked. The main output from the UN's review, *Growth of the World's Urban and Rural Population, 1920-2000* (UN, 1969), was followed by a paper entitled 'Statistical definitions of urban population and their uses in applied demography', published as the theme chapter in the 1972 Demographic Yearbook (UN, 1973). The latter, in somewhat modified form, then formed the lead paper in the first of the three Working Papers prepared by the IUSSP Committee (Goldstein and Sly, 1975a). Its author, John V. Grauman of the UN's Population Division, made such an impression on the Committee that, on his death in 1976, the latter dedicated its final publication (Goldstein and Sly, 1977) to his memory. We look, in turn, at the main features of the two reviews.

The UN's Review by John Grauman

Although the main part of John Grauman's *Growth* report (UN, 1969) was devoted to a description of the results of compiling and processing data from past censuses and making forward projections, it also contains a penetrating discussion of the methodological issues faced in this work. In particular, it fully acknowledges the deficiencies in the data on which the estimates are based. It admits that many arbitrary decisions have had to be taken in the endeavor to arrive at comparable estimates: '...these are debatable and leave much scope for improvement on the basis of more detailed research' (UN, 1969, p.iii). It goes on to refer to the 'drastic alterations in settlement patterns' and stresses that, 'In such changing conditions, conventional terms become less adequate as descriptions of the environment' (UN,

1969, p.1). Drawing on a brief survey of the historic development of the 'urban' concept, it concludes that, 'A definition of "urban" places cannot be devised which has unvarying relevance throughout the changes in time and diversity of local conditions' (UN, 1969, p.1). Indeed, it is reckoned that, 'With the increase in number of urban attributes and their wider diffusion, it is doubtful whether the historic twofold "urban" and "rural" distinction will retain its relevance much longer' (UN, 1969, p.3).

Going beyond this opening gambit, the report (UN, 1969, pp.1-6) then makes a range of important points that can be summarized as follows:

- The urban phenomenon is associated with numerous aspects and, furthermore, these aspects can coincide or overlap to a varied extent and not all are necessarily present at the same time. 'Urbanization, consequently, will not be confined to any single definition ....'
- Instead of making a direct opposition between urban and rural, 'a distinction may have to be drawn in new terms, as when a locality of intermediate characteristics develops differences from the type in which it is classified, or there may be a continuum of types within a group, according to the scale of measurement'.
- Though in many parts of the world the primary distinction of settlements as urban or rural serves most purposes of study, '... this simple scale is not adequate for entire regions with dominantly urban interdependent functions'.
- 'The increasing scale and complexity of the interplay of urban functions are enlarging the recognizable units of urban dominance'.
- Much more needs to be known about world-wide variations in the degrees of population concentration and dispersal in the *rural habitat*. The scattered information now existing with respect to the world's rural habitat 'is woefully inadequate for an assessment of basic problems affecting the welfare of man'.

In terms of the data presented in the *Growth* report, there was not much that could be done to take these considerations on board, given the form in which national census results had already been compiled. In terms of 'The world's urban and rural population in 1950 and 1960, as nationally defined' (the title of Chapter I), the report is merely able to highlight the range of approaches used by the 123 countries with censuses taken between 1955 and 1963, as outlined in the previous section. Somewhat more discretion, however, was possible in Chapter II on 'World urbanization trends as measured in agglomerations'. In particular, a distinction was drawn between agglomerations of 20,000 inhabitants or more and smaller ones. Attention is concentrated on the former, which were labelled 'city and big-town population', firstly because such places are almost universally treated as urban and secondly because it was reckoned that the population of these places was generally more accurately recorded than for smaller places. Indeed, the Population Division itself invested much effort in using the more detailed statistics provided by national statistical offices to try and achieve maximum consistency in the delineation of these larger agglomerations.

Nevertheless, mindful of remaining deficiencies and wishing to provide a fairly simple overview, the report contented itself with a five-fold breakdown in this size range, using a multiplier of five, as follows:

- Super-conurbations: 12,500,000 or more inhabitants.
- Multimillion cities: 2,500,000 or more inhabitants.
- Big-city population: 500,000 or more inhabitants.
- City population: 100,000 or more inhabitants.
- Agglomerated population: 20,000 or more inhabitants.

The terms are inclusive, in that each one includes all the larger size categories, but for most of the statistical reporting the data is for the separate size bands 20,000-99,999 and so on, with the top two usually amalgamated into a single category of 2.5 million and over. Beyond this, the report lumps together the populations of all 'urban' localities with under 20,000 residents under the heading 'small-town population'. This is principally in recognition of the fact that the lower size limit of 'small towns' varies widely between countries, though it also reflects the feeling that the reported data for these localities will suffer from wider errors than for larger places. Hence the report's plea for more information about the so-called 'rural habitat', mentioned above.

Finally, the *Growth* report provides two other pointers to the way forward, even though neither was implemented in the report itself. One was to consider moving away from the simple urban-rural dichotomy. Grauman quotes the 1964 recommendation of the Conference of European Statisticians that an intermediate category of 'semi-urban' be introduced for settlements of between 2,000 and 9,999 inhabitants (UN, 1969, p.8). The other concerns the point that the simple urban-rural scale is not adequate for entire regions with dominantly urban interdependent functions, linked to the recognition that the increasing scale and complexity of the interplay of urban functions are enlarging the recognizable units of urban dominance. Grauman makes clear that this is by no means a new trend, referring to the coalescence of separate towns into 'conurbations' from the 1860s onwards. He goes on to cite the emergence of even less compact urban forms such as 'metropolitan areas' and 'megalopolis', whose integrity is based primarily on functional criteria that lays stress on the high degree of mutual interdependence of activities. In such cases, it is suggested that, 'A fourfold classification, separating "urban" and "rural" areas both within and outside the larger regions of urban dominance, might provide a more relevant framework of analysis' (UN, 1969, p.3).

This point is taken further in the paper in the 1972 Demographic Yearbook in conjunction with the idea that the urban phenomenon comprises more than a single aspect. The paper also introduces the user perspective, notably the idea that population and related data is needed for a range of different applications, each involving a different definition. These can be reduced to three types: economic, administrative, ecological. Each of these three types have their separate practical uses: 'Roughly speaking, ... for three types of government policy or program: economic plans, social measures, and physical projects' (UN, 1973, p.9).

The following extracts (taken from UN, 1973, p.12) give a flavor of the main conclusions that the author draws:

> On the whole, it would seem that, for different purposes, we need different sets of criteria in the statistical measurement of urban populations.
>
> For policies whose implementation depends on organs of local administration, there is a continuing need for data and estimates of 'administratively urban' population.
>
> For plans and programmes which have to be detailed according to forms of the settlement pattern, it is most useful to describe urban settlements as areas inhabited, or built up, at high density and having, within the contours of densely settled terrain, at least some minimum number of inhabitants.
>
> Considering the conditions already existing in the most industrialized countries and likely to emerge soon or later in the rest of the world, the focus of interest seems to shift partly to regional concepts. But the metropolitan regions which have become recognized are territorially more extensive than the corresponding more strictly urbanized areas.
>
> Looking into the future …one can incline to the view that a fourfold classification may eventually serve most practical purposes the best. It may become necessary to distinguish urbanized and non-urbanized areas (i.e., mainly areas settled at high, and at lower densities) both within the metropolitan regions and outside such regions, resulting in four categories such as metropolitan urbanized, metropolitan rural, non-metropolitan urbanized, and non-metropolitan rural populations.
>
> With urbanization and urban regionalization now progressing so fast, there is the possibility that various concepts, which are still useful today, will soon become outmoded.

These, then, were the thoughts emanating from the UN in the early 1970s. As mentioned above, they were submitted to the IUSSP Committee and were reproduced in virtually identical form as the first article in the Committee's Working Paper 1 (UN Secretariat, 1975). They proved to have a strong influence on that Committee's final recommendations, as we shall now see.

Conclusions and Recommendations of the IUSSP's Goldstein Committee

The IUSSP's Committee on Urbanization and Population Redistribution, set up in 1971 under the chairmanship of Sidney Goldstein, had two primary goals: first, a review of basic data needed for the study of urbanization and, second, a review of methods for measuring urbanization and projecting urban population. The results of the former were published in Working Paper 1 (Goldstein and Sly, 1975a), those of the latter in Working Paper 2 (Goldstein and Sly, 1975b). In addition, the Committee sponsored a series of country case studies which were published in two volumes as Working Paper 3 (Goldstein and Sly, 1977).

This set of publications provides a wealth of insights into what was then considered to be 'state of the art' in the study of urbanization, but for present purposes the main interest is in what was seen as the priorities for improving the situation. To this end, we present a summary of the Committee's 'major conclusions and recommendations' (Goldstein and Sly, 1975a, pp.12-13):

- The lack of basic data needed for the study of urbanization was seen to arise primarily from the unavailability of skills, funds, and/or the failure to realize the relevance and importance of an item for the study of urban phenomena.
- The problems of international comparability of urban data arose primarily from differences among countries with respect to three factors: 1) nomenclature, 2) the definition and demarcation of boundaries, and 3) the criteria and/or the value of criteria needed to attain urban status.
- The basic data needed for the study of urbanization could be obtained through either population censuses or sample surveys but, as the demand for the scope of urban data broadened, it would become increasingly necessary to supplement census data with information from various types of record systems and sample surveys.
- Although it was recognized to be difficult, given the diverse national settings within which urbanization occurs and the importance of preserving the integrity of individual nations, it was considered imperative that standard procedures and criteria be developed which would facilitate the comparative study of urbanization.
- A single statistical definition of urban which would be applicable to all countries of the world probably was seen as being unrealistic at that time or in the near future.
- Similarly, it did not appear that a single definition of urban could meet all the needs of individual countries. Therefore, it was recommended that, for the immediate future, efforts should be directed at the development of standard procedures and definitions for groups of countries at similar levels of development with similar cultural backgrounds. Longer-range planning, however, should be undertaken at the same time with the goal of developing a standard set of statistical procedures to differentiate urban from rural areas in all countries of the world.
- In many countries of the world, the conventional urban-rural dichotomy appeared to the Committee as being outdated. It was therefore recommended that three- or four-fold classification systems be considered for adoption.
- The development of standard definitions and procedures should attempt to narrow international differences along the three dimensions mentioned above: 1) Nomenclature. It was noted that international recommendations tended towards 'localities' defined as agglomerations of dense settlement without regard to political boundaries. While the Committee supported this approach, it also emphasized the importance of also having data for administrative areas because of the number of programs and projects based on administrative units. 2) Boundaries. It was seen as good sense for countries to report data for administrative subdivisions as well as for areas which contain population

agglomerations demarcated in a manner such as that used in Japan for the densely inhabited district, or the United States for the urbanized areas.

3) Criteria. Having noted that problems of international comparability arise from using different values of criteria like population size and density, the Committee suggested that these would be reduced if all countries were to tabulate data for localities according to the same population size groups.

This, then, can be seen as the situation reached over a quarter of a century ago as a result of both the UN's own assessment and the deliberations of the Goldstein Committee's eight members and the many others that participated in their meetings and contributed their country case studies. Key among these points is a perceived lack of resources going into this area, despite the rapid pace of urbanization and major changes in settlement patterns. It was clearly felt that more effort needed to go into both resolving problems of international data comparability and obtaining data from more sources. Particular emphasis should be put into achieving greater harmonization for groups of similar countries, even if worldwide consistency may have to be left as a long-term goal. Also, there should be a much stronger push towards collecting and reporting data for other than administrative areas, not least because the international diversity of these areas forms a major obstacle to harmonization. While it was seen as important still to have data for administrative areas for certain purposes, an increasing number of applications were needing data on the basis of alternative spatial units like urban agglomerations and metropolitan regions. Finally, and most closely related to the thrust of this book, already then it was seen as important to move beyond the urban-rural dichotomy and consider more sophisticated classifications of places and people.

Concluding Comments and Questions

Arising out of this review, the bottom line is that the world's main source of published international comparative statistics on urban population – the United Nations – is still using essentially the same 'spectacles' as it adopted half a century ago. Its basic output continues to relate to two types of geographical concept: an urban-rural dichotomy, and agglomerations for which statistics for around half the world's countries are still being presented for cities according to administrative boundaries (the so-called 'city proper') rather than the properly defined built-up area. On the credit side, following the lead given by John Grauman's pioneering report of 1969, the UN Population Division now publishes not just estimates of past and current urban population but also projections, the preparation of which entails a huge amount of work on the two-year cycle of *World Urbanization Prospects*. Yet, what is the value of all this forward-looking effort? As is clear from the above review, the definitions on which this work is based still seem to be as problematic now as they were 50 years ago. Moreover, the whole underlying conceptual basis was already being challenged – both within and outside the UN – over a quarter of a century ago and will likely have less relevance in 2030, when the projection period ends.

If progress is to be made in this area, it is important that the following sorts of questions be addressed:

- What needs to be done to achieve greater consistency between countries in defining and delineating their human settlement patterns?
- How far has there been an attempt to follow the Goldstein Committee's recommendation that efforts should be made to develop standard procedures and definitions for groups of countries at similar levels of development and cultural backgrounds?
- Why has it not seemed possible to implement the Goldstein Committee's recommendation that 'three- or four-fold classifications' be adopted? (By which presumably is meant the idea of including a 'semi-urban' category so as to get away from the simple urban-rural dichotomy, and John Grauman's suggestion of a four-way split of national territory based on the dual frameworks of urbanized area and metropolitan area definitions.)
- Could more be done to go beyond the urban-rural dichotomy in reporting demographic data, most obviously by presenting data for urban settlements grouped by population size, even if a relatively high cutoff of 20,000 residents needs to be used in order to achieve comprehensive coverage across the world?
- Finally, in relation to the lower levels of the settlement system, what has come of Grauman's plea for better intelligence on the 'rural habitat', including the nature of settlement outside urban centres of 20,000 or more inhabitants?

The list of questions could go on. Surely, given the rapid pace of change in human settlement systems and the great attention given to the problems arising, the apparent lack of progress in the international definition and measurement of urbanization does not lie in any lack of interest in these types of intelligence. Has the effort made by the UN been too weak? Or have national statistical offices clung to old conventions because they may have other priorities or because, out of inertia, they have been able to ignore changes in their own settlement systems? Or perhaps, have national agencies developed alternative and better approaches to measurement, but not reported the resulting data to the UN because the latter are not consistent with the UN's current specifications?

A central goal must be to provide explanations for the stagnation that has prevailed so far and make suggestions about how best to move forward. Better ways of conceptualizing and measuring human settlement systems need to be devised. Furthermore, if such ways are developed, the issue of how best to disseminate them so that they are actually implemented needs to be considered. The subsequent chapters in this book provide a good basis for addressing these challenges.

Chapter 3

World Urbanization: Trends and Prospects

Hania Zlotnik[1]

Following the previous chapter's review of the approach used by the United Nations (UN) to compile its global set of statistics on urbanization, this chapter presents the main results of the most recent set of estimates and projections of world urbanization produced by the United Nations Population Division, namely the *2001 Revision* of *World Urbanization Prospects* (UN, 2003). As in the full report produced by the United Nations, the chapter focuses first on the estimated dynamics of urbanization during 1950-2000 and those projected for 2000-2030. The estimates and projections are based on the reported proportions of persons living in urban areas. Information is presented for the world as a whole; for its constituent More Developed Regions (MDRs) and Less Developed Regions (LDRs); and for six mutually exclusive major areas. Some comments are also made about the 228 countries and areas covered by the report. In addition, the principal findings concerning the population of cities are presented, including trends during 1975-2015 in the distribution of the urban population across five city-size groups ranging from those under 0.5 million inhabitants to 'megacities' of at least 10 million.

The chapter has two primary objectives. The first is to set the scene for the chapters that follow, especially those in Part II. Thus, it documents the truly remarkable increases in the number of people living in urban areas since 1950 and the consequent rise of the world's overall level of urbanization, which is expected to reach the 50 per cent mark in 2007. It also documents the great unevenness of experience between different parts of the world, not just between MDRs and LDRs, but also the great diversity among countries within each major area. Secondly, the chapter prompts the reader to reassess the value of using the urban-rural dichotomy to assess urbanization levels and trends. Although *World Urbanization Prospects* constitutes the only comprehensive set of estimates and projections of urban and large-city populations available – figures that have the advantage of being based on a uniform methodology – the limitations of the insights those data can provide about the complexities of urbanization occurring in different countries are clear. Both researchers and policy makers require more

---

[1] The views and opinions expressed in this chapter are those of the author and do not necessarily reflect those of the United Nations.

detailed information about the process of urbanization which, as the data presented in this chapter amply demonstrate, constitutes one of the major transformations that humanity has undergone during its life on earth.

Urbanization Trends

The Dynamics of Urbanization at the World Level

Over the course of the twentieth century, the population of the world grew more rapidly than ever before in human history, passing from 1.65 billion in 1900 to 6.06 billion in 2000. Yet, even this fourfold increase in the overall population pales compared to the thirteenfold rise in the number of persons living in urban areas that occurred during the same period. From just 220 million in 1900, the urban population of the world rose to 2.9 billion by 2000 (Figure 3.1). Clearly, the bulk of the urban population growth achieved to date took place during that century.

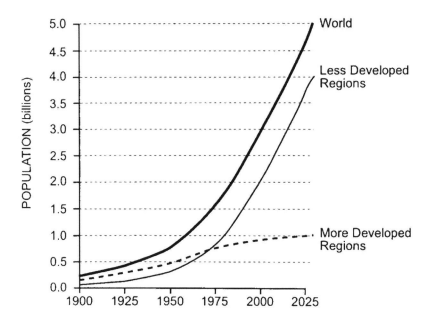

Figure 3.1   Urban population, 1900-2030

The acceleration of urban population growth started in the more developed countries, where the process of industrialization led to the concentration of population in urban centers and resulted in an early increase in the number of urban dwellers. As a result, by 1900 the urban population of the MDRs as a whole was nearly double that of the LDRs (150 million vs. 70 million) and even by 1950 the number of urban dwellers in the MDRs still surpassed that in the LDRs by a substantial margin (447 million vs. 304 million). However, the rest of the twentieth century would witness the rapid urbanization of the LDRs. By 1975 the urban population of the LDRs, having grown at an annual average rate of 3.9 per cent over the previous 25 years, had surpassed that of the MDRs. By the end of the century, the LDRs had nearly 2 billion urban dwellers, more than double the number in the MDRs, and it is expected that by 2030 the ratio will be nearly 4 to 1 (Figure 3.1).

In contrast with the rapid rise of the urban population, the growth of the world's rural population has been slower (Figure 3.2). In contrast to the thirteenfold increase in urban population during the twentieth century, the number of rural dwellers just about doubled, passing from 1.4 billion in 1900 to 3.2 billion in 2000. Most of this growth took place in the second half of the century. Moreover, all the latter was accounted for by the LDRs, as the rural population of the MDRs had in aggregate peaked by 1925. According to the projections published in the *2001 Revision*, the LDR's rural population will peak by 2025. As a result, the world's overall rural population growth during 2000-2030 will be minimal, with the number of rural inhabitants in 2030 likely to be only 94 million higher than in 2000.

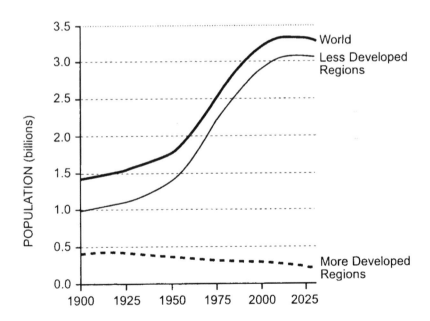

Figure 3.2   Rural population, 1900-2030

The projected stagnation of the world's rural population implies that virtually all the population increase expected during 2000-2030 will be absorbed by urban areas. The urban areas of the LDRs will account for the vast majority of that increase, continuing a long-term trend (Table 3.1). Even in 1950-1955 they accounted for 55 per cent of the annual increment of the world's urban population, but by 1995-2000 their share had risen to 93 per cent, with 96 per cent expected by 2025-2030. In terms of absolute gains, the average annual increment of the LDRs' urban population rose from 13 million in 1950-1955 to 55 million in 1995-2000, whereas in the MDRs it declined from 11 million to just 4.4 million. These trends are expected to continue in the future so that, by 2025-2030, 71 million inhabitants are expected to be added annually to the urban population of the LDRs compared to just 2.8 million added to that of the MDRs (Table 3.1).

Table 3.1   Annual increment of the world's total population and the urban populations of the world, more developed and less developed regions, 1950-2030

Period	Increment in world's total population (millions)	Urban population increment (millions)			LDRs' urban increment as percentage of:	
		World	MDRs	LDRs	World urban increment	World total increment
1950-55	47.0	24.3	11.0	13.4	54.9	28.4
1955-60	53.1	28.9	12.3	16.7	57.6	31.4
1960-65	62.7	33.4	12.5	21.0	62.7	33.4
1965-70	71.4	34.4	11.4	23.0	66.7	32.2
1970-75	74.9	37.2	10.3	26.9	72.3	35.9
1975-80	72.8	42.7	8.0	34.7	81.3	47.7
1980-85	79.0	48.9	7.2	41.7	85.3	52.8
1985-90	86.1	57.0	7.3	49.7	87.2	57.7
1990-95	81.4	55.9	5.8	50.1	89.6	61.6
1995-2000	79.0	59.3	4.4	54.8	92.5	69.4
2000-05	76.9	63.0	3.7	59.3	94.1	77.2
2005-10	76.9	67.4	3.8	63.6	94.4	82.6
2010-15	76.3	71.2	3.8	67.3	94.6	88.2
2015-20	74.4	73.5	3.8	69.7	94.8	93.7
2020-25	71.5	74.7	3.5	71.2	95.3	99.6
2025-30	66.7	74.0	2.8	71.3	96.3	106.9

Source: UN (2003).

Not only are the urban areas of the LDRs absorbing most of the population growth occurring in urban areas worldwide, they are also increasingly absorbing most of the growth of the total world population (Table 3.1, final column). Thus, whereas in 1950-1955 the increase in the population of the urban areas of the LDRs accounted for 28 per cent of the total increment to the world population, by 1995-2000 that increase accounted for 69 per cent of the annual increment to the world population and by 2020-2025 it is expected to surpass the latter. Such an outcome in 2020-2025 is consistent with a net transfer of population from rural to urban areas in the LDRs, either through migration or as a result of the territorial expansion of urban settlements and the transformation of rural villages into cities.

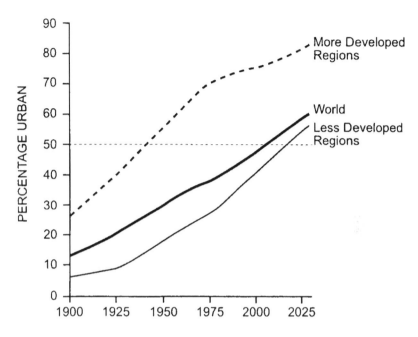

Figure 3.3   Percentage of population in urban areas, 1900-2030

In relative terms, by contrast, the LDRs have been and remain considerably less urbanized than the MDRs (Figure 3.3). In 1900, only 6.5 per cent of the population of the LDRs lived in urban areas, whereas 26.1 per cent of the population in the MDRs did so. By 1950 the difference between the two levels had widened considerably and continued to grow until 1975, when the proportion urban reached 70 per cent in the MDRs compared to under 27 per cent in the LDRs. Since 1975 the gap has been narrowing, so that by 2000 the former had 75 per cent of the population in urban areas and the latter 40 per cent. During 2000-2030, the proportion urban is expected to increase by 1.1 percentage points per year in the LDRs, more than thrice as rapidly as in the MDRs (0.3 points). As a result, by 2030 their respective levels of urbanization are expected to be 56.4 per cent and 82.6 per cent – a difference of barely 26 points, lower than at any time since 1925.

The Dynamics of Urbanization for Major Areas

Levels and trends of urbanization also vary considerably among the world's major areas. Europe and Northern America, being part of the more developed world, exhibit high levels of urbanization and slowing rates of urban population growth.

In the less developed world, Africa and Asia remain largely rural, whereas Latin America and the Caribbean, considered jointly, have a high proportion of their population living in cities. Oceania, which straddles the more and less developed world, is also highly urbanized.

High levels of urbanization do not, however, imply equally high numbers of urban dwellers. Asia, despite having the second lowest level of urbanization in the world (37.5 per cent of its population lived in urban areas in 2000), had the highest number of persons living in urban areas in 2000 (1.4 billion). Europe (with 534 million urban dwellers), Latin America and the Caribbean (391 million) and Africa (295 million) follow as the major areas with the second to fourth largest urban populations in 2000 (Figure 3.4).

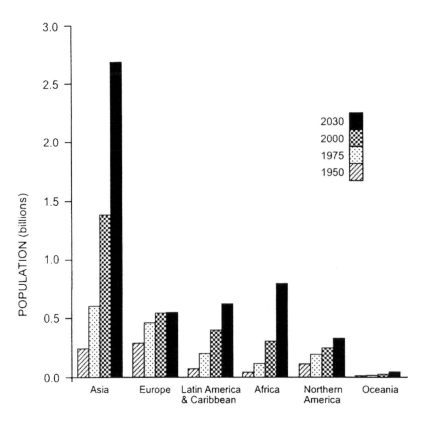

Figure 3.4   Urban population by major area, 1950-2030

Because the urban populations of Africa and Asia are expected to continue experiencing rapid growth during 2000-2030, by 2030 they will rank first and second in terms of the number of urban dwellers: 2.7 billion in Asia and 0.8 billion in Africa (Figure 3.4). In addition, their rates of urbanization will continue to be among the highest in the world (Table 3.2). Thus, the proportion urban in Asia is expected to rise 1.23 per cent per year during 2000-2030, to reach 54 per cent by 2030. For Africa, the projected rate of urbanization is 1.17 per cent per year, leading to a 2030 population that is 53 per cent urban. However, even with this rapid rise in the proportion urban, Africa and Asia will still be the least urbanized major areas of the world in 2030, although their populations are likely to have become more urban than rural by that time (Table 3.2).

Table 3.2   Level and rate of urbanization for the world and major areas, 1950-2030 (per cent)

Year/period	World	Africa	Asia	LA & C	N Amer	Europe	Oceania
Level							
1950	29.8	14.7	17.4	41.9	63.9	52.4	61.6
1975	37.9	25.2	24.7	61.4	73.8	67.3	72.2
2000	47.2	37.2	37.5	75.4	77.4	73.4	74.1
2030	60.2	52.9	54.1	84.0	84.5	80.5	77.3
Rate							
1950-1975	0.97	2.17	1.39	1.53	0.58	1.00	0.64
1975-2000	0.88	1.55	1.67	0.82	0.19	0.35	0.10
2000-2030	0.81	1.17	1.23	0.36	0.30	0.31	0.14

Level refers to the proportion of total population that is urban, rate to the annual average percentage change in level. LA & C Latin America and the Caribbean, N Amer Northern America.

Source: UN (2003).

Latin America and the Caribbean has been the most highly urbanized major area of the less developed world since 1900. During 1950-2000, its proportion urban grew at a rate of 1.2 per cent per year, more rapidly than those of Europe or Northern America. Consequently, by 2000, it had become just as urbanized as the major areas of the more developed world, with three out of every four inhabitants living in cities (Table 3.2). Over the next thirty years, however, its rate of urbanization is expected to be lower than in the past, at 0.36 per cent per year, mainly because the proportion urban is already high. That rate of urbanization would yield by 2030 a population that is 84 per cent urban, making this the second most urbanized major area of the world after Northern America, and give it the third largest urban population in the world, with 0.6 billion urban dwellers.

Europe, Northern America and Oceania were the most urbanized major areas of the world in 1950, but since then the urbanization rates of these major areas have been considerably lower than those exhibited by the major areas of the less developed world (Table 3.2). Over the second half of the twentieth century, Europe, the least urbanized major area of the developed world in 1950, experienced the highest rate of urbanization, averaging 0.7 per cent per year, about double that experienced by Northern America and Oceania (0.4 per cent per year).

The urbanization rates for the major areas of the developed world were particularly low or even negative during 1965-1985. This period is particularly associated with the phenomenon known as 'counterurbanization', which entailed shifts in population distribution down the urban hierarchy and into rural areas (Korcelli, 1984; Champion, 1998). Particularly low rates of urbanization were recorded in Northern America and Oceania during 1970-1990: 0.11 per cent annually in Northern America and –0.03 per cent annually in Oceania. Although urbanization rates were higher in Europe at this time (at 0.55 per cent per year), they dropped to 0.18 per cent per year during 1990-2000. In contrast, those for Northern America and Oceania increased somewhat during the 1990s, to 0.26 per cent per year in Northern America and 0.45 per cent per year in Oceania. During 2000-2030, all three of these major areas are expected to experience fairly low urbanization rates, at about 0.3 per cent per year for Europe and Northern America and 0.1 per cent per year for Oceania (Table 3.2).

The flip side of urbanization is the reduction of rural population growth rate. In the major areas of the less developed world, two different trends are discernible. In Latin America and the Caribbean the rural growth rate declined steadily from 1.31 per cent per year in 1950-1955 to -0.04 in 1985-1990 and is projected to remain negative during the whole projection period, averaging -0.33 during 2000-2030. In contrast, Africa's annual rate of rural growth increased steadily from 1.75 per cent in 1950-1955 to 2.27 in 1980-1985 and, although it has declined since then, it was still a high 1.57 in 1995-2000. In Asia, the rural growth rate stood at 1.54 per cent per year in 1950-1955, reached a peak of 2.16 in 1965-1970 and has been declining since then, reaching 0.57 in 1995-2000. During 2000-2030, the rural population of Asia is expected to decline at an average annual rate of -0.4 per cent, but Africa's rural population will likely continue to increase at a robust rate of 1.14. Consequently, whereas the rural population of Asia will remain virtually unchanged at 2.3 billion during 2000-2030, that of Africa will increase substantially, passing from 0.5 to 0.7 billion persons during the same period (Figure 3.5).

Turning to the more developed world, the rural population of Europe has been declining steadily since 1950 and is projected to experience an acceleration of the rate of decline, from -0.54 per cent per year during 1975-2000 to -1.31 per cent during 2000-2030. As a result, the number of rural inhabitants in Europe will drop sharply, passing from 193 million in 2000 to 131 million in 2030 (Figure 3.5). Neither Northern America nor Oceania has experienced such a sustained reduction of the rural population. In Northern America, the rural population grew at a low rate of 0.11 per cent per year during 1950-1975 but experienced a spurt of faster growth after 1975, maintaining an average annual rate of 0.44 per cent during 1975-2000. Although the rural population of Northern America is projected to decline during 2000-2030, its rate of decline is expected to be more moderate than that for Europe (-0.5 per cent per year). As a result, the rural population of Northern America is expected to be 61 million in 2030, down from 71 million in 2000. In Oceania the rural population has increased steadily since 1950, experiencing a faster rate of growth during 1975-2000 than in 1950-1975, and it is projected to continue increasing at a rate of 0.6 per cent per year during 2000-2030.

By 2030, Oceania is expected to have 10 million people living in rural areas (Figure 3.5).

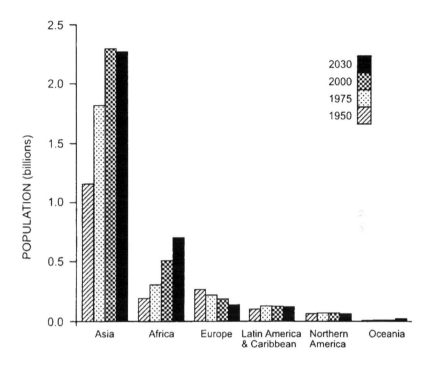

Figure 3.5 Rural population by major area, 1950-2030

In sum, there is still considerable heterogeneity in the levels of urbanization reached by the world's major areas and in the trends they expect to follow in the future. Europe, Latin America and the Caribbean, Northern America and Oceania are already highly urbanized, with more than 70 per cent of their populations living in urban areas in 2000. They are expected to experience lower urban population growth than in the past and, with the exception of Oceania, to see their levels of urbanization rise above 80 per cent. However, their urban populations taken jointly, which in 1975 surpassed that of Asia, are already about 0.2 billion below that of Asia and will not catch up with the latter during 2000-2030. In contrast to these highly urbanized areas, both Africa and Asia are still largely rural. Although their levels of urbanization are expected to increase at rates higher than one per cent per year during 2000-2030, their urban populations will constitute barely more than half their respective total populations by 2030. But despite their low levels of urbanization, Africa and Asia are projected to have the largest urban populations in 2030. In addition, the rural population of Africa will also continue to grow, thus maintaining the potential for high urban population growth beyond 2030.

Urbanization at the Country Level

So far, the discussion of urbanization trends has focused on regional aggregates
that do not reveal the diversity of experience at the country level. The changes that
have occurred in individual countries are as remarkable as those observed at higher
levels of aggregation. In 1950, for instance, only a quarter of the 228 countries or
areas in the world had more than 46 per cent of their population living in urban
areas. By 2000 half had 57 per cent or more of their population living in urban
areas and according to current projections, by 2030 over three-quarters of all
countries or areas are likely to have over half of their population living in urban
areas.

Important differences in the distribution of countries by level of urbanization
are noticeable when countries in the LDRs and the MDRs are considered
separately. As Table 3.3 shows, the upper extreme of the distributions of countries
in both groups has been 100 per cent since 1950, indicating that there is at least one
country or area in both these parts of the world that is totally urban.

**Table 3.3   Distribution of countries or areas in more developed and less
developed regions by level of urbanization, 1950-2030**

	1950	1975	2000	2030
Less developed regions				
Lower extreme	0.0	0.0	0.0	0.0
Lower quartile	9.5	19.6	32.1	49.1
Median	23.5	35.6	49.4	64.4
Upper quartile	37.9	54.0	69.5	79.3
Upper extreme	100.0	100.0	100.0	100.0
More developed regions				
Lower extreme	13.7	20.0	21.4	34.6
Lower quartile	27.8	51.5	59.7	70.7
Median	50.1	66.4	72.9	80.5
Upper quartile	69.9	81.5	88.5	91.8
Upper extreme	100.0	100.0	100.0	100.0

The upper and lower extremes indicate the highest and lowest values observed; the lower
quartile, the median and the upper quartile divide the distribution into four parts, each with
the same number of observations. Consequently, half of the observations lie between upper
and lower quartiles.

Source: UN (2003).

Except for this upper extreme, the other indicators of the distribution of
countries according to the level of urbanization are lower in value for the LDRs
than the MDRs, confirming that the process of urbanization is more advanced
among the latter (Table 3.3). In 1950, for instance, three-quarters of all LDR

countries were no more than 37.9 per cent urban, whereas three-quarters of all MDR countries were at least 27.8 per cent urban. By 1975 half of all LDR countries had a level of urbanization of 35.6 per cent or higher and 25 years later three-quarters had reached levels above 32 per cent. By 2030 three-quarters are expected to have urbanization levels of 49.1 per cent or higher. The overall trend towards urbanization is clear from the upward movement of the central part of the distribution lying between lower and upper quartiles (Table 3.3, upper panel).

Similarly in the MDRs, the central part of the distribution has shown a tendency to move upward. There, however, the main feature is the narrowing of the interquartile range, as an increasing number of countries reach very high urbanization levels (Table 3.3, lower panel). By 1975, over three-quarters of all developed countries were more than 50 per cent urban and over a quarter had urbanization levels above 80 per cent. By 2030, at least three-quarters of all developed countries are projected to be more than 70 per cent urban and at least half are expected to have 80 per cent or more of their populations living in urban areas. At the top of the distribution, at least of quarter of all developed countries will likely be more than 91 per cent urban.

High levels of urbanization are more common among small countries than among larger ones, but this is changing, as is the MDR/LDR balance of countries containing the largest numbers of urban people. Table 3.4 presents the countries accounting for 75 per cent of the world's urban population in 1950, 2000 and 2050, listing them according to level of urbanization at each of those points in time. As the world's urban population increases, so does the number of countries accounting for three-quarters of it: 17 in 1950, 25 in 2000 and 28 in 2030. In 1950, most of the countries with large urban populations were to be found in the MDRs (11 out of 17), but by 2000 the contribution of the latter was down to 9 out of 25 and by 2030 that number is projected to diminish to seven out of 28.

As Table 3.4 indicates, in 1950 the developed countries in the list tended to have considerably higher levels of urbanization than most developing countries with large urban populations. Among the latter, Argentina had the highest proportion urban, at 65.3 per cent. The other developing countries with large urban populations had levels of urbanization ranging from 12.4 per cent in Indonesia to 42.7 per cent in Mexico. China, the country with the second largest urban population in 1950, had just 12.5 per cent of its population living in urban areas at the time.

By 2000, although the developed countries with large urban populations still had levels of urbanization in the high range, several developing countries were displaying similarly high levels. Argentina, Brazil and the Republic of Korea had become more than 80 per cent urban, and Colombia and Mexico were about 75 per cent urban. The rest of the developing countries with large urban populations had levels of urbanization below 66 per cent. China and India, the countries with the largest urban populations at the time, had levels of urbanization that were still low, at 36 per cent and 28 per cent, respectively.

Table 3.4   Countries accounting for 75 per cent of the world's urban population in 1950, 2000 and 2030, ordered by level of urbanization

	1950				2000				2030		
Rank	Country	% urban	Millions urban	Rank	Country	% urban	Millions urban	Rank	Country	% urban	Millions urban
1	UK	84.2	43	1	UK	89.5	53	1	Saudi Arabia	92.6	41
2	Netherlands	82.7	8	2	Argentina	88.2	33	2	UK	92.4	57
3	Germany	71.9	49	3	Germany	87.5	72	3	Argentina	91.9	45
4	Argentina	65.3	11	4	Rep. of Korea	81.8	38	4	Venezuela	91.8	34
5	USA	64.2	101	5	Brazil	81.2	138	5	Germany	91.7	71
6	France	56.2	23	6	Japan	78.8	100	6	Republic of Korea	90.5	48
7	Italy	54.3	26	7	Spain	77.6	31	7	Brazil	90.5	205
8	Spain	51.9	15	8	USA	77.2	219	8	Colombia	84.9	53
9	Japan	50.3	42	9	France	75.4	45	9	Japan	84.8	103
10	Russian Fed.	44.7	46	10	Colombia	75.0	32	10	USA	84.5	303
11	Mexico	42.7	12	11	Mexico	74.4	74	11	France	82.2	52
12	Ukraine	39.2	15	12	Russian Fed.	72.9	106	12	Mexico	81.9	110
13	Poland	38.7	10	13	Ukraine	67.9	34	13	Iran (Islamic Rep. of)	78.8	82
14	Brazil	36.5	20	14	Italy	66.9	39	14	Russian Fed.	77.9	95
15	India	17.3	62	15	Turkey	65.8	44	15	Turkey	77.0	69
16	China	12.5	70	16	Iran (Islamic Rep. of)	64.0	45	16	Italy	76.1	39
17	Indonesia	12.4	10	17	Philippines	58.6	44	17	Philippines	75.1	85
				18	South Africa	56.9	25	18	Indonesia	63.7	180
				19	Nigeria	44.1	50	19	Nigeria	63.6	140
				20	Egypt	42.7	29	20	China	59.5	883
				21	Indonesia	41.0	87	21	United Rep. of Tanzania	55.4	36
				22	China	35.8	456	22	Egypt	54.4	54
				23	Pakistan	33.1	47	23	Dem. Rep. of the Congo	49.1	65
				24	India	27.7	279	24	Pakistan	48.9	133
				25	Bangladesh	25.0	34	25	Bangladesh	44.3	99
								26	Viet Nam	41.3	45
								27	India	40.9	576
								28	Ethiopia	31.0	39

Source: UN (2003).

In 2030, Argentina, Brazil, the Republic of Korea, Saudi Arabia and Venezuela are all expected to have populations that are more than 90 per cent urban and that are among the largest in the world. In addition, Colombia, Mexico, the Philippines and Turkey are all expected to have more that 75 per cent of their population living in urban areas. Among all countries with large urban populations, Ethiopia is expected to be the least urbanized, with 31 per cent of its population in urban areas. China is expected to have become nearly 60 per cent urban and India nearly 41 per cent. Both countries will continue to have the largest urban populations in the world, accounting jointly for 29 per cent of all urban dwellers on earth. Among the developed countries, the United States, which in 1950 had the world's largest urban population, had moved to third place on this criterion by 2000 and is expected to maintain that place through 2030. The other highly urbanized developed countries that will remain in the list of those with large urban populations in 2030 include France, Germany, Italy, Japan, the Russian Federation and the United Kingdom.

Both the unprecedented population growth that many countries experienced during the second half of the twentieth century and their rising levels of urbanization have resulted in increasing numbers of people living in urban areas. Even so, in the majority of the 228 countries or areas of the world, the number of urban dwellers remains modest. In 2000, for instance, their median urban population size was just 2.1 million persons and three-quarters of them contained no more than 7 million urban dwellers. Although the urban populations of most countries are expected to grow significantly in the future, by 2030 three-quarters of them are expected to have urban populations below 14 million, half of them below 4 million and a quarter below 378,000. In terms of monitoring world urbanization trends and urban population growth, however, it is not surprising that most attention is focused on the relatively small number of countries with large urban populations, where an individual city can contain more urban people than many entire countries, as we shall now see.

The Urban Hierarchy

We focus now on the distribution of the urban population across a hierarchy of urban agglomerations classified by population size. Analysis of this distribution sheds light on a different process of population concentration than the shift from rural to urban areas discussed above. Consideration of the urban population distributed across the urban hierarchy allows an assessment of trends toward an increasing concentration of the urban population into larger cities (Champion, 1989). *World Urbanization Prospects* produces estimates of the urban population in five urban size categories: fewer than 500,000 inhabitants; 500,000 to under 1 million; 1 million to under 5 million; 5 million to under 10 million, and 10 million or more inhabitants. Urban agglomerations belonging to the last category are labelled 'mega-cities'. Although there are no specific names for cities in other categories, those in the first category we shall call 'small cities' and those in the middle three categories 'medium-sized cities'. We begin with an examination of the distribution of the urban population across this hierarchy over the period 1975-2015, before having a closer look at the two ends of the spectrum.

The Distribution of the Urban Population

Figure 3.6 shows how the world's population divides up between the five city-size categories and the rural areas in 1975, 2000 and 2015, according to the estimates and projections reported in the *2001 Revision*.

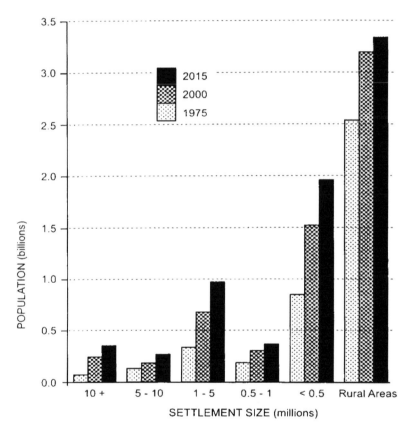

Figure 3.6   World population according to population size of settlement, 1975-2015

Between 1975 and 2000, the population living in mega-cities more than trebled, passing from 68 million to almost 225 million, and is projected to reach 340 million by 2015. By contrast, the population in cities with 5-10 million inhabitants, though larger than that in mega-cities in 1975, grew more slowly to 2000, partly because several of its original members became mega-cities over that period. The

1-5 million category accounted for more people than the two larger categories put together and displayed strong growth over the whole period. Such growth reflects the rapid population increases of individual members, plus the recruitment of members from the class below offsetting the loss of some members to the class above. The category of cities with 0.5-1 million inhabitants appears remarkably small and static compared to those above and below it. Perhaps the most remarkable feature of the distribution is the large share of cities with less than half a million inhabitants and their rapid growth, from 844 million in 1975 to 1.5 billion in 2000 and an expected 1.95 billion in 2015. Only the rural areas, among the settlement types shown in Figure 3.6, exceed the small-city category in population size and, while their absolute growth closely matched that of small cities during 1975-2000, it is not expected to do so during 2000-2015.

The top panel of Table 3.5 shows how the urban population of the world is distributed between the five size classes of cities and how these shares have shifted between 1975, 2000 and 2015. In relative terms, the small-city category is slipping somewhat over time, yet even in 2015 it is still expected to house one in two of the world's urban dwellers. Cities with 0.2-1 million inhabitants appear to be steadily losing out, with the result that cities with at least 1 million inhabitants are seeing their combined share move up progressively: from 33.8 per cent of all urbanites in 1975 to 37.3 per cent in 2000 and 40.4 per cent projected for 2015. Among the three categories of cities with a million or more inhabitants, the megacity category is clearly the main gainer, fully doubling its share between 1975 and 2015, although most of this rise was achieved before2000. During 1975-2015, the 1-5 million category experiences a steady rise in its share steadily, but the 5-10 million category oscillates, partly because of the relatively small number of cities that belong to it and the effects of their passage to megacity status.

This overall pattern is duplicated very closely by both MDRs and LDRs, as is shown by the next two panels in Table 3.5. At the upper end of the urban hierarchy, both recorded big gains between 1975 and 2000 in the share of their urban populations accounted for by megacities, partly at the expense of the 5-10 million category as some of these became megacities. In addition, both MDRs and LDRs display a progressive contraction of the proportions of their urban population living in the bottom two categories (with less than one million inhabitants) and a progressive increase in the proportions living in cities with 1-5 million people. Perhaps the most impressive feature of this comparison, however, is the relatively strong similarity in the urban distributions of the MDRs and LDRs, with both having around half their urban populations living in small cities, a quarter in the 1-5 million category, and about a tenth or less in each of the other three categories.

Somewhat greater diversity, however, is evident when disaggregating by major area of the world, especially at the two extremes of the city-size range (Table 3.5, lower panels). In Europe, for instance, there is a stronger concentration at the lower end of the hierarchy, with a majority of the urban population (at least 63 per cent) living in cities with fewer than 500,000 inhabitants. In addition, cities with 1-5 million inhabitants have attracted a growing share of the population to the detriment of larger cities. In 1975 there were five urban agglomerations in Europe with more than 5 million inhabitants, but by 2000 only four agglomerations remained in that size class.

Table 3.5   **Distribution of urban population between five population size groups, by world area, 1975-2015 (per cent)**

Area/year	Fewer than 500,000	500,000 to 1 million	1 to 5 million	5 to 10 million	10 million or more
World					
1975	54.7	11.4	21.5	7.9	4.4
2000	52.5	10.1	23.6	5.9	7.9
2015	50.4	9.2	24.8	6.8	8.8
More Developed Regions					
1975	57.5	9.3	19.8	8.5	4.9
2000	55.4	8.6	24.1	4.4	7.5
2015	54.6	7.8	25.4	4.8	7.4
Less Developed Regions					
1975	52.2	13.3	23.0	7.4	4.0
2000	51.2	10.8	23.3	6.6	8.0
2015	49.0	9.6	24.6	7.5	9.3
Africa					
1975	68.4	13.8	11.8	5.9	0.0
2000	61.4	9.0	21.8	7.9	0.0
2015	57.2	8.6	24.5	4.2	5.5
Asia					
1975	48.8	13.2	24.9	7.9	5.3
2000	49.8	10.9	23.1	6.3	9.9
2015	48.4	9.6	23.3	8.1	10.7
Europe					
1975	63.8	10.1	18.1	8.1	0.0
2000	64.3	9.1	20.5	6.0	0.0
2015	63.5	8.9	21.7	6.0	0.0
Latin America & Caribbean					
1975	55.3	9.0	16.4	8.7	10.6
2000	48.1	10.0	21.9	5.0	15.0
2015	43.5	8.9	27.5	7.0	13.1
Northern America					
1975	44.1	9.5	28.9	8.7	8.8
2000	39.1	10.8	34.9	2.9	12.3
2015	41.2	8.1	34.8	4.6	11.2
Oceania					
1975	42.6	21.5	35.9	0.0	0.0
2000	45.8	0.0	54.2	0.0	0.0
2015	44.2	4.2	51.7	0.0	0.0

Rows may not sum to exactly 100.0 because of rounding.

Source: UN (2003).

In comparison with Europe, Northern America saw the share of its population in small cities decline between 1975 and 2000. Meanwhile, its proportion living in medium-sized cities (particularly those with 1-5 million inhabitants) and in megacities has risen. However, during 2000-2015, Northern America is expected to experience a slight increase in the proportion of the population living in cities with fewer than 500,000 inhabitants and declines in the proportions living in cities with 0.5-1 million inhabitants and in megacities.

In Latin America and the Caribbean there has been a tendency for the population to become more concentrated in medium-sized cities with populations of 1-5 million, as well as in megacities. There the trend towards somewhat greater concentration of the population in the upper echelons of the urban hierarchy is expected to continue until 2015 with the exception of megacities. In Asia there is a slight trend towards a greater concentration of the population in larger cities, particularly megacities, but the distribution of the urban population by size class is expected to change little though 2015. In Africa the tendency towards a greater concentration in larger cities is clearer, especially if one considers jointly the proportion of the population living in urban agglomerations with 1 million inhabitants or more. The overall proportion in those agglomerations has increased from 4.5 per cent in 1975 to 11 per cent in 2000 and is projected to reach 15.5 per cent in 2015.

For Oceania, the changes in the distribution of its urban population by size class are difficult to interpret because they are affected by its small number of cities and the discontinuities associated with the transfer of cities from one category to the next. In 1975 its two largest cities had populations in the range of 1-5 million inhabitants. By 2000 the number of cities in that category had increased to 6 and that number is expected to remain unchanged until 2015. Those six cities accounted in 2000 for 54.2 per cent of the area's total urban population and, while this share is expected to drop to 51.7 per cent in 2015, it is indicative of the high level of concentration in just a few agglomerations.

The Important Role of Small Cities

Given that the majority of the world's urban dwellers live in cities with fewer than 500,000 inhabitants, it is important to examine the role of these settlements in terms of the whole population and, in particular, by comparison with the rural population from which small cities draw much of their dynamism through migration and territorial reclassification. While the small cities' share of the urban population is expected to be a little lower in 2015 than 2000, their share of total population is continuing to rise as the balance between rural and urban components shifts further towards the latter. Table 3.6 compares the proportion of population accounted for by small cities and rural areas in 1975, 2000 and 2015.

In both the MDRs and the LDRs, the proportion of persons living in cities with fewer than 500,000 inhabitants has been rising, reaching 41.8 per cent in the MDRs and 20.7 per cent in the LDRs by 2000 (Table 3.6). Because that trend is expected to continue, by 2015 more than a quarter of the world population is expected to live in these small cities. In the MDRs, small cities have accounted for a greater population share than rural areas since at least 1975 and by 2015 the ratio is expected to rise to two to one. In the LDRs, by contrast, rural areas will likely remain the main areas of residence, accounting for 51.4 per cent of their total population in 2015, more than double the population of the small cities there at that time.

At the level of the major areas (Table 3.6, lower panel), today a higher percentage of the population lives in small cities than in rural areas in Europe, Northern America, Oceania and Latin America and the Caribbean. By contrast, in Asia and Africa the share of the small-city population is lower than that of the rural population by a very considerable margin. Between 2000 and 2015 the absolute number of people living in small cities is expected to increase in all major areas except Europe, with the highest growth rates projected for Africa (3 per cent per year) and Asia (2.3 per cent per year). Despite such rapid growth in the population of small cities, by 2015 the number of rural dwellers in Africa and Asia is expected to remain much larger than the number of residents of small cities.

Table 3.6   Proportion of total population living in small cities and rural areas, by world area, 1975-2015 (per cent)

Area	1975		2000		2015	
	Small cities	Rural areas	Small cities	Rural areas	Small cities	Rural areas
World	20.8	62.1	24.8	52.8	27.1	46.3
More Developed Regions	40.3	30.0	41.8	24.6	43.0	21.4
Less Developed Regions	14.0	73.2	20.7	59.6	23.8	51.4
Africa	17.3	74.8	22.9	62.8	25.9	54.7
Asia	12.0	75.3	18.6	62.5	22.2	54.1
Europe	42.9	32.7	47.2	26.6	48.4	23.7
Latin America & Caribbean	34.0	38.6	36.3	24.6	35.0	19.5
Northern America	32.6	26.2	30.3	22.6	33.4	18.9
Oceania	30.7	27.8	33.9	25.9	33.6	23.9

Small cities are those with fewer than 500,000 inhabitants.

Source: UN (2003).

The World's Megacities

The emergence of the megacity was one of the distinctive traits of the process of urbanization during the latter half of the twentieth century. In 1950 there was just one city with at least 10 million inhabitants, namely New York with 12.3 million people. By 1975 the number had grown to five, and by 2000 to 16, while the numbers living in megacities more than tripled between 1975 and 2000. The *2001*

Revision projections indicate that the number of megacities will likely increase to 21 by 2015 and their population rise by 115 million (Table 3.7). Yet, despite their huge size and undoubted importance, megacities still account for only a small share of both the world population (3.7 per cent) and the urban population (7.9 per cent). Moreover, on the basis of these figures, it would appear that their growth is diminishing rapidly, with their shares of the world's total and urban population not expected to rise markedly by 2015, when they are projected to be 4.7 and 8.8 per cent respectively.

Table 3.7   Number (N) and population (Pop, in millions) of megacities, by world area, 1975-2015

Area	1975		2000		2015	
	N	Pop	N	Pop	N	Pop
World	5	68.1	16	225.0	21	340.5
More Developed Regions	2	35.7	4	67.4	4	70.6
Less Developed Regions	3	32.5	12	157.6	17	269.9
Africa	0	0.0	0	0.0	2	27.5
Asia	2	31.2	10	136.3	13	214.3
Europe	0	0.0	0	0.0	0	0.0
Latin America & Caribbean	2	21.0	4	58.7	4	66.4
Northern America	1	15.9	2	29.9	2	32.4
Oceania	0	0.0	0	0.0	0	0.0

Source: UN (2003).

The megacities are not uniformly distributed among major areas, nor are they more likely to exist in the most highly urbanized regions (Table 3.7). Even in 1975, only one megacity was located in Northern America and there were none in Europe. At that time Asia and Latin America had two each, while Asia had the largest number of inhabitants living in megacities. By 2000, Asia had come to dominate the megacity picture, with ten of the world's 16 megacities and 136 out of their 225 million combined population. Latin America had four megacities and Northern America had two. Neither Africa nor Europe had any megacities in 2000. During 2000-2015 Asia is expected to continue having more megacities than any other major area and to have the largest population living in megacities. Yet the relative concentration of the population in megacities is and will continue to be considerably lower in Asia than in most of the other major areas. Thus, just 3.7 per cent of the population of Asia is estimated to be living in megacities in 2000, much lower than the 11.3 per cent of Latin America and the Caribbean and the 9.5 per cent of Northern America.

Because megacities attract considerable attention from the media, policymakers and the public at large, there seems to be a perception that they absorb a large share of urban growth and tend to grow very rapidly. In fact, the opposite is true. An analysis of past and future growth rates of the populations of megacities has indicated that, as a city's population rises, its growth rate tends to decline (UN, 2003). So, although some of today's megacities experienced very high rates of population growth when they were still medium-sized urban centers, their rates of growth appear to moderate considerably as they approach the 10 million mark.

Table 3.8   Growth rates, 1950-2015, of 21 urban agglomerations that are expected to be megacities in 2015

Urban agglomeration	Country	Population 2015 (000s)	Average annual growth rate (%)		
			1950-75	1975-2000	2000-15
Tokyo	Japan	27,190	4.2	1.2	0.2
Dhaka	Bangladesh	22,766	6.6	7.0	4.0
Mumbai	India	22,577	3.6	3.1	2.3
São Paulo	Brazil	21,229	5.6	2.2	1.1
Delhi	India	20,884	4.6	4.1	3.5
Mexico City	Mexico	20,434	5.2	2.1	0.8
New York	USA	17,944	1.0	0.2	0.5
Jakarta	Indonesia	17,268	4.8	3.3	3.0
Calcutta	India	16,747	2.3	2.0	1.7
Karachi	Pakistan	16,197	5.4	3.7	3.2
Lagos	Nigeria	15,966	7.5	6.1	4.1
Los Angeles	USA	14,494	3.2	1.6	0.6
Shanghai	China	13,598	3.1	0.5	0.4
Buenos Aires	Argentina	13,185	2.4	1.1	0.6
Metro Manila	Philippines	12,579	4.7	2.8	1.6
Beijing	China	11,671	3.1	1.0	0.5
Rio de Janeiro	Brazil	11,543	4.0	1.2	0.5
Cairo	Egypt	11,531	3.7	1.8	1.3
Istanbul	Turkey	11,362	4.8	3.6	1.6
Osaka	Japan	11,013	3.5	0.4	0.0
Tianjin	China	10,319	3.8	1.6	0.8

Urban agglomerations are listed in order of population size in 2015.

Source: UN (2003).

This point is well illustrated by the growth rate statistics presented in Table 3.8 for the 21 agglomerations that are expected to be megacities in 2015. Having become the first megacity by 1950, New York had the lowest rate of growth among all future megacities, averaging just one per cent per year during 1950-1975. During 1975-2000, four of the megacities or future megacities (Beijing, New York,

Osaka and Shanghai, all with populations of at least 8 million inhabitants) had rates of growth lower or equal to one per cent. Furthermore, Tokyo, by then the world's largest urban agglomeration, grew at just 1.2 per cent per year during 1975-2000. Over the period 2000 to 2015, 10 of the 21 megacities of 2015 are expected to grow at rates lower than one per cent per year and, among them, Osaka is expected to exhibit zero population growth. Nevertheless, there are some cities that maintain high growth rates even when their populations have already soared. Although a moderation of the rates of growth of all current megacities is expected during 2000-2015, cities such as Delhi, Dhaka, Jakarta, Karachi and Lagos are projected to grow at annual rates of at least 3 per cent.

A similar conclusion can be drawn if attention is broadened to cover all cities that are expected to reach the 5 million mark by 2015. As a group, these have exhibited relatively low rates of growth since 1950 and are expected to grow very slowly, if at all, during 2000-2015 (UN, 2003). It thus appears that very rapid population growth is more likely to occur in smaller urban agglomerations that in those with several million inhabitants. It is for that reason that, in a world that is still mostly rural, the continued urbanization of rural settlements and the growth of smaller cities is expected to contribute a more important share of future urban growth than the population increases of the world's largest cities.

Conclusion

This chapter has presented an overview of past trends in urban, rural and city growth and outlined the urbanization trends expected in the future. Clearly, world urbanization levels increased markedly during the twentieth century and will continue to rise. For the LDRs, whose population is estimated to have been just 6.5 per cent urban in 1900, the move to a 40 per cent level in 2000 and an expected 56 per cent level in 2030 constitutes a major transformation in society and way of life. For the MDRs, the rise from barely a quarter urban in 1900 to three-quarters at the end of the twentieth century is no less impressive, but it is clear that the rate of increase has declined markedly since the 1970s. Moreover, in both the MDRs and the LDRs, nearly half of the urban population growth has been absorbed by urban centers with fewer than 500,000 inhabitants, a pattern which looks set to continue. Megacities, while attracting much attention owing to their individual size and well-publicized problems, accounted for less than 4 per cent of world population in 2000 and their share of the urban population seems likely to remain low for the foreseeable future.

The overview of world urbanization trends presented above is based on the most comprehensive set of data available on the subject. However, it must be stressed that both the accuracy of past estimates and the adequacy of projected figures depend on the quality of the basic statistics available at the country level and, as other chapters in this book discuss, there are numerous limitations in the data available. Lack of comparable definitions of urban areas between countries, inconsistencies in the measurement of urban agglomerations over time within countries, missing data, problems in coping with the natural geographical expansion of the urbanized territory, all impinge on the quality and reliability of the data presented here. Nevertheless, despite these shortcomings, the main trends reflected by the estimates and projections prepared by the United Nations Population Division are widely recognized as sound and the general conclusions derived from the global analysis presented in this chapter would probably change little if perfect data underpinned them.

Perhaps a more important drawback of the present analysis is that it cannot go any further. Given the limitations of the only data commonly available, there are many key questions regarding the urbanization process that cannot be answered even at the regional level. For instance, to what extent is the urban population increasing because of an extension of the urbanized territory around already established urban centers rather than by the development of new urban centers in formerly rural areas? To answer this question, it is necessary not only to obtain more detailed data on the population of localities but also information on the territorial location and extension of those localities. The development of guidelines on how to identify and delimit localities, and on how to encapsulate key geographical information about them in a parsimonious way amenable to comparative analysis over time and between countries, is a key task that is only just beginning to be addressed. Given the crucial impact that urbanization has already had in the social, economic and cultural fabric of societies all over the world, it is high time that the instruments needed to study its dynamics in depth were developed. This book makes a major contribution in that direction.

PART II
REGIONAL PERSPECTIVES ON
SETTLEMENT CHANGE

Chapter 4

The Fading of City-Suburb and Metro-Nonmetro Distinctions in the United States

William H. Frey[1]

In the USA and other developed countries, a strong correspondence once existed between metropolitan morphology, demographic changes, and population characteristics associated with the twin distinctions between central city and suburb within metropolitan areas and between metropolitan and nonmetro areas (Frey and Speare, 1988; Fuguitt *et al.*, 1989). This correspondence served to validate the use of metropolitan areas and subareas in both academic and policy analyses which used this classification as an underlying paradigm for assessing spatial demographic issues (e.g. policies targeted to city or rural poverty, suburban homeownership for minority populations, public transportation to permit 'reverse commuting' of city residents without cars to suburban jobs). Indeed, when the basis for the current US metropolitan definition was formulated in the late 1940s, there was a high degree of correspondence between a region's labor market area, its housing market area and local activity space. This area also tended to take on a common physical form where a highly dense core area served both integrative and distributive functions for a less dense, largely residential hinterland.

It is the contention of this chapter, however, that these synergies between function, form, and demographic attributes no longer hold, for three reasons. First, changes in transportation, communication, production technologies and the organization of production, as well as nationwide industrial shifts, have led to a decoupling of these functional and physical spaces. The expansion of existing areas and creation of new areas in a low density mode have led to a diversity of physical configurations for the daily activity space of community residents, including areas that have no discernible cores. As a consequence, the city-suburb distinction cannot be defined in a clearcut manner, and the populations within each of these two categories have become increasingly heterogeneous. Second, since the original

[1] This chapter draws from discussions and collaborations with Kao-Lee Liaw, Kenneth Johnson, Douglas Geverdt, Zachary Zimmer and especially the late Alden Speare, Jr. The author is also grateful to Cathy Sun of the University of Michigan Population Studies Center for extensive data preparation work, and to Brookings Institution in Washington DC, for research support.

metropolitan concept was defined in the late 1940s, the country's rural territory has become more strongly integrated into the national economy. Some portions of this nonmetro space have become closely tied to specific metropolitan areas, while others stand relatively isolated from metropolitan influence. Nonmetro areas that are attractive to retirees and exurbanites have grown and taken on demographic attributes different from those of declining nonmetro areas that have lost workforces associated with their earlier agricultural, manufacturing, and mining economic bases. Third, the classic role of central cities as destinations for immigrants and rural-urban migrants, who later move to the suburbs, is much less relevant today for a large number of metro areas, as entire metro areas are serving distinctly different roles as receivers of immigrants or domestic migrants.

This chapter documents the changing demographic dynamics and characteristics of US metropolitan areas and subareas, and how they no longer conform to the patterns that existed when these metro categories were originally developed. It begins with an account of how the new 'South to North' immigration to selected port-of-entry metros is changing the context for metropolitan population dynamics by creating distinctions between different types of metro areas based on their metropolitanwide migration dynamics. There follows a discussion of the changing demographic distinctions emerging within metro areas, and how they are rendering the city-suburb distinction obsolete as a means of classifying demographically distinct populations. Next, the changing demographic characteristics of metropolitan and nonmetro areas are examined in the context of the national settlement system and in light of the varied economic functions of nonmetro areas. The final section summarizes these shifts and discusses their implications for metropolitan classifications as a means of identifying areas with similar demographic dynamics and characteristics.

Distinguishing Metro Areas by Immigration and Domestic Migration

An important new demographic dynamic affecting metropolitan populations is the tendency for immigrant flows and domestic migration flows to dominate growth in different metro areas and regions. Like many other developed countries, the US has begun to experience a significant 'South to North' immigration. The destinations of these immigrants are very unevenly distributed, being concentrated primarily in large port-of-entry metro areas. Many of these areas are losing domestic migrants who are more likely to relocate in faster-growing, but smaller, metro areas, as well as in nonmetro territory. Because the immigrant flows tend to have younger age structures and higher fertility levels than the US native population, the 'port-of-entry' areas are becoming demographically distinct from the parts of the urban system that are attracting mostly domestic migrants. In the US, 'South-to-North' migration is thus impacting settlement systems in ways that further isolate immigrant groups from longstanding residents.

While immigration to the US has always been high, it has changed in both magnitude and character as the result of revisions in immigration legislation in the mid-1960s which were further modified in 1986 and in 1990 (Martin and Midgley,

1994). The increasing number of immigrants from Latin America and Asia has tended to accentuate their concentration into port-of-entry areas where there are like race-ethnic and nationality populations who can provide social and economic support as well as information about employment in the informal economy. The most recent immigrant cohorts tend to comprise a disproportionate number of labor force aged persons with at most high school educations who are best suited for lower-level service kinds of employment (Briggs, 1992). The port-of-entry cities, in their turn, tend to be characterized by a net migration loss of domestic migrants (Champion, 1994) who relocate to other types of growing metro areas. A number of explanations for this phenomenon have been put forward (Frey, 1996).

A Metropolitan Area Typology

The emergence of distinctive destinations for immigrants and domestic migrants can be seen in the typology of metropolitan areas[2] presented in Table 4.1. The nine with the largest net gain of international migration in 1990-1999, termed High Immigration metros, are sustaining all or most of their migration-related growth from immigration rather than from domestic migration. These are quite distinct from the metros classed as High Domestic Migration – those with the highest gains on domestic migration – and High Outmigration – those with the largest domestic losses – as their internal migration balance is greater than that of international migration. High Domestic Migration metros such as Atlanta, Seattle, Raleigh-Durham, and Charlotte are among the fast-rising national or regional 'command and control' corporate or banking centers with significant advanced service components to their economies. Also on this list are places like Las Vegas,

---

[2] First used in the 1950 census, the metropolitan area in the US is a functionally-based concept designed to approximate a socially and economically integrated community. As originally defined, individual metropolitan areas included a central city nucleus with a population of at least 50,000 along with adjacent counties (or towns in the New England states) that were economically and socially integrated with that nucleus, as determined by commuting data, population density, and measures of economic activity. Since that time it has been commonplace to make a distinction between the 'central city' and 'suburban' (or residual) portions of the metropolitan area. While most of the US's present metropolitan areas can still be characterized by this concept, minor modifications to the definition have been implemented to account for special cases and more complex urbanization patterns. Current metropolitan areas are designated as Metropolitan Statistical Areas (MSA), stand-alone areas; or Consolidated Metropolitan Statistical Areas (CMSAs), combinations of smaller metropolitan units (Primary Metropolitan Statistical Areas) which show commuting relationships with other such units. In 2000, there were 280 metropolitan areas (MSAs and CMSAs) which housed approximately 80 per cent of the US population; the residual 20 per cent was defined as 'nonmetropolitan'. The present analysis will follow the conventional definitions of metropolitan and nonmetropolitan with one minor exception. In the six New England states, where metropolitan definitions based on towns preclude the availability of some population data, county-based New England County Metropolitan Areas (NECMAs) are used. See Chapter 18 for further details of the US approach to metropolitan classification.

Phoenix, and Orlando – noted retirement and recreation centers which are also attracting an increasing working-age population lured by new job growth in these areas. At the other extreme, Detroit, Cleveland, and other High Outmigration metros are losing internal migrants due to more sluggish economies.

Table 4.1   Metropolitan areas classed by international and domestic migration contributions to population change, 1990-1999

Metropolitan categories/ rank		Contribution to 1990-1999 change	
		Net international migration	Net domestic migration
High Immigration			
1	New York	1,408,543	-1,913,850
2	Los Angeles	1,257,925	-1,589,222
3	San Francisco	494,189	-373,187
4	Miami	420,488	-84,884
5	Chicago	363,662	-516,854
6	Washington DC	267,175	-172,425
7	Houston	214,262	85,537
8	Dallas	173,500	235,611
9	San Diego	159,691	-139,649
High Domestic Migration			
1	Atlanta	81,037	498,283
2	Phoenix	60,800	396,092
3	Las Vegas	35,506	394,331
4	Dallas	173,500	235,611
5	Denver	50,089	200,658
6	Portland	55,583	198,896
7	Austin	27,114	168,817
8	Orlando	44,244	167,120
9	Tampa	42,088	157,209
10	Charlotte	14,719	154,320
11	Raleigh	16,269	154,049
12	Seattle	90,492	153,946
High Outmigration			
1	Philadelphia	106,951	-269,874
2	Detroit	76,185	-238,994
3	Boston	137,634	-199,506
4	Cleveland	19,705	-103,945
5	Buffalo	8,927	-82,174
6	Hartford	24,028	-79,177
7	Pittsburgh	8,681	-73,980
8	St. Louis	24,828	-71,014
9	Milwaukee	11,883	-70,223
10	New Orleans	14,128	-70,036

Metro areas are CMSAs, MSAs, and (in New England) NECMAs, as defined by OMB in June 2000; names are abbreviated. See text for definition of categories.

Source: Author's analysis of US Census Bureau County Estimates Censuses.

In contrast to these two categories, the High Immigration metros are distinct in a number of respects. First, most of them can be thought of as either global cities or national corporate headquarters and trade centers. Not only do they attract large numbers of immigrants from Latin America and Asia, but they are also centers of finance and corporate decisionmaking at a national or worldwide level. Second, it is plain that there is a strong net outmigration of domestic migrants from most of these areas and especially from those that are among the largest 'world cities'. This suggests that these areas are taking on a dual-city character (Sassen, 1991; Waldinger, 1996) in that their economic and labor force structures will become highly bifurcated between mostly well-off native-born professionals and lower-level service workers comprised largely of immigrants.

Although the metros in each category have somewhat distinct patterns, there is some overlap. Dallas appears on both High Immigration and High Domestic Migration metro lists, drawing on both sources because of the strength of the Texas economy in the 1990s and its continued attraction for migrants from Mexico and elsewhere. Secondly, two High Outmigration metros, Philadelphia and Boston, show modestly high levels of immigration which serve to cushion the effects of high net domestic outmigration. Finally, evidence from the 2000 Census suggests some dispersal of the foreign-born population to several of the High Domestic Migration metros, likely due to the employment opportunities generated in low skilled retail, service and construction industries (Frey, 2002). Among the metro areas showing greatest 1990-2000 increases in foreign-born shares are Las Vegas, Dallas, Phoenix, Atlanta, Raleigh-Durham and Denver.

The metros in each category tend to have distinct regional locations. High Domestic Migration metros are located in the South and the West regions, whereas the High Outmigration ones are located primarily in the North, New Orleans being the exception. This reflects the greater generation of employment in the 'Sunbelt' (South and West) regions than the more heavily industrialized 'Rustbelt' (North) zone, which has characterized the nation's redistribution shifts for several decades. High Immigration metros are located in each of the nation's major regions, but most are located in coastal states that have served historic roles as immigrant entry points. The role of 'chain migration', along with continued establishment of race and ethnic communities in these areas, allowed them to draw international migrants during periods, such as the 1990s, when most of the domestic migrants moved to more economically robust areas.

Race-Ethnic and Age Distinctions

The selective immigration and domestic migration patterns are shaping distinct demographies across these three types of metro areas. Table 4.2 shows that, with a few exceptions, High Immigration metros have a larger percentage of Hispanics and Asians than those in the other two categories. Aside from Washington DC and Chicago, Hispanics and Asians account for over one-quarter of the populations in each of these areas, and fully one-half in the case of Los Angeles. Moreover, consistent with the domestic migration patterns observed in Table 4.1, these metros show declines in their combined white and black populations, with six out of nine down by 9 percentage points or more.

Table 4.2   2000 race-ethnic and age characteristics, and 1990-2000 changes for categories of large metro areas

Metropolitan categories/ rank	% Hispanics and Asians		% Whites and Blacks		% Aged Under 18		% Aged 65 and Over	
	2000	Change	2000	Change	2000	Change	2000	Change
High Immigration								
1 New York	25.0	5.9	72.4	-8.2	24.7	1.7	12.7	-0.5
2 Los Angeles	50.9	9.2	46.3	-11.4	28.5	1.9	9.9	0.1
3 San Francisco	38.6	8.8	57.7	-11.9	23.6	0.6	11.1	0.1
4 Miami	42.1	7.6	55.7	-9.5	24.3	1.6	14.5	-2.2
5 Chicago	20.6	6.7	77.8	-8.1	26.9	0.8	10.9	-0.5
6 Washington DC	11.7	4.2	86.0	-6.2	25.3	1.4	10.1	0.3
7 Houston	33.8	9.6	64.6	-10.8	29.0	0.2	7.7	0.4
8 Dallas	25.3	9.9	72.9	-11.2	28.0	0.8	8.1	-0.3
9 San Diego	35.9	8.1	60.5	-10.9	25.7	1.3	11.2	0.2
High Domestic Migration								
1 Atlanta	9.9	6.2	88.5	-7.6	26.6	0.7	7.6	-0.5
2 Phoenix	27.3	8.8	69.3	-10.2	26.8	0.5	11.9	-0.6
3 Las Vegas	25.8	12.4	70.9	-14.6	25.3	1.0	11.8	0.2
4 Dallas	25.3	9.9	72.9	-11.2	28.0	0.8	8.1	-0.3
5 Denver	21.4	6.4	76.3	-8.1	25.7	-0.1	8.9	-0.3
6 Portland	13.1	6.0	83.4	-8.6	25.7	-0.1	10.7	-1.7
7 Austin	29.8	6.7	68.4	-8.1	25.4	-0.2	7.3	-0.5
8 Orlando	19.3	9.4	78.4	-11.4	24.8	1.0	12.4	-0.5
9 Tampa	12.3	4.5	85.9	-6.0	21.9	1.5	19.2	-2.4
10 Charlotte	7.1	5.2	91.6	-6.2	25.4	0.7	10.2	-0.7
11 Raleigh	9.0	6.1	89.4	-7.5	24.2	1.5	8.6	-0.9
12 Seattle	13.7	4.8	81.6	-8.3	24.8	-0.2	10.3	-0.4
High Outmigration								
1 Philadelphia	8.9	3.1	89.6	-4.4	25.3	1.0	13.5	0.0
2 Detroit	5.2	1.8	92.5	-3.6	26.4	0.3	11.7	0.2
3 Boston	9.9	3.3	87.6	-5.3	24.0	1.3	12.7	-0.1
4 Cleveland	4.1	1.2	94.4	-2.5	25.3	0.3	14.3	0.3
5 Buffalo	4.2	1.3	94.0	-2.4	24.3	0.8	15.8	0.6
6 Hartford	11.7	3.3	86.5	-4.9	24.2	1.7	14.0	0.6
7 Pittsburgh	1.9	0.6	97.1	-1.5	22.3	0.2	17.7	0.6
8 St. Louis	3.0	1.0	95.6	-2.2	26.3	0.0	12.9	0.1
9 Milwaukee	8.4	3.5	89.8	-4.6	26.5	0.0	12.5	0.1
10 New Orleans	6.5	0.7	92.0	-1.9	26.8	-1.3	11.4	0.4

Change is percentage point change since 1990.

Source: Author's analysis of 1990 and 2000 US decennial census data.

Still, there is a surprisingly large Hispanic presence in several High Domestic Migration metros. This is evident in Phoenix, Las Vegas and Orlando, which each increased their Hispanic and Asian share by over 8 percentage points in the 1990s. These three also saw their combined white and black shares fall by over 10 per cent. High Outmigration metros tend to have smaller Hispanic/Asian percentages, and 1990-2000 increases are less than in other metro categories. These metro areas are at least 85 per cent white and black in their race-ethnic composition.

Age comparisons are somewhat less distinct across these three categories. High Outmigration Migration metros have lower shares of their populations under 18 and higher shares over age 65 than the other two, while High Immigration metros tend to have the youngest age structures. Among all of the metros listed in Table 4.2, three High Immigration metros – Los Angeles, Houston, and Dallas – lead the rest, with more than 28 per cent of their populations under 18. High Domestic Migration metros also exhibit relatively young age structures, although, in some cases, this is moderated by the aging of their large middle-aged baby boom populations (Frey, 2001b).

In sum, distinct demographic processes are at work for the three categories of metros. High Immigration metros dominate with respect to attracting immigrants, and in their concentration of Hispanic and Asian immigrant minority groups. These metros will also become more distinct in terms of their socioeconomic attributes as well, displaying more bipolar distributions on income and employment attributes (Frey, 1996).

The Fading of Demographic Distinctions between Central City and Suburb

The new migration processes that are shaping these *metropolitan-level* changes overlay significant morphological and demographic dynamics that have affected *internal* shifts within individual metro areas. These changes are rendering the city-suburb dichotomy obsolete as an indicator of distinctly different demographic attributes. The 'suburbs', especially, have become much less homogeneous, serving to support our contention that the latter dichotomy is fading.

The Outward Spread of Population and Jobs

Over the last three decades of the twentieth century, the expansion of metropolitan population was the result of the continued lateral spread of population into new territory around existing metro areas, together with the establishment of new metro areas in less densely populated parts of the country (Long and DeAre, 1988). From 1970, too, employment deconcentration grew in pace and widened in scope. It was during the 1970s that the balance of metropolitan jobs shifted from the central city to the suburbs in many older metro areas. Also then, for the first time, nonmanufacturing jobs suburbanized faster than manufacturing jobs in these older areas (Frey and Speare, 1988). These included many white-collar office and service-industry jobs, heralding the beginning of the 'suburban office boom' (Cervero, 1986).

Hartshorn and Muller (1986) characterized the 1970-80 decade as a period of 'catalytic growth' for suburban downtowns (following the pre-1960 'bedroom community' and 1960-70 'independence' stages). During this stage, suburban employment clustered in various types of places classed as: suburban freeway corridors, retail strip corridors, high-technology corridors, regional mall centers, diversified office centers, large-scale mixed use centers, old town centers, and suburban specialty centers. Although there had been some development of regional shopping centers, industrial parks and office parks in the 1960s, the widespread growth of these suburban employment sites accelerated during the 1970s. Stanback (1991) contends that these suburban employment changes are associated with a new era of metropolitan economic development wherein suburban employment centers have begun to compete with historical central cities and are becoming more independent.

Two investigations of suburban employment patterns suggest that local labor markets can be recognized within the broad expanse of suburbia. Cervero (1989) identified 57 Suburban Employment Centers (SECs), each of which had more than one million square feet of office floor space and two thousand or more workers and was located more than five miles from the historic CBD. Areas with the greatest concentrations of jobs were classed as 'office growth corridors', 'subcities' and 'large mixed-use developments', with an average of 234,000, 33,500 and 27,500 jobs, respectively. Secondly, Garreau (1991), who labels his suburban centers 'edge cities', found 203 such areas within the boundaries of 36 major metros. These are characterized by more than 5 million square feet of leasable office space, more than 600,000 square feet of retail space and a high employment/population ratio. They are also perceived locally as a single-end destination for mixed use (jobs, shopping, entertainment) and over the last 30 years have transformed from residential or rural in nature to mixed use.

Heterogeneous Suburban Populations

These contemporary redistribution shifts render the city-suburb model less useful for distinguishing socioeconomic and demographic settlement patterns. The 1950s distinction between suburban populations oriented toward 'familism' and a more heterogeneous central-city population has broken down, as suburban populations have taken on much more of an 'urban' character. Particularly in the last three decades, the portions of metro areas lying outside their central cities have become markedly more heterogeneous in terms of age structure, racial-ethnic composition, social status and household type.

Of course, within this broad spatial category, one finds the usual clustering of population characteristics across smaller communities (Muller, 1981). Yet even these configurations do not conform to the kinds of distance-based or sectoral models that urban sociologists and geographers showed to be consistent with core-hinterland development in earlier times (Schnore, 1965; Hawley, 1971; Johnston, 1971). Detailed examinations of tract cluster variations on a range of 1980 population and housing characteristics, in selected metro areas, indicate that neither central-city/ring nor urbanized-area/ring dichotomies are ideal ways of distinguishing intrametropolitan differences (Treadway, 1990, 1991).

Indeed, the social geography of many settlement areas has now evolved to a situation where it is the central city rather than suburbia that is more homogeneous in its sociodemographic makeup. This characterization is most applicable to large older industrial central cities that have served, historically, as destinations for immigrants from abroad or the black rural-to-urban migrants. These central cities, whose physical configurations most closely approximate the classic model, have been sustaining race- and class-based population declines for decades. As a result, these cities' social and demographic compositions are decidedly unrepresentative of the broader metro area.

An Alternative Classification of Subareas within Metro Areas

The evidence cited above argues for a classification scheme that recognizes analytically meaningful categories within the broad expanse of territory classed simply as 'noncity part of the metro areas' under the present statistical system. One approach is an 'extended suburban typology' that distinguishes employment centers from residential suburbs on the basis of their employment/residence ratios, and residential suburbs from low-density areas on the basis of their population sizes and densities (Speare, 1993; Frey and Speare 1995). This differentiates six community types, with definitions as follows:

1. *Major city* (referred to below as 'city'). The largest city; or pair of cities if the second city has at least 25,000 people and is at least one third of the population size of the largest city and is adjacent to it.
2. *Inner employment centers* (or 'inner center'). Places located within 10 miles of the center of the major city, either with at least 25,000 people and more people working in the place than living there, or with at least 10,000 people and more than 40 per cent of the resident workers working in the place.
3. *Outer employment centers* (or 'outer center'). Same criteria as 'inner center' apart from being located 10 miles or more from the center of the major city.
4. *Inner residential suburbs* (or 'inner suburb'). Areas located within 10 miles of the center of the major city, and with populations of 10,000 or more, that have densities of at least 1,000 per square mile but do not meet the criteria for 'inner center'.
5. *Outer residential suburbs* (or 'outer suburb'). Areas located 10 miles or more from the center of the major city, and with populations of 10,000 or more, that have densities of at least 1,000 per square mile but do not meet the criteria for 'outer center'.
6. *Low density areas.* Places with fewer than 1,000 persons per square mile, the residual parts of counties outside places of 10,000 or more, or whole counties with no places over 10,000 people.

The discriminatory power of this classification scheme can be demonstrated by an examination of population characteristics for the different types of suburban communities in selected metro areas. For this purpose, one metro is chosen from each of the three categories shown in Tables 4.1 and 4.2: Los Angeles as the example of a High Immigration metro, Atlanta for the High Domestic Migration metro, and Detroit as the High Outmigration metro. In a sense, Los Angeles and Detroit represent two extremes. Los Angeles features the most ethnically diverse population in the US, high levels of immigration and a growth history that does not sharply distinguish between central city and suburbs. By contrast, Detroit is known to be one of the most racially segregated metropolitan areas in the country, with

sharp distinctions between a largely black central city and largely white suburbs
(Frey and Farley, 1996). In between, Atlanta is an area gaining mostly from
domestic migration and attracting large numbers of middle-class blacks. While an
older city, which was also highly segregated, it has shown recent tendencies toward
greater minority suburbanization and reduced neighborhood segregation.

The six zones of the extended suburban typology for these three metros are
depicted in Figure 4.1. One feature of Los Angeles is its large number of outer
centers and outer suburbs, representing the larger commuting fields in this low-
density metro area. A distinguishing feature of Atlanta is that a large part of its
population lies in the low-density category, reflecting its new outward spread from
an initially concentrated inner-city area. Detroit, on the other hand, has a larger
number of inner centers and inner suburbs than the other two, although there is also
growth in its periphery.

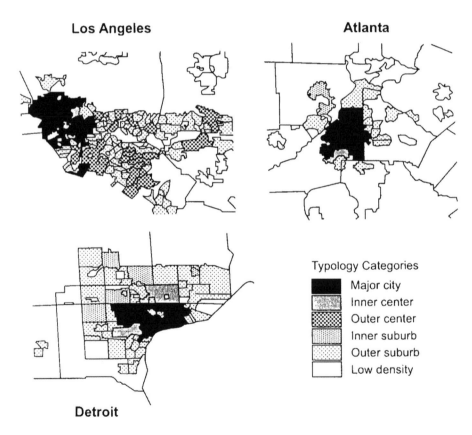

Figure 4.1   Extended suburban typology

Population Growth and Race Characteristics

Table 4.3 presents information on the relative growth and racial profiles for each zone of the metropolitan areas, using metropolitan CMSA and MSA definitions and data from the 1990 Census because the relevant data were not available from the 2000 Census at the time of writing. In general, for all three metros, it is the outer centers, outer suburbs, and low-density areas that show the highest rates of growth or, in the case of Detroit, lowest rate of decline. An important departure from this pattern is shown by Los Angeles' significant growth in its major city, with immigrants tending to reside there. Indeed, immigration contributed to all of Los Angeles' zones, especially in the outer centers and outer suburbs. In contrast, Atlanta showed population decline in its city and inner center, whereas its other zones all showed growth, with the highest rate of increase in the low-density zone. Detroit showed a decentralization pattern as well, but with negative metro-wide growth over the 1980-1990 period, there were declines in three of the five suburban zones. The strongest growth in Detroit is in its low-density zone, as well.

Table 4.3 **Population change and immigration 1980-1990 and race-ethnic composition 1990: extended suburban typology of Los Angeles, Atlanta and Detroit**

Metro/ Zone type	% change 1980-1990	% 1980-1990 immigration	Race-ethnic composition 1990 (%)			
			White	Black	Hispanic	Asian
Los Angeles	26	18	50	8	32	9
City	18	25	37	13	39	9
Inner Center	3	11	72	4	17	7
Outer Center	24	18	49	9	32	10
Inner Suburb	0	11	66	24	7	3
Outer Suburb	27	16	51	5	33	10
Residual	60	9	69	5	21	4
Atlanta	33	3	70	26	2	2
City	-7	2	30	67	2	1
Inner Center	-8	2	31	66	2	1
Outer Center	24	3	66	29	2	1
Inner Suburb	5	4	58	37	3	2
Outer Suburb	26	4	88	7	2	3
Residual	52	3	79	17	2	2
Detroit	-2	1	76	21	2	1
City	-15	1	21	75	3	1
Inner Center	-9	3	90	7	1	1
Outer Center	-1	3	80	14	3	3
Inner Suburb	-8	2	89	7	2	1
Outer Suburb	1	1	92	4	2	1
Residual	12	1	94	3	1	1

Source: Author's analysis of US census data.

The race-ethnic data in Table 4.3 make plain that significant segregation does exist across suburban communities. Race-ethnic concentration is least evident in multiethnic Los Angeles. Significant Hispanic populations are found in all its zones except for the inner suburbs which have a high representation of blacks. Asians are also somewhat evenly distributed across these zones except for a lower percentage in the inner suburbs. The only black concentration – in the inner suburb zone – should not be interpreted as representing widespread integration, as in fact the only two inner suburbs are 87 and 3 per cent black respectively. Still, these minorities are more spread across suburban zones in this High Immigration metro than is the case for our examples of the other two metro types.

Both Atlanta and Detroit show more highly segregated race patterns where blacks are the dominant minority. Blacks in Atlanta are more widely spread, with high proportions extending beyond the major city across inner centers, outer centers and inner suburbs. Still, this is somewhat deceptive, for while blacks comprise 37 per cent of Atlanta's inner-suburb residents, the figure is over 70 per cent for three communities and less than 5 per cent for the other two. Meanwhile, consistent with other findings for High Outmigration metro areas (Frey and Farley, 1996), Detroit shows high levels of segregation where the plurality of African Americans live in the city of Detroit itself, with some additional concentrations in outer centers. The fact that the city of Detroit is only 21 per cent white, and four of the five suburban zone types are at least 89 per cent white indicates the very high level of segregation that still exists within this metro area.

Household and Socioeconomic Characteristics

From what has just been shown on the basis of this extended suburban typology, between-zone distinctions can also be expected on family and household attributes. In particular, outer suburbs and low-density areas contain communities that are largely residential and attractive to families, while cities and suburban employment centers should be more likely to house nonfamily households and less well-off female-headed families. As shown in Table 4.4, these distinctions are apparent for both Los Angeles and Atlanta, where the representation of married-couple-with-child families is highest at the periphery. In contrast, Detroit's main distinction is between the city and all of the other zones. The city of Detroit has a higher percentage of female-headed families than married-couple families. The remaining zones do not show very great distinctions except for the residual low-density zone where about one-third of the households are traditional families.

It is also apparent from Table 4.4 that the 'familism' of the suburbs transcends ethnic divisions. For instance, the black families that reside in Detroit's suburban zones are more likely to be traditional families, these outnumbering female-headed families by a particularly wide margin in low-density, residual areas. Similar city-suburb distinctions for blacks are evident for Los Angeles and Atlanta. The Hispanic population's family patterns tend to be more 'traditional' overall, but again for all three exemplar metros it is the outer suburbs and low-density areas that are the zones with the highest proportions of married-couple families.

Table 4.4   Total, black and Hispanic household composition: extended suburban typology of Los Angeles, Atlanta and Detroit

Metro/ Zone type	% of total households			% of black households		% of Hispanic households	
	H-W w/child	Female head w/child	Non-family	H-W w/child	Female head w/child	H-W w/child	Female head w/child
Los Angeles	27	6	30	18	16	43	10
City	22	7	37	12	15	39	11
Inner Center	16	4	48	11	12	31	9
Outer Center	26	7	31	20	19	44	10
Inner Suburb	7	2	67	16	4	13	4
Outer Suburb	32	6	24	25	17	45	10
Residual	31	6	24	28	15	47	9
Atlanta	27	7	29	21	18	36	6
City	11	14	44	11	22	19	8
Inner Center	19	13	33	22	20	22	16
Outer Center	17	8	42	17	19	27	8
Inner Suburb	18	7	42	22	18	31	3
Outer Suburb	23	4	33	19	16	33	3
Residual	33	6	24	29	16	42	6
Detroit	25	8	29	14	23	31	13
City	14	19	34	13	24	23	20
Inner Center	22	5	31	8	20	25	8
Outer Center	24	7	34	20	18	31	11
Inner Suburb	22	6	32	21	19	33	12
Outer Suburb	26	5	29	18	20	35	10
Residual	32	5	24	25	15	40	8

H-W, husband and wife; w/child, with one or more own children under 18 years old.

Source: Author's analysis of US census data.

Socioeconomic differentiation exhibits a similar pattern, with status generally lowest in the central city and highest in the outer suburbs and low-density areas according to the indicators presented in Table 4.5. This is somewhat less clearcut for educational attainment, because some zones contain universities and colleges and also because young singles associated with 'city lights' tend to be more highly educated than older cohorts in more wealthy communities. In Detroit, for example, the highest percentage of college graduates exists in the outer center, where the University of Michigan at Ann Arbor is located. As regards poverty and home ownership, there is a particularly clear distinction for Atlanta and Detroit between the city, on the one hand, and the outer two zones, on the other. Here the only significant departure is the relatively low level of owners in their outer centers. For Los Angeles, the picture is broadly similar, except for the low levels of poverty in its inner center and of owners in its inner suburbs and the high poverty level in its outer center.

Mean household incomes vary across zones, as well, with the distinction being sharper within Atlanta and Detroit than for Los Angeles (Table 4.5). In the former two areas, the highest incomes are reported in the outer suburbs and low-density zones and (in Detroit) the outer centers. In contrast, Los Angeles shows general variability on this mean measure across zones with highest incomes in the inner center. Although the average incomes of each racial group differ from each other as discussed earlier, race and ethnic incomes vary across zones according to the pattern of the total population. Hence, the highest earning whites, blacks, Hispanics and Asians in Los Angeles reside in the inner center. One exception is the income of city whites in Los Angeles, who, unlike the other race and ethnic groups there, have incomes well above the metro mean. This anomaly is also the case in Atlanta where well-off whites live in the same zone as relatively low-income blacks.

Table 4.5	Socioeconomic attributes for race-ethnic groups, 1990: extended suburban typology of Los Angeles, Atlanta and Detroit

Metro/ Zone type	% college graduates	% in poverty	% owners	Mean household income ($1,000s)				
				Total	White	Black	Hispanic	Asian
Los Angeles	19	15	54	48.1	54.9	33.9	34.8	49.8
City	20	22	39	45.7	60.3	29.0	29.5	41.9
Inner Center	33	9	40	59.4	62.9	40.4	41.0	60.7
Outer Center	19	16	51	45.9	50.8	34.6	35.0	48.6
Inner Suburb	37	10	32	42.0	39.6	54.0	34.0	49.7
Outer Suburb	18	12	61	50.1	55.3	39.4	38.0	54.4
Residual	17	12	70	49.2	51.9	39.0	36.8	59.7
Atlanta	27	10	62	45.0	50.4	28.8	40.7	43.9
City	26	27	43	37.9	62.3	22.3	37.7	31.9
Inner Center	18	17	50	32.1	32.2	31.6	45.0	50.0
Outer Center	23	17	36	32.7	37.1	21.0	24.1	31.3
Inner Suburb	30	12	50	38.5	43.3	29.2	31.9	32.9
Outer Suburb	44	5	62	66.3	69.5	32.4	51.8	55.2
Residual	25	7	69	46.2	48.7	34.2	42.4	46.1
Detroit	19	13	69	42.4	46.2	27.1	37.5	55.7
City	9	32	53	25.6	27.4	25.1	24.4	26.3
Inner Center	13	11	73	38.2	39.7	19.0	36.9	38.4
Outer Center	31	12	61	45.6	47.0	38.2	36.9	47.5
Inner Suburb	17	10	74	38.6	38.8	35.7	33.4	45.8
Outer Suburb	18	7	72	44.8	45.4	30.7	43.6	53.4
Residual	22	5	78	52.2	52.1	44.0	53.1	82.2

Source: Author's analysis of US census data.

This analysis of these three metro areas suggests that an extended suburban classification is necessary to reflect their evolving internal population patterns. This kind of framework, when melded with other important jurisdictional boundaries such as suburban school districts, can drive other studies of intra-suburban race and ethnic population change and migration, in a way that improves upon current studies which treat 'the suburbs' as an undifferentiated entity. At the same time, the findings point up certain distinctions in demographic characteristics that can be linked to the different ways in which these exemplar metro areas are being affected by immigration and domestic migration flows.

The Metro-Nonmetro Dimension of the Evolving Demographic Dynamics

The other major aspect of the US settlement system that has changed in recent decades is the nature of the areas that lie outside metro areas, as currently defined. This territory is now less predominantly rural and more integrated into the national economy than was the case in the 1940s (Fuguitt *et al.*, 1989). Also, around 1970, residential and employment activities began to deconcentrate around many small and moderate-sized places, following a pattern that had previously existed only in metro areas. More recently, the concentrated clustering of immigration in certain metro areas has added a further impulse toward the deconcentration of the native-born population toward smaller metro areas and nonmetro territory (Champion, 1989, 1994; Long and DeAre, 1988; Frey and Speare, 1988, 1992; Fuguitt *et al.*, 1989; Johnson and Beale, 1995). These developments are affecting migration dynamics and population profiles in a way that further blurs the metro-nonmetro dimensions of the US settlement system.

Migration Dynamics in the Settlement System

Table 4.6 uses a nine-fold grid to examine immigration's differential impact across the settlement system and the extent to which domestic migration disperses across it. This grids crosses the three broad regions – North (combining Northeast and Midwest), South, and West – with a metro-status trichotomy comprising the High Immigration metros, all other metro areas and nonmetro territory. This confirms the point made earlier about the degree to which immigration is concentrated on a few port-of-entry areas. Nationally, 65 per cent of all the 1990s immigrants located in the nine High Immigration metros, which house less than 28 per cent of the total US population, with this source of population growth more than compensating for their high level of net loss through domestic migration.

In regional terms, were it not for attracting one-quarter of US immigration in the 1990s, the High Immigration metros in the North would have sustained overall population decline because their domestic migration loss was greater than their natural increase (Table 4.6). The even higher immigration rate of the West's High Immigration metros replaced their domestic migration losses almost exactly, and while those in the South recorded a small gain through domestic migration, again here immigration was by far the most important source of migration growth. In contrast, other metro areas in both the South and the West sustained greater growth from domestic migration than from immigration, and the nonmetro areas there even more so. This pattern is consistent with the scenario where domestic migrants are locating in less congested, lower-cost metros at the same time that immigration concentrates in a few larger metros.

Table 4.6   Demographic components of change by metropolitan status and region, 1990-1999

| Region/ | % change 1990-1999 | | | | % share of US | |
Metro status	Total	IM	DM	NI	2000 Pop	1990-1999 IM
United States						
High Immig metros	9.8	6.9	-5.7	8.6	27.7	65.1
Other metros	9.5	1.7	1.7	6.1	52.7	30.3
Nonmetro	7.1	0.7	3.4	3.0	19.7	4.6
North						
High Immig metros	5.1	6.4	-8.0	6.7	10.8	24.3
Other metros	3.8	1.2	-2.2	4.8	23.2	9.7
Nonmetro	3.8	0.3	1.4	2.1	8.0	0.9
South						
High Immig metros	15.2	6.1	0.3	8.8	7.6	14.7
Other metros	12.4	1.6	4.5	6.3	19.3	10.2
Nonmetro	7.6	0.6	4.1	2.9	8.8	1.8
West						
High Immig metros	10.8	8.2	-8.0	10.6	9.3	26.2
Other metros	18.1	3.4	5.5	9.2	10.2	10.4
Nonmetro	15.2	1.9	6.9	6.4	3.0	1.8

The two 'share' columns each sum to 100.0 for the US. IM International migration, DM Domestic migration, NI Natural increase, Pop population.

Source: Author's analysis of US Census Bureau County Estimates.

Race-Ethnic and Age Distinctions

The new immigrant minority groups are far more concentrated in the nation's nine High Immigration metros than the general population. The impact of this concentrated immigration is reflected in their race-ethnic compositions compared with other categories of areas in the settlement system. As shown in Table 4.7, nearly one-third of the population of the High Immigration metros, nationally, is comprised of Hispanics and Asians in contrast to less than 12 per cent of other metros and less than 7 per cent for nonmetro areas.

The patterns differ regionally such that all three categories in the West exhibit a greater Hispanic/Asian presence than those in the North and South (Table 4.7). This suggests some dispersal of the new immigrant minority groups away from the High Immigration metros in the West toward smaller places and some nonmetro counties. Recent findings from the US Census suggest this dispersal involves both recent immigrants to the US and the secondary migration of longer-term residents among the foreign-born (Frey, 2002). In the North and South, however, there is a sharp difference between the race-ethnic compositions of High Immigration metros and the other metros. Nonmetro areas in the North and South and other metro areas in the South are over 90 per cent white/black. At the other extreme, barely half of the population of High Immigration metros in the West is white/black.

Table 4.7 2000 race-ethnic and age characteristics, and 1990-2000 changes, by metropolitan status and region

Region/ Metropolitan status	% Hispanics and Asians		% whites and blacks		% aged under 18		% aged 65 and over	
	2000	Change	2000	Change	2000	Change	2000	Change
United States	16.3	4.5	81.2	-6.2	25.7	0.1	12.4	-0.1
High Immigration	31.7	7.5	65.8	-9.5	26.2	1.4	10.9	-0.3
Other metros	11.8	3.8	85.8	-5.5	25.6	0.0	12.4	0.0
Nonmetro	6.5	2.0	90.4	-3.3	25.3	-1.5	14.7	-0.2
North								
High Immigration	23.7	6.1	74.0	-8.1	25.4	1.5	12.1	-0.5
Other metros	5.9	2.2	92.2	-3.7	25.2	0.2	13.0	0.1
Nonmetro	2.9	1.2	95.2	-2.3	24.9	-1.3	15.4	-0.1
South								
High Immigration	25.3	7.7	72.7	-9.3	26.6	1.1	9.9	-0.3
Other metros	12.0	3.7	86.1	-5.1	25.3	-0.2	12.4	0.0
Nonmetro	6.4	2.2	91.3	-3.3	25.0	-1.5	14.5	-0.2
West								
High Immigration	46.0	9.0	50.9	-11.5	26.9	1.5	10.4	0.1
Other metros	24.9	5.5	70.8	-8.4	26.9	-0.2	10.9	-0.1
Nonmetro	16.5	2.3	75.0	-4.7	27.2	-2.0	13.0	0.1

Change is percentage point change since 1990.

Source: Author's analysis of 1990 and 2000 US decennial census data.

The differences in age structure among the nine categories of the settlement system are less sharp than those for race and ethnicity (Table 4.7). The 'youngest' areas of the system include High Immigration metros in the South and West, along with the West's other metros and nonmetro areas. Nonmetro areas in each region have the highest shares of people aged 65 and over, resulting from their sustained outmigration of younger people and gains of retirees together with relatively little infusion of young immigrants.

In sum, this section has shown that immigration over the 1990s tended to be concentrated within the High Immigration metros, although there has been some dispersal of Hispanic/Asian people to other metropolitan and nonmetro areas, especially in the West. The 'other metro areas' in the South and West have also shown large gains in their white/black populations. As a result, there is a greater representation of immigrant groups in the High Immigration metros in all three regions.

Selective Population Gains in Nonmetropolitan Areas[3]

Nonmetro areas have experienced something of a roller coaster ride, since the 'rural renaissance' was identified in the 1970s, with their overall population growth rate falling back from 13.3 per cent then to only 2.6 per cent in the 1980s. With their overall rate bouncing back to 10.2 per cent for the 1990s, nonmetro growth has again become widespread, but there is much selectivity between types of area. In particular, there is a major contrast between nonmetro counties that are adjacent to metro areas and the rest, and retirement and recreation areas are also especially dynamic.

The importance of metro-area adjacency is clear from the 1990-1995 analysis shown in Table 4.8. Adjacent nonmetro counties saw growth as strong as metro areas at this time, more than 85 per cent of them gained population then, and their net migration gain exceeded that in metro areas by a substantial margin. Even among more remote nonmetro counties, recent population gains were much greater than during the 1980s. Growth occurred in 68 per cent of these counties in the early 1990s, compared to 36 per cent during the 1980s. Overall, these remoter counties recorded a further migration turnaround, with their net gain of 2.3 per cent for 1990-95 contrasting markedly with their net loss of 5.2 per cent for the 1980s.

Table 4.8   Population change, net migration and natural increase, by 1993 metropolitan status and adjacency, 1990-1995

	Nonmetro			Metro	Total US
	All	Nonadjacent	Adjacent		
Number of counties	2,304	1,297	1,007	837	3,141
Population 1990 (000s)	50,820	22,669	28,151	197,893	248,718
Population change					
000s	2,580	989	1,591	11,456	14,037
% change	5.1	4.4	5.7	5.8	5.6
% counties gaining	75.3	67.5	85.4	90.7	79.4
Net migration					
000s	1,555	529	1,026	2,873	4,429
% change	3.1	2.3	3.6	1.5	1.8
% counties gaining	66.8	59.4	76.4	73.7	68.6
Natural increase					
000s	1,025	460	565	8,583	9,608
% change	2.0	2.0	2.0	4.3	3.9
% counties gaining	74.3	67.2	83.4	96.3	80.1

Net migration combines international and domestic.

Source: Frey and Johnson (1996).

---

[3] Much of this section is taken from Frey and Johnson (1996), based on more extensive work by Johnson and Beale (1995).

The other categories of nonmetro counties that grew substantially were destinations for retirement-age migrants and centers of recreation. All 190 nonmetro counties classed as retirement destinations gained population and almost all had net inmigration between 1990 and 1995 (Table 4.9). Such areas are located in the Sunbelt, coastal regions, and parts of the West and in the Upper Great Lakes. They were attracting retirees, while retaining their existing population (Fuguitt and Heaton, 1993). Population gains also occurred then in 92 per cent of the 285 recreational counties, with a large majority receiving net inmigration. Such counties had been prominent growth nodes during the 1970s and 1980s and this trend clearly persisted in the early 1990s (Johnson and Beale, 1995).

Table 4.9   Population change, net migration and natural increase, by type of nonmetropolitan county, 1990-1995

Nonmetro county type	Number of counties	Population change		Net migration		Natural increase	
		% change	% gaining	% change	% gaining	% change	% gaining
All non-metro	2304	5.1	75	3.1	67	2.0	74
Retirement	190	13.8	100	12.2	98	1.6	64
Federal lands	269	12.1	94	8.8	87	3.3	84
Recreational	285	9.7	92	7.6	88	2.2	79
Manufacturing	506	4.6	90	2.6	76	2.0	91
Commuting	381	6.9	90	5.0	85	1.9	83
Government	242	5.4	88	1.8	74	3.6	83
Service	323	7.3	85	5.6	76	1.7	74
Nonspecialized	484	5.2	81	3.7	75	1.5	74
Transfer	381	4.8	77	3.6	71	1.3	66
Poverty	535	4.3	75	1.6	60	2.7	83
Mining	146	2.7	64	0.4	53	2.3	82
Low density	407	5.9	54	2.8	46	3.1	64
Farming	556	3.2	50	1.6	46	1.6	54

1993 metropolitan definition. 14 previously metro counties are excluded. % change is aggregate change for all cases in category. % gaining is share of counties gaining population from the specified component. Recreational counties as defined by Johnson and Beale (1995). Low density counties contain fewer than six persons per square mile in 1990. All other types as defined by Cook and Mizer (1994). Each county is classified into one economic type (Farming, Mining, Manufacturing, Government, Service, Nonspecialized), but other types are not mutually exclusive.

Source: Frey and Johnson (1996).

What is noteworthy with the 1990s is that nonmetro population gains were also widespread in government-dependent counties and those with concentrations of manufacturing jobs. Evidence of increasing nonmetro diversification is reflected in the fact that much of the recent growth in these manufacturing counties appears to have been fueled by jobs in sectors other than manufacturing. While population gains in manufacturing and government-dependent counties have been smaller than for recreational and retirement counties and have been more evenly balanced between natural increase and net migration, this diversification portends a longer-term growth prospects. Other county types with high growth rates fueled by net migration include those with a large proportion of their workforce commuting to jobs in other counties (including but not exclusively the metro-adjacent counties) and those with economies dominated by service-sector jobs.

At the other extreme on the ranking of nonmetro types by 1990-1995 population growth in Table 4.9 are counties dependent on farming. Only 50 per cent of these grew then and only 46 per cent had net inmigration. Natural decrease was also more common in farming-dependent counties than elsewhere. Population gains were more widespread in mining counties, but the magnitude of the gains was quite small. Migration gains occurred in only slightly over half of the mining counties. The smaller than average population gains and widespread out migration from mining and farming dependent counties of the early 1990s represents a continuation of the trends of the 1980s. Counties with histories of persistent poverty also had low growth rates during the early 1990s and, as in the case of the mining and farming counties, what growth there was came from natural increase.

On the whole, therefore, the greatest gains among nonmetro counties were largely due to their 'consumer' functions – as 'exurban' places of residence within close commuting distance of metro areas; or as places for retirement or recreation with high natural amenities. The sociodemographic attributes of these counties have more in common with metropolitan suburban populations than with slow-growing or declining counties with the more traditional rural 'production' functions of agriculture, mining, and manufacturing.

Conclusions

As with many developed countries, metropolitan and nonmetro areas in the United States have undergone significant morphological and demographic changes since 1950 when its system of metropolitan-area standards was instituted. Metropolitan growth has deconcentrated markedly within the older parts of the country, and has spread into less developed, less dense areas and regions. The outward, suburban, spread of more diverse population groups and economic activities has created the need for new settlement categories pertaining to new activity spaces and local labor market areas. Nonmetro areas have become more diversified economically and more fully integrated into the national economy.

Moreover, in recent decades the US and other developed countries have increasingly become destinations of large flows of 'South to North' international migration, and these flows have been directed primarily to only a subset of metropolitan areas. Within these new port-of-entry metro areas, the impact of immigrant populations is borne in both the city and suburban subareas. At the same time, the native-born population, led by new domestic migration flows, has been directed to different metro areas, tending to deconcentrate to smaller metros and to nonmetro areas.

As a result, with respect to spatial differences in demographic characteristics, an analogy can be drawn between the city-suburb dichotomy of the past and a new distinction between High Immigration metros and the rest of the settlement system. In the past, central cities were recipients of new immigrant groups as well as rural-to-urban migrants, while more assimilated urban and native-born groups advanced to the suburbs. Thus, the dichotomy between central city and suburb also represented a distinction between minority, foreign-born, and new inmigrant populations, on the one hand, and more middle-class native-born households on the other. In the present context, immigrant populations are directed to both cities and suburbs in port-of-entry metro areas, while the native-born and more assimilated minority populations are gravitating to different metropolitan and nonmetro areas. Hence the distinction between the High Immigration metros and the other metro-nonmetro categories also represents a distinction between different sets of population attributes.

Further, for each of these broad metropolitan contexts, demographic attributes vary considerably within metro areas. The extended suburban typology described earlier on the chapter, based on function, density and distance from city center, helps to delineate differences in demographic characteristics that were previously associated with 'urban' or 'suburban' parts of the metro area, but now distribute themselves in clusters across all parts of the area. While the overall demographic context of the metro area is becoming shaped by the macro migration forces of immigration and domestic migration, this typology provides a useful means of identifying homogenous population spaces within metro areas.

Finally, the analysis has shown that the sharp distinction that used to exist between metro and nonmetro areas, with respect to demographic attributes, has also become blurred as some nonmetro areas have become 'exurban' extensions to expanding metros and are taking on residential attributes that were previously associated with the suburbs. The growing leisure and 'footloose' retiree populations are also finding certain nonmetro areas attractive places to vacation or live, thereby also altering the more rural population characteristics of the past. At the same time, a larger number of still rural areas are losing their traditional economic bases, and house aging and declining populations that differ sharply from before. Between these extremes, there are growing nonmetro communities, also in attractive locales, which are gaining residents in new 'footloose' industries which are made possible by the telecommunications revolution.

These changes in demographic dynamics and metropolitan structure over the past 50 years would appear to render obsolete the correspondence between metropolitan area morphology, function, and demographic structure. Indeed, one proposal for altering the basis for metropolitan area definitions in the US called for ignoring both morphology and demographic characteristics, and delineating 'community areas' simply on the basis of function, using commuting clusters, as their basis of definition (Frey and Speare, 1995). While this approach may be a radical departure from the past, it does reflect the reality that settlement systems in twenty-first century postindustrial societies may no longer correspond closely to either physical form or demographic attributes. Some vestiges in the past, such as the extended suburban community typology proposed above, can be used to maintain some semblance of the old correspondences. But there is a clear need to distinguish between separate functional aspects of the settlement system from its morphology and demography, in ways that can still be useful to planners, policymakers and scholars who need to focus on space-based demographic issues.

Chapter 5

Population Dynamics and Urbanization in Latin America: Concepts and Data Limitations

Alfredo E. Lattes, Jorge Rodríguez and Miguel Villa[1]

This chapter analyzes and discusses the urbanization process in Latin America and the Caribbean countries within the framework of development and population dynamics. It points to a number of deficiencies in our current knowledge regarding the evolution of the settlement systems there. It goes on to suggest that this situation is highly related to the use of inadequate concepts and to limitations in the available population data collection systems. The chapter is organized in four parts. The first one comprises a historical overview of population dynamics and urbanization trends in Latin America as a whole (including the Caribbean) and in comparison with those of other world regions. Then, in the second part, the different experiences of the region's 22 largest countries are analyzed, as far as available statistics permit. The third part reviews and critiques current rural-urban classification schemes, with selected examples from national case studies. Finally, the chapter presents suggestions for the improvement of the production and dissemination of data for settlement analysis in the region.

Urbanization in Latin America: An Overview

Human development is a complex and heterogeneous social process among different societies and within them. Population dynamics and settlement systems are components of the development process. The evolution of settlement systems involves not only population changes but also the changing infrastructure that supports the population as well as the economic, political and cultural activities of the society. The different territorial areas interact through the movement of goods, capital, information and population. In other words, the study of the settlement process and, particularly, the urban system contributes to a better understanding of social change (Bourne, 1992). In this overview section, we briefly outline the

[1] The authors wish to acknowledge the assistance of Pablo Comelatto, CENEP, Buenos Aires. The chapter includes an update of Lattes (1998), using estimates and projections from UN (2001b).

historical evolution of the region's settlement and development and then go on to examine its overall urbanization experience within the global context.

Historical Background

Most of the major cities of the Latin American region were established by the sixteenth century (Eisenstadt and Shachar, 1987). The traditional colonial capitals of Mexico City and Lima reached maturity in the sixteenth and seventeenth centuries and Buenos Aires, Caracas, Santiago de Chile and Montevideo in the eighteenth century (Socolow and Johnson, 1981). More recent developments, such as the large-scale international migration during the nineteenth and early twentieth centuries, shaped the settlement system and the urbanization process of the eight more urbanized countries of the region. However, the most direct causes of the rapid urbanization and the concentration of population in a few cities of Latin America are more recent, and are linked to the two most important structural transformations undergone by societies in the region as a result of severe international crises.

The first one coincided with the prevalence of a development model based on government support for industrialization which rested on an import substitution policy that took place in the region during the 1930-1970 period and was responsible for, among other things, the rapid modernization of most Latin American countries and, also, for extraordinary rural-urban population shifts. During that period the region experienced its highest urban growth and high rates of industrialization which, combined with the increase in economic and territorial integration, led to the introduction of capitalist modes of production in rural areas.

The Keynesian model began to show signs of exhaustion and in the 1970s the collapse of the economy became evident. Latin America experienced a serious and prolonged economic recession during 'the lost decade' of the 1980s. The Economic Commission for Latin America and the Caribbean (ECLAC, 1990) estimates that at the end of the 1980s the GNP per capita in the region was the same as 13 years earlier.

The second and present phase of urbanization in the region is associated with the new development model directed at opening the national economies and promoting structural adjustment. The changes have involved, among other things, deindustrialization, growth of informal sectors, increases in urban poverty, and a decline in the attraction of people to large metropolitan areas. However, these trends do not necessarily indicate a reversal of urban concentration, as De Mattos (1998) pointed out.

Latin American Urbanization in a Global Context

As a world region, Latin America is distinctive in its combination of high level of urbanization now and its rapid pace of change over the past 75 years. In 1925 its level of urbanization was halfway between that of the developed world and that of the other developing regions (Table 5.1). Since then, the region has urbanized at a more rapid pace than Northern America and Europe, though somewhat less rapidly than Africa and Asia. While it took 75 years (from 1925 to 2000) for the level of urbanization in Northern America to rise from 53.8 to 77.2 per cent, Latin America covered the same ground in only half the time. Projections indicate that, by 2025, 82 per cent of the region's population will live in urban areas – on a par with the 83 and 81 per cent of Northern America and Europe respectively, but much higher than the 51-52 per cent of Africa and Asia.

Table 5.1   Urbanization levels for selected major world regions, 1925-2025

Region	Level of urbanization (%)				
	1925	1950	1975	2000	2025
World	20.5	29.7	37.9	47.0	58.0
More developed regions	40.1	54.9	70.0	76.0	82.3
Less developed regions	9.3	17.8	26.8	39.9	53.5
Northern America	53.8	63.9	73.8	77.2	83.3
Latin America	25.0	41.4	61.2	75.3	82.2
Europe	37.9	52.4	67.3	74.8	81.3
Oceania	48.5	61.6	71.8	70.2	73.3
Africa	8.0	14.7	25.2	37.9	51.8
Asia	9.5	17.4	24.7	36.7	50.6

Regions are ordered by level of urbanization in 2000.

Source: 1925 estimated from Hauser and Gardner (1982); 1950 to 2025 from UN (2001b).

The rapid urbanization of Latin America took place in a context of very strong demographic growth to which it was necessarily related. In particular, as shown in Table 5.2, between 1925 and 1975 the rate of total population growth was the highest of all world regions and the urban population grew more rapidly than every region except Africa. At the same time, a remarkable change occurred between the two halves of this period. Urban growth showed only slight increase between 1925-50 and 1950-75, whereas the urbanization rate decreased. Clearly enough, this is the result of the total growth rate rising faster than that of urban growth, possibly due to the effect of a slowdown in rural-urban migration being offset by increased international immigration and higher natural growth. Both these latter factors also contribute to a total population higher growth.

On the other hand, the current similarity of Latin America's urbanization level with that of the most developed regions of the world does not imply that other social and economic changes have also been achieved. Although urbanization and the concentration of population in large cities may be a prerequisite of development, it is not a sufficient condition. A recent report (ECLAC, 2000) shows that the number of poor people continues growing, particularly in the urban areas of the region, rising from 122 million in 1990 to 130 million in 1999. While in

1970 only 37 per cent of the poor people were urban residents, by 1999 the figure had increased to 62 per cent.

Table 5.2   Annual growth rates for total and urban populations, and rate of urbanization, by major world region, 1925-2025 (per cent)

Region	1925-1950	1950-1975	1975-2000	2000-2025
World				
Total growth rate	1.0	1.9	1.6	1.0
Urban growth rate	2.5	2.9	2.4	1.9
Urbanization rate	1.5	1.0	0.9	0.8
Northern America				
Total growth rate	1.3	1.4	1.0	0.6
Urban growth rate	2.0	2.0	1.1	0.9
Urbanization rate	0.7	0.6	0.2	0.3
Latin America				
Total growth rate	2.1	2.6	1.9	1.2
Urban growth rate	4.1	4.2	2.7	1.5
Urbanization rate	2.0	1.6	0.8	0.4
Europe				
Total growth rate	0.3	0.8	0.3	-0.1
Urban growth rate	1.6	1.8	0.7	0.2
Urbanization rate	1.3	1.0	0.4	0.3
Oceania				
Total growth rate	1.3	2.1	1.4	1.1
Urban growth rate	2.2	2.7	1.3	1.2
Urbanization rate	1.0	0.6	-0.1	0.2
Africa				
Total growth rate	1.5	2.4	2.6	2.0
Urban growth rate	3.9	4.6	4.3	3.3
Urbanization rate	2.4	2.2	1.6	1.3
Asia				
Total growth rate	1.1	2.2	1.7	1.0
Urban growth rate	3.5	3.6	3.3	2.3
Urbanization rate	2.4	1.4	1.6	1.3

Regions are ordered by level of urbanization in 2000.

Source: 1925 estimated from Hauser and Gardner (1982); 1950 to 2025 from UN (2001b).

Diversity among 22 Latin American Countries

This section examines the varied experiences of urbanization across Latin America, focusing on the 22 countries with at least two million inhabitants in 2000. The major part of this review deals with similarities and differences in national urbanization levels and trends since 1950, the demographic components involved in this, and the structure of the settlement system in terms of the role of large cities, based primarily on the international data sets provided by the UN. After this, a variety of studies are drawn upon to examine the expansion of metropolitan areas and the population changes taking place in rural areas and in the demographic frontier zones.

Urbanization

There is a high level of diversity in urbanization levels and trends, largely reflecting differences between nations in the degree and nature of development. In 1950 only three of the 22 countries studied here (Uruguay, Argentina and Chile) had more than 50 per cent of their total population residing in urban areas, whereas 18 did so by 2000 (Table 5.3).

Table 5.3   Urbanization level for 22 countries of Latin America, 1950-2030

Country	Level of urbanization (%)								
	1950	1960	1970	1980	1990	2000	2010	2020	2030
Uruguay	78.0	80.1	82.1	85.2	88.7	91.2	93.0	94.1	94.7
Argentina	65.3	73.6	78.4	82.9	86.5	89.9	92.0	93.1	93.9
Venezuela	46.8	61.2	71.6	79.4	84.0	86.9	89.1	90.7	91.8
Chile	58.4	67.8	75.2	81.2	83.3	85.7	87.8	89.5	90.7
Brazil	36.0	44.9	55.8	66.2	74.7	81.3	85.2	87.3	88.9
Cuba	49.4	54.9	60.2	68.1	73.6	75.3	77.3	79.7	82.3
Puerto Rico	40.6	44.5	58.3	66.9	71.3	75.2	78.5	81.3	83.6
Mexico	42.7	50.8	59.0	66.3	72.5	74.4	76.7	79.3	81.9
Colombia	37.1	48.2	57.2	63.9	69.5	73.9	77.6	80.5	83.0
Peru	35.5	46.3	57.4	64.6	68.9	72.8	76.3	79.3	81.9
Ecuador	28.3	34.4	39.5	47.0	55.1	65.3	73.1	77.8	80.6
Dominican R.	23.8	30.2	40.3	50.5	58.3	65.1	70.5	74.5	77.7
Bolivia	37.8	39.3	40.7	45.5	55.6	62.5	67.8	72.1	75.7
Panama	35.8	41.3	47.7	50.5	53.7	56.2	59.6	64.0	68.6
Nicaragua	34.9	39.6	47.0	50.3	53.1	56.1	60.3	65.1	69.5
Jamaica	26.7	33.8	41.5	46.8	51.5	56.1	61.0	65.9	70.3
Paraguay	34.5	35.6	37.1	41.7	48.7	56.0	62.3	67.3	71.5
Honduras	17.6	22.8	28.9	34.9	41.8	52.7	61.2	66.7	71.0
Costa Rica	33.5	36.6	39.7	43.1	45.8	47.8	51.2	56.0	61.4
El Salvador	36.5	38.4	39.4	41.6	43.9	46.6	51.0	56.6	62.0
Guatemala	29.5	32.5	35.5	37.4	38.1	39.7	43.5	49.4	55.4
Haiti	12.2	15.6	19.8	23.7	29.5	35.7	42.3	48.8	54.9
Total	41.4	49.3	57.5	65.0	71.1	75.4	78.6	81.1	83.3

Countries are ordered by level of urbanization in 2000.

Source: UN (2001b).

The high concentration of population and the positive association between population size and urbanization level means that the trends of the region overall have been, to a great extent, the trends of a small number of countries. In 2000, more than 80 per cent of total population and more than 85 per cent of urban population of Latin America was accounted for by just eight countries: Argentina, Brazil, Chile, Colombia, Ecuador, Mexico, Peru and Venezuela. These, together with Uruguay, Cuba and Puerto Rico, are also the most urbanized countries of the region.

If we classify the 22 countries into five categories based on their urbanization level in 2000, we obtain the groups shown in Figure 5.1. This figure clearly exposes a double convergence, both among the groups of countries as well as within each group. It is also observed that such convergence will continue in the next three decades.

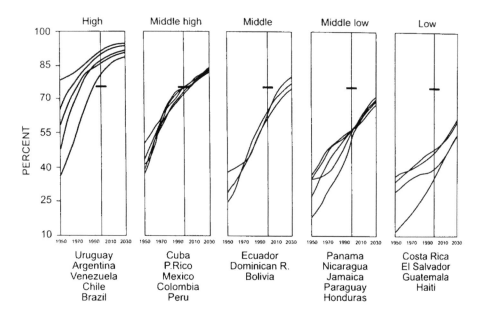

The vertical line in each panel represents the year 2000 and the short horizontal bar represents the urbanization level of the whole region in that year. Countries are listed in declining order of urbanization level in 2000, from left to right and within each panel.

Figure 5.1   Latin America: groups of countries by level of urbanization, 1950-2030

Urban Growth and the Role of Migration

Migration has played different and changing roles as a demographic component of urban growth, of city growth and of urbanization. From the estimates shown in Table 5.4 we can conclude that, in general, the net rural-urban transference of population (including the effect of international migration and the reclassification of localities) has been diminishing over time in its contribution to the urban growth of the 22 countries. During the 1950s it accounted for 46.4 per cent of the urban growth of the region, whereas by the 1990s the proportion had fallen to 38.4 per cent.[2] At the same time, the role of migration varies greatly between countries. During the 1990s, for instance, it ranged from 8.8 per cent in Guatemala to 51.7 per cent in Honduras. The cases of Mexico (-7.9) and Cuba (-5.4) are explained by the fact that their urban migration balances were affected by international migration.

Table 5.4   Rural-urban net transference as a component of urban growth and urbanization, for 22 countries of Latin America, 1950-2000

Country	Urban growth due to rural-urban net transference (%)					Relationship between rural-urban net transference and urbanization (%)				
	1950-1960	1960-1970	1970-1980	1980-1990	1990-2000	1950-1960	1960-1970	1970-1980	1980-1990	1990-2000
Uruguay	27.8	9.0	-42.2	25.9	24.2	155.1	46.1	-84.0	67.3	85.4
Argentina	51.0	37.9	31.1	30.2	27.6	128.8	128.1	119.7	132.6	120.0
Venezuela	56.9	39.4	43.2	22.1	13.7	136.4	124.0	182.0	122.4	98.1
Chile	41.3	33.6	30.2	11.8	16.3	102.0	104.6	92.8	88.6	101.9
Brazil	49.7	51.6	49.9	42.8	34.5	113.7	115.3	117.3	111.3	91.2
Cuba	39.2	16.7	43.9	45.7	-5.4	104.6	52.0	89.9	98.3	-18.2
Puerto Rico	-85.1	52.2	47.6	21.2	36.3	-141.6	78.3	103.4	53.8	99.1
Mexico	-40.9	36.1	32.1	21.6	-7.9	106.4	109.4	109.9	72.2	-59.9
Colombia	50.5	37.6	36.6	33.0	30.8	104.5	99.7	111.8	114.1	125.1
Peru	56.8	50.9	37.6	26.2	14.8	110.6	115.7	123.3	114.1	61.5
Ecuador	48.2	39.0	46.7	48.3	50.5	112.1	120.8	122.5	123.5	111.0
Dominican R.	50.2	53.3	51.5	41.9	35.3	113.3	108.5	106.9	105.6	91.7
Bolivia	8.2	11.1	34.7	48.3	36.2	52.5	79.6	109.6	96.3	108.3
Panama	36.6	36.6	23.0	25.3	20.4	104.4	109.0	126.8	108.6	97.8
Nicaragua	31.5	39.8	17.7	1.0	10.3	107.4	111.3	100.9	5.6	61.6
Jamaica	35.4	19.1	15.8	15.1	12.0	57.3	31.5	33.0	31.7	24.0
Paraguay	-62.2	-14.4	37.0	45.7	42.2	-517.2	-98.6	124.1	132.7	120.4
Honduras	53.3	48.3	44.1	45.5	51.7	115.9	109.0	116.7	121.2	112.6
Costa Rica	23.3	26.1	35.1	35.8	42.9	118.0	131.5	150.6	208.7	305.9
El Salvador	10.2	13.0	1.2	-52.2	16.0	66.8	172.2	6.5	-156.1	69.6
Guatemala	28.5	26.1	5.9	-10.9	8.8	113.8	105.4	35.8	-157.6	65.8
Haiti	62.6	58.5	52.6	61.1	50.1	100.2	99.7	105.6	125.6	94.3
Total	46.4	45.8	42.3	41.6	38.4	115.3	123.8	123.5	133.6	145.9

Countries are ordered by level of urbanization in 2000.

Source: UN (2001b).

The direct contribution of the net transference of rural-urban population to urbanization has been so important that it fully accounts for the *tempo* of the urbanization of the region and of most of the countries. The right part of Table 5.4 very clearly shows that, in all five decades, the rural-urban net transference in the

---

[2] Based on a reduced number of countries and with a more refined calculation procedure, UN (1981) estimated that in the 1950s and 1960s the contributions of rural-urban migration to urban growth would have been 39 and 35 per cent respectively.

region as a whole reached values that explain more than 100 per cent of the urbanization rate.[3] The variation among the countries is again very wide.

There is now much evidence regarding the growing weight of urban-to-urban migration among the different internal population movements that occur in most of the countries of the region. The trend was already observed in the 1970s and became much more apparent in the 1980s and 1990s. For instance, in 1987-1992, almost half of the movements of residents among the states of Mexico had urban origins and destinations.[4] A similar case was verified in Brazil, where it is estimated that more than 60 per cent of the 26.9 million intermunicipal migrants in 1981-91 moved among cities (Baeninger, 1997). The phenomenon has not been fully acknowledged by politicians and technicians who still regard rural-urban migration as the main cause of 'urban problems', nor – it is fair to say – has it yet been incorporated as a central issue in migration analyses (CONAPO, 1998; Martinez, 1999; Tuiran, 2000).

The increasing prevalence of urban-urban migration has important consequences for the urban areas involved as well as for the migrants and their perception by the recipient society. In fact, there are many similarities between the characteristics of the localities of origin and destination, opposite to the pattern in rural-urban migration. Today it is thought that there is much more interurban population interchange, though it is very difficult to quantify it. Migrants are not as different from nonmigrants as they were in the past. For instance, analysis for Greater Santiago (Chile) for 1987-1992 shows that nonmigrant heads of families had 8.8 years of education on average, whereas the inmigrants had 10.7 years and outmigrants 10.6 years (ECLAC/ILPES, 2000). The data for Mexico City are even clearer and indicate a change: in 1965-1970, when rural migration to Mexico City was very intense, the migrants had a lower average level of education than the nonmigrants, but in 1992-1997 the opposite was the case (CONAPO, 1998, p.69).

On the other hand, many jobs originating in the 'new rural economy' do not lead to the permanent settlement of workers, but instead attract those residing in the cities on a seasonal or daily basis, as is the case of the floriculture in the savanna of Bogota, Colombia. This may be because these jobs demand a type of qualification usually lacking for country people, as illustrated by the expansion of forestry in Chile (Carrasco *et al.*, 1997). In turn, many residential movements from the cities to their rural surroundings constitute, rather than a return to the country, a new form of mobility among urban localities.

International migration, often forgotten in the analyses of the spatial distribution of population, has gained great importance in recent decades. This is particularly the case in Central America, where international migration is mixed with the traditional internal migration and has produced important effects on urban as well as rural communities (Lungo, 1993). Until the mid 1970s, Central

---

[3] The net rural-urban transference rate surpasses the urbanization rate due to the fact that the urban-rural differential of natural growth does not contribute, or negatively contributes, to the progress of urbanization.

[4] These data refer only to people moving to localities of 20,000 or more inhabitants and exclude the intra-metropolitan movements (CONAPO, 2001).

American migration was almost exclusively limited to intranational or intraregional movements. These were caused by localized and temporary labor demands that, with the mechanization of agriculture and the creation of an incipient industry in the urban zones, stimulated some economic sectors (Castillo and Palma, 1999). Given this scenario, many of the migrants were peasants with poor training. Since the mid 1970s this pattern has undergone a profound change: while movements related to the frontier labor markets still persist, those outside the region (especially to the United States) have gained more importance.

In the English- and French-speaking Caribbean countries, international migration is often the most relevant component of the demographic dynamics, with remarkable effects on the social and demographic urban structure. Recent studies indicate that a very high proportion of Caribbean people have relatives and friends that emigrated (predominantly to the United States or to the metropolis of the former colonies), facilitating the development of support networks as well as the promotion of migratory expectations (Thomas-Hope, 1999). The fluctuations of the urbanization level in some countries of the Caribbean mainly reflect the effect of international emigration.

In South America, three international migration dynamics have affected the patterns of the population settlement. Firstly, the massive European immigration has contributed to populating different zones of some countries (principally Argentina and Uruguay, but also Brazil and Chile) and promoted the consolidation and growth of most urban centers. Secondly, the migratory exchange among the countries of the region has impacted on the urban system. This exchange, of cultural and historical roots, is explained by the unequal levels of development of the countries and is very sensitive to economic and political conditions. Seasonal movements of agricultural laborers, occupation of the internal frontiers, expansion of some economic activities (such as tourism, commerce, transportation and services) and development of large infrastructure projects are some of the most important stimuli of trans-frontier migration. Finally, emigration outside the region, initially propelled by political upheavals (such as those occurring in the Southern Cone in the 1970s), has gained great importance in recent years. In general, this emigration goes to the developed countries and originates in the cities, with their abundant supplies of qualified personnel.

The Changing Structure of the Urban Population

The urban systems of Latin America stand out for their huge cities: in 2000, with only 8.5 per cent of total world population and 13.7 per cent of total urban population, the region had four of the 15 most populated cities of the world (São Paulo, Mexico, Buenos Aires and Rio de Janeiro). On the other hand, the degree to which the region's urban population is concentrated in the largest cities peaked about 1960 according to the UN data (Table 5.5). At that time, almost two in five urban people (38.7 per cent) were living in the 25 largest cities, and just five cities accounted for well over half of this proportion. By 2000, the share of the largest 25 had dropped to only slightly more one in three (34.1 per cent), with that of the top five falling even more steeply, down by almost a quarter over the 40 years.

Table 5.5   **Proportion of urban population living in the 5, 10, 15, 20 and 25 largest cities, Latin America, 1950-2000**

Number of cities	Proportion of urban population (%)					
	1950	1960	1970	1980	1990	2000
5	21.4	22.5	22.0	21.3	18.4	17.2
10	28.2	29.0	28.7	27.7	24.5	23.4
15	31.8	33.4	33.0	32.1	28.7	27.7
20	34.6	36.4	36.1	35.2	32.1	31.4
25	36.7	38.7	38.5	37.6	34.5	34.1

Source: UN (2001b).

Alternatively, using as our indicator of concentration the proportion of urban population residing in the largest city of each of the 22 selected countries (Table 5.6), we can observe that the regional total has been falling continuously since 1950, down from 28.7 to 24.6 per cent. Looking at the individual countries, however, the pattern is quite varied. In only four countries (Argentina, Cuba, Uruguay and Venezuela) has the concentration of urban population followed the regional trend of being at its highest in 1950. For several countries, the reduction set in around 1970: Puerto Rico, Ecuador, Bolivia, Nicaragua, Paraguay, Costa Rica. The year 1980 marked the high point for two others (Brazil and Mexico), and 1990 for two more (Colombia and Honduras). For the eight remaining countries (Chile, Guatemala, El Salvador, Haiti, Honduras, Panama, Peru and Dominican Republic), the degree to which their urban population is concentrated in their largest city still seems to be growing, albeit at very different paces. Among them, four countries (Panama, Guatemala, Dominican Republic and Haiti) stand out for their extraordinary urban concentration. From the above set of figures, it may be concluded that the degree of urban concentration in the largest city has diminished since 1970 in most of the countries of the region.

The observation that some large cities of the region have decreased their urban predominance is not new. It has been pointed out as early as two decades ago by many authors (Alberts, 1977; Landstreet and Mundigo, 1981; Urzúa *et al.*, 1982; Lattes, 1984; Gatica, 1980). Now, however, we verify that this phenomenon has reached a larger number of countries and, also, that the decrease has become more noticeable.

The reduction in the degree of urban concentration in the largest cities of the region and of its individual countries has been interpreted in at least four different ways, for example:

- as a consequence of changes in the development pattern (opening to global economic and sociocultural exchange, reduction in state spending, etc.), that strongly value primary activities at the expense of those prevailing in cities;

- as a consequence of the economic crisis of the 1980s (the 'lost decade') which affected large cities more intensely due to the fiscal restrictions that limited the action of the State, traditionally concentrated on them;
- as a long-term response to the 'urban problems' quite evident before the economic crisis of the 1980s (ECLAC, 2001a; Chant, 1999; Rodríguez and Villa, 1997); and
- as a temporary situation whereby the larger cities, in the longer term and due to the 'new economy', will regain strength and be again a concentrating pole (ECLAC/ILPES, 2000).

Table 5.6   Proportion of urban population living in the biggest city, for 22 countries of Latin America, 1950-2000

Country	Proportion of urban population (%)					
	1950	1960	1970	1980	1990	2000
Uruguay	65.3	56.8	50.7	48.8	45.3	40.6
Argentina	45.0	44.6	44.8	42.6	39.7	37.7
Venezuela	28.3	27.6	26.8	21.5	17.5	15.0
Chile	37.5	39.4	39.7	41.0	41.9	42.5
Brazil	14.8	15.0	15.0	15.5	13.6	12.8
Cuba	39.7	36.9	34.0	28.9	26.9	26.7
Puerto Rico	44.2	52.1	52.1	50.7	48.7	47.5
Mexico	24.4	28.9	30.4	31.0	25.1	24.7
Colombia	14.5	16.0	18.4	20.0	20.5	20.1
Peru	35.9	36.7	38.7	39.3	39.2	39.9
Ecuador	26.4	29.4	29.7	28.8	26.4	27.8
Dominican R.	39.2	45.6	47.1	49.6	58.6	65.1
Bolivia	25.9	28.1	30.1	29.9	28.6	28.4
Panama	55.5	60.9	63.4	62.3	65.8	73.0
Nicaragua	27.8	32.6	37.9	35.8	35.0	33.7
Jamaica*	-	-	-	-	-	-
Paraguay	43.4	47.2	51.9	51.7	45.2	41.0
Honduras	30.5	29.7	29.8	32.8	34.9	27.8
Costa Rica	63.3	62.6	63.8	61.0	55.6	51.3
El Salvador	22.8	25.0	36.9	39.5	46.2	48.1
Guatemala	48.9	41.4	35.4	29.4	50.3	71.8
Haiti	36.3	43.3	51.6	54.2	55.6	60.3
Total	28.7	28.5	28.1	27.3	25.0	24.6

Countries are ordered by level of urbanization in 2000.

* Biggest city of Jamaica, Kingston, is not included since only the cities with more than 750,000 inhabitants in 1995 are included in the data source.

Source: UN (2001b).

Another factor that has helped to lessen the attraction of large cities is the 'connectivity explosion' of the last two decades. With the construction of large highways connecting big metropolitan areas and main cities, access to the downtown areas of large cities was facilitated, and the transfer time between nodes of the urban system reduced. There was also an important expansion of medium- and long-distance transportation, most notably domestic air services. Towards the end of the 1980s, the telecommunications revolution significantly reduced the distance cost, in some cases making it practically disappear. As a result of these transformations, many activities that used to require location in the largest cities can now be done in their outskirts. In some cases this has involved the deconcentration of both activities and population, while in others it solely resulted in a bigger dispersion around a center that continues maintaining its predominance.

Much evidence exists on the impact of improved transportation infrastructure. Interconnection corridors are serving to bind Mexico City with Toluca-Lerma, Puebla-Tlaxcala, Cuernavaca, Pachuca-Tizayuca and Querétaro-San Juan in both productive and demographic terms (Hiernaux, 1998; Asuad, 2000; see also Chapter 8 of this book). In Santiago (Chile) the construction of new roads has opened traditionally agricultural valleys to residential occupation by wealthy population groups, sometimes as weekend houses but also as permanent homes for people working in Santiago. Something similar has occurred in Buenos Aires where many 'close urbanizations' and 'countries', built close to the main highways, have left behind the previous pattern of suburbanization developing exclusively within the reach of suburban trains (Torres, 2001, p.47).

Finally, the reduction of the concentrating impetus of the large cities can also be linked to the consolidation of intermediate cities (CELADE, 2001; ECLAC/ HABITAT, 2001; Chant, 1999; Jordan and Simioni, 1998). Cities with between 50,000 and 1 million inhabitants retained a high degree of demographic dynamism in the 1980s and 1990s, albeit somewhat lower than that registered in previous decades. Consequently, their role within national urban systems increased. At the same time, however, it should be remarked that several of those cities have begun to reproduce the problems of the large cities. This suggests that the intermediate city is not necessarily assured of a promising future. Their viability depends on the basis of their economies including their degree of integration to the global scenario, on the type of articulation that links them to the regional and national urban systems, and on the ability to use their comparative advantages regarding production, supply of services, availability of infrastructure, generation of knowledge and information, and living conditions.

Expansion of Metropolitan Areas

The metropolitan areas of the region are characterized by much stronger growth in their peripheral zone than in their central nucleus. This results in a great expansion of their area and leads to the need for substantial investment in infrastructure. At least three forces appear to be fuelling this process.

One is the movement of poor people who, when faced with the lack of housing and land in central zones, progressively occupy the external rings which later become integrated into the agglomeration. Contrary to what happened previously, in the 1980s and 1990s this land occupation in the periphery was State-promoted. Several studies of peripheral expansion related to the movements of poor people remark that these generated many disadvantages for them: notably, the high cost of daily movement, both in terms of money and time; the great physical and social distance from the nucleus of power and of higher socioeconomic dynamism; the difficulties of obtaining services, due either to their remoteness or to the scant resources of the peripheral local administrations; and a higher environmental vulnerability (ECLAC/HABITAT, 2001).

A second force underlying the expansion of the metropolitan periphery is the suburbanization of high-status groups – a longstanding phenomenon but one that only became really visible in the 1990s. Such groups move their residence to rural environments that possess urban facilities and are self-sufficient and exclusive. While this originated in the desire to avoid 'urban problems', it is also explained by the emergence of a young and 'winning' segment, socioeconomically speaking, that looked for spacious and safe housing alternatives. The car, along with new road axes, has strongly increased the connectivity of these wealthy suburbs with the center of the city. Argentina is one of the countries where this phenomenon is most evident. In other countries, the spatial redistribution of people adopted a 'spreading' towards semi-rural zones close to their traditional areas of residence. Such is the case of several condominiums in Chile.

The ever-widening peripheral spaces are another propelling force of physical metropolitan expansion (Aguilar, 2000; Rodríguez and Villa, 1997). De Mattos (2001), among others, reckon this to be a distinctive feature of 'postindustrial' cities that may imply the gradual adoption of a spatial model ('urban archipelago') similar to that of Los Angeles. This new configuration of 'extended metropolitan areas' is neither limited to the incorporation of new territories nor to the enlargement of the road network: it also implies industry and other activities moving to the periphery. Thus, in the areas surrounding the central city, a number of subcenters with relatively autonomous social dynamics are generated, receiving daily flows of workers from the central city. This phenomenon combines the effects of the operations of the market agents with those related to the processes of unregulated settlement. The term 'concentrated deconcentration' has been used to denote this shift of growth to towns and secondary cities located within the wider metropolitan region but at some distance from the main center (Champion, 1997).

An image of how the temporal-spatial dynamics of the populations of the three largest metropolitan areas of Latin America have developed is provided by Rodríguez and Villa (1997). From Table 5.7, we can see that:

- In Mexico and Buenos Aires, the peripheral areas are more dynamic than the agglomerated municipalities, though in 1990 they still represented a smaller proportion of the extended metro area than the latter in both cases. In São Paulo the periphery, while less dynamic, was accommodating a slightly higher proportion of the population of the extended area by then (27.6 per cent as opposed to 27.2 per cent).

• Inside all three metropolitan areas, the agglomerated municipalities are by far the more dynamic part, with the central administrative district diminishing in its relative importance in the extended metro area in each case, as expected. Buenos Aires Autonomous City shows the highest relative fall, down from 55.8 in 1950 to 23.1 per cent in 1990.

Table 5.7   Population of the extended metropolitan areas of Mexico City, São Paulo and Buenos Aires, by area components, 1950-1990

Area components	Millions			Percentage		
	1950	1970	1990	1950	1970	1990
Mexico City						
Metropolitan area	3.4	9.0	15.0	83.8	86.6	80.7
Federal district	3.1	6.9	8.2	76.0	66.0	44.2
Agglomerated municipalities	0.3	2.1	6.8	7.8	20.6	36.5
Peripheral areas	0.6	1.4	3.6	16.2	13.4	19.3
Adjacent to metro area	0.2	0.5	1.5	5.6	4.8	7.9
Non-adjacent	0.4	0.9	2.1	10.6	8.6	11.4
Extended metropolitan area	4.0	10.4	18.7	100.0	100.0	100.0
São Paulo						
Greater São Paulo	2.6	8.1	15.2	70.5	76.2	72.4
São Paulo municipality	2.1	5.9	9.5	58.3	55.6	45.1
Agglomerated municipalities	0.4	2.2	5.7	12.2	20.6	27.2
Peripheral areas	1.0	2.5	5.8	29.5	23.8	27.6
Adjacent to metro area	0.4	1.0	2.0	12.2	9.7	9.6
Non-adjacent	0.6	1.5	3.8	17.3	14.0	18.0
Extended metropolitan area	3.6	10.6	21.0	100.0	100.0	100.0
Buenos Aires						
Greater Buenos Aires	4.7	8.4	11.0	88.3	87.1	85.2
Buenos Aires autonomous city	3.0	3.0	3.0	55.8	31.0	23.1
Agglomerated municipalities	1.7	5.4	8.0	32.6	56.1	62.1
Peripheral areas	0.6	1.3	1.9	11.7	12.9	14.8
Adjacent to metro area	0.4	0.9	1.4	7.8	9.1	10.8
Non-adjacent	0.2	0.4	0.5	3.9	3.8	4.0
Extended metropolitan area	5.3	9.6	12.9	100.0	100.0	100.0

Source: Rodríguez and Villa (1997).

A type of population mobility that has gained great importance in recent decades within the metro areas themselves (Katzman, 2001) is that generated by the expulsion of poor people from 'exclusive' residential zones or by public action that builds social housing in the least expensive land of the urban periphery. On the other hand, the evidence on different countries and cities is hardly consistent. While some studies note that residential segregation can be attenuated (Sabatini, 1999; Godard, 1994), others show that intraurban movements can worsen this phenomenon. Despite the lack of appropriate data to determine a predominant trend, it is very clear that residential segregation in large cities is a disturbing issue (Clichevsky, 2000). Further to the tension caused by such contrasting usage of the land, it increases the vulnerability of settlements in the face of natural disasters and infectious diseases.

Expansion of Demographic Frontiers and Rural Dispersion

Besides the spreading of metropolitan areas, there are other forces that make people occupy less populated spaces. This expansion of the demographic frontier is very clear in the Amazon and Orinoco basins, in zones of the Argentinean and Chilean Patagonia, in the North of Mexico and along Central America coasts. This process, which continues after the reduction or disappearance of official colonization programs, responds to the attraction of abundant resources in those spaces (ECLAC/HABITAT, 2001).

Moreover, thanks to mechanisms fostering market integration, some of these zones have become active binational spaces where economic complementarity generates employment and labor movements (CELADE, 2002; ECLAC, 2001a; ECLAC/ILPES, 2000). The Paraguayan East, centered on the department of Alto Parana that doubled its share of national population from 1950 to 1990, illustrates the attraction of the international frontier which exploits natural sources of energy. Similarly, along the extended frontier between the US and Mexico, Baja California stands out due to an urban subsystem – articulated by Tijuana, Mexicali and Ensenada – that is structurally linked to the US economy and has attracted huge migrant flows from the rest of the Mexican territory (CONAPO, 1997; Cosio-Zabala, 1998). In the case of Quintana Roo, that registers even higher immigration rates than Baja California, tourism is the cause of the spectacular growth of the City of Cancun, which has risen from being a small fishing village in 1950 to an international seaside resort of more than 300,000 inhabitants.

From its rural demographic predominance of 50 years ago to its high urban level of today, Latin America is a clear example of a deep and rapid process of population redistribution. Although in many countries of the region the recent opening of national economies has brought about the revalorization of primary production for export, their effects do not seem to diminish the rural-urban shift. On the contrary, there are clear indications that the productive specialization, the incorporation of capital-intensive technology and the expansion of agricultural-industrial centers have all deepened the segmentation of the economic units and generated population movements towards the urban environment. Furthermore, the seasonal hiring of labor and the qualification profile demanded by new agricultural occupations increase the mobility of workers (even daily) from urban localities, or force them to maintain double residence both in the country and in the city (Ortega, 1992, 1998; Ramírez, 1998).

Nevertheless, about 130 million of the region's people live in rural areas. Approximately a third of them are settled in zones close to urban localities and therefore maintain intense interaction with the latter. A lower proportion is disseminated in vast 'peripheral' territories and in a multiplicity of small places, which are characterized by relatively low density, considerable distance from more populated areas, and lack of roads, transportation and communications. It is not surprising that, in these contexts of great dispersion, social interaction is minimal, and there is an acute failure to meet people's basic needs.[5] Not surprisingly, the proportion of poor people is significantly higher in rural than in urban areas (ECLAC, 2001a, 2001b), although, as mentioned above, it is the urban areas that contain the majority of the region's poor.

Definitions of Urban Population in Latin America Countries

From the outset, the UN's recommendations have left it to the individual countries to choose their criteria for defining 'urban' (UN, 1950; see also Chapter 2 in this book). It is therefore perhaps not surprising that the censuses of Latin American countries differ greatly in their definitions of urban population. They differ both in the kind of spatial unit in which the population is recorded (mainly, agglomerations or localities, and minor political-administrative areas), and in the criteria used to classify a particular spatial unit as urban, rural, etc. This becomes especially clear when looking at national definitions in detail, as we do later on for three countries.

Variety of Criteria Used by 17 Countries

The variety is clearly shown in Table 5.8, which presents a classification of countries according to the criteria used, including any changes between 1940 and 1996. Countries that have used a sole criterion at any particular time are in the diagonal of Table 5.8, whereas countries lying outside the diagonal are those that combine two or more criteria. A date following the name of the country means that the criterion has changed through time and therefore the country name appears more than once. The main points that we can draw from Table 5.8 are as follows:

- at least five types of criteria have been in use (population size, presence of particular services, proportion of nonagricultural activities, landscape and political status);
- two criteria prevail: population size and political status of the territorial unit;
- two or more criteria are combined in several countries;
- in some countries the criteria utilized for censuses carried out at different times have been modified, thus affecting the comparison of results from one date to another.

These observations confirm the need to exercise great care in the comparative analysis of the urbanization processes, both among Latin American countries and also when comparing them with countries elsewhere.

---

[5] A study on Mexico (CONAPO, 2001) reports that a third of the populated localities of the country were in an isolation status in 2000. These 65,000 rural localities had an average size of barely 70 inhabitants.

Table 5.8 Criteria used for defining urban population in some Latin American censuses, 1940-1996

Main Criterion \ Secondary Criterion	Population size	Presence of particular services	Proportion of nonagricultural activities	Landscape	Political status
Population size	Argentina Bolivia Mexico Puerto Rico Venezuela	Chile (1970) Cuba (1970, 1981) Guatemala (1950) Honduras (1961, 1974, 1988) Nicaragua (1963, 1971, 1995) Panama	Chile (1992) Nicaragua (1963, 1971)		Colombia (1964, 1973) Nicaragua (1995) Peru (1972, 1981, 1993)
Presence of particular services	Cuba (1953)				
Proportion of nonagricultural activities					
Landscape	Chile (1982)	Chile (1960)	Chile (1952)		
Political status	Peru (1940)	Costa Rica Paraguay (1962) Peru (1961)			Brazil Colombia (1951, 1985, 1993) Dominican Republic Ecuador El Salvador Guatemala (1964, 1973, 1981, 1994) Haiti Honduras (1950) Jamaica Nicaragua (1950) Paraguay (1950, 1972, 1982, 1992) Uruguay

Source: CELADE (2001).

One way of testing the robustness of definitions is to use a range of alternative criteria. A simple example is presented in Table 5.9, where the results of applying three different population-size cutoffs are compared with the census figures on level of urbanization for 17 countries. If the threshold is set at 2,000, the resulting picture is very similar to that derived from the national definitions, with similar levels of urbanization shown, a correlation of 0.9925 and only one difference in the country ranking, for 1990. If set at 20,000, the urbanization level diminishes in every country, as expected, as they do further when the threshold is lifted to 100,000. The degree of reduction varies between countries, being generally greatest for smaller countries like Guatemala and least for countries where most of the urban population lives in one or more large cities like Bolivia. At these higher cutoffs, there is also a greater displacement of country ranks compared with that based on national definitions, but even so the distributions are not so far different, as reflected in the still very high correlations.

Table 5.9   Urbanization level according to national definitions and the proportion of population in cities of different size, for 17 countries of Latin America, census years around 1990

Country*	Urbanization level according to national definitions (1990)		Population in cities of:					
			2,000 and more		20,000 and more		100,000 and more	
	%	Rank	%	Rank	%	Rank	%	Rank
Uruguay	88.7	(1)	84.7	(3)	74.3	(2)	50.3	(4)
Argentina	86.5	(2)	86.5	(1)	74.7	(1)	61.7	(1)
Venezuela	84.0	(3)	85.7	(2)	71.3	(4)	59.5	(3)
Chile	83.3	(4)	84.6	(4)	72.3	(3)	60.7	(2)
Brazil	74.7	(5)	75.9	(5)	61.7	(5)	47.6	(6)
Mexico	72.5	(6)	74.4	(6)	56.4	(7)	46.5	(7)
Colombia	69.5	(7)	69.9	(7)	59.2	(6)	49.3	(5)
Peru	68.9	(8)	66.0	(8)	55.2	(8)	45.9	(8)
Dominican Rep.	58.3	(9)	55.9	(10)	45.2	(12)	32.2	(12)
Bolivia	55.6	(10)	57.5	(9)	49.6	(9)	41.3	(9)
Ecuador	55.1	(11)	55.3	(11)	48.0	(10)	36.4	(10)
Panama	53.7	(12)	53.5	(12)	46.8	(11)	36.3	(11)
Nicaragua	53.1	(13)	53.4	(13)	41.0	(13)	22.7	(15)
Paraguay	48.7	(14)	48.5	(14)	39.0	(14)	31.6	(13)
El Salvador	43.9	(15)	47.6	(15)	35.9	(15)	28.0	(14)
Honduras	41.8	(16)	39.1	(16)	28.0	(16)	19.1	(17)
Guatemala	38.1	(17)	38.1	(17)	24.3	(17)	19.6	(16)
Range	50.6		48.8		50.4		42.6	
Mean	63.3		63.3		51.9		40.5	
Median	58.3		57.5		49.6		41.3	
Correlation coefficient	-		0.9925		0.9842		0.9367	

Countries are ordered by their level of urbanization in 1990 according to national definitions.

Source: UN (2001b) and CELADE (2001).

This simple exercise shows that, in fact, controlling the differences in size of urban areas does not modify greatly the overall distribution of countries on the basis of urbanization level at a single time point. From another perspective, one can say that the level of urbanization is a very general measure, not sensitive to different definitions as long as population size of the areas classified as urban is controlled. Nevertheless, something very different would result if the matter under study was change in the urban structure of countries over time. Also, it is of course impossible to control the territorial unit used in each case, so the comparability of the three criteria is only apparent.

Three National Case Studies

A clearer sense of the problems of comparability can be obtained through looking in more detail at the definitions of urban (and rural) population of three countries of the region: Argentina, Chile and Ecuador. They are very different from each other and their approaches all suffer from important limitations for measuring their own urbanization, let alone comparing one case with another.

Argentina From the outset, the definition of locality was based on a physical criterion. In the first two national censuses (1869 and 1895) the criterion was implicit and the locality was named 'populated center'. In those censuses, all the population residing in a populated center (regardless of its size) was considered urban. The 1914 census introduced a threshold of 2,000 inhabitants to classify a populated center as urban. The population residing outside those centers was classified in a residual category: rural population.

Since the 1960 census the name 'populated center' was replaced by that of 'locality'. This change has persisted, although the concept of 'locality' did not become fully operational until the 1991 census, which says: 'Locality is a portion or several interrelated portions of the surface of the earth delimited by a surrounding line and formed as a mosaic of built and nonbuilt areas' (INDEC, 1998).

In order to apply the prior definition and identify the localities by proper names, it is also necessary to take into account the territorial boundaries stipulated by law. Argentina has three orders of jurisdictions; the first being the provinces and the Federal District; the second, the departments; and the third, local governments (municipalities, communes, etc.). The latter do not exhaust the whole territory, though they do so in several provinces. The first- and second-order jurisdictions do cover the whole country and are delimited on statistical and political-administrative criteria. All of them are subdivided into territorial units known as 'fractions', 'radios' and 'segments', used exclusively for census purposes. In short, the Argentine territory is divided without residue and in a mutually exclusive way in two orders of political-administrative jurisdictions (provinces and departments) which are also statistical units, and in three additional orders of statistical areas (for census purposes).

On the other hand, the localities or 'agglomerations' comprise a separate set of spatially delimited areas. They fall into two types: the *simple locality*, which is fully extended over a unique political jurisdiction, and *the integrated locality* which extends over two or more political-administrative areas, either jurisdictions of first or second order, or areas of local government. The principal example of the latter is the Buenos Aires Agglomeration, with more than 11 million people in the 1991 census. This comprises, either totally or partially, 27 different political jurisdictions: the Federal District or Autonomous City of Buenos Aires, a first-order jurisdiction, plus 26 departments of the Province of Buenos Aires. Of the latter, nine have all their census radios within the Agglomeration, another nine have either integrated or nonintegrated radios, and the remaining eight have either totally comprised radios, partially comprised radios as well as radios not integrated in the Agglomeration.

Altogether, 3,058 localities were identified in the 1991 census, of which the Buenos Aires Agglomeration is just one. The patterning of Argentina's settlement provided by this agglomeration approach is summarized in Table 5.10, using a size-based classification that goes down to 50 inhabitants. This threshold cuts out 187 smaller localities, whose population is included with the 'dispersed population' living outside localities. This leaves an 'agglomerated population' that makes up 90.6 per cent of national population. Just 3.5 per cent of national population lives in a further 2,086 localities of under 2,000 inhabitants, giving an urbanization level of 87.1 per cent on the basis of this latter threshold size.

Table 5.10   Agglomerated population (by size of the agglomeration) and dispersed population, Argentina, 1991

Agglomerated population by size and dispersed population	Number of localities*	Population (thousands)	Percentage distribution
Agglomerated population	2,871	29,564	90.6
Buenos Aires Agglomeration	1	11,298	34.6
From 1,000,000 to 1,999,999	2	2,327	7.1
From 500,000 to 999,999	4	2,551	7.8
From 250,000 to 499,999	7	2,204	6.8
From 100,000 to 249,999	12	1,839	5.6
From 50,000 to 99,999	29	2,007	6.2
From 20,000 to 49,999	74	2,221	6.8
From 10,000 to 19,999	110	1,574	4.8
From 5,000 to 9,999	181	1,255	3.8
From 2,000 to 4,999	365	1,162	3.6
From 1,000 to 1,999	354	507	1.6
From 500 to 999	471	338	1.0
From 50 to 499	1,261	281	0.9
Dispersed population	-	3,052	9.4
TOTAL	2,871	32,616	100.0

Source: Vapñarsky (1999).

How much confidence can be placed in figures like these? In fact, post-census evaluations can reveal discrepancies in the assignment of population figures to agglomerations that are very important at the individual locality level. For instance, according to the evaluation and adjustment carried out by Vapñarsky (1968) after the 1960 census, in some localities the differences surpassed 50 per cent and there were even a few whole provinces where the urbanization level was out by around 25 per cent. Even then, however, for the country as a whole, the difference was very small, with the census figure of 73.8 per cent being only slightly higher than Vapñarsky's adjusted figure of 72 per cent. Moreover, these kinds of problems were solved, to a large extent, as of the 1991 census.

Finally, it is worth pointing out that, for the analysis of the urbanization process in Argentina, an important limitation is the lack of correspondence between the localities used in the census to identify the urban areas and those spatial units utilized for the production of other basic demographic data (vital, health, education, change of address statistics, etc.).

Chile The urban definition applied here in the last five censuses has altered significantly, as shown above in Table 5.8. The 1952 and 1960 censuses used the presence of particular services and the political-administrative status. In 1970 those criteria were combined with a peculiarly low threshold of 40 grouped houses. In 1982 the threshold was raised to 60 grouped houses and a demographic variable was added: not less than 301 persons. The presence of 'urban characteristics' included a series of exceptional cases and *sui generis* situations which had to be evaluated by the local people in charge of the census, though they lacked the specific criteria to do so. In the census of 1992, the threshold of number of houses was used just for exceptional situations, but the main change was a big rise in the population threshold to 2,001 or more inhabitants. But places with a population of between 1,001 and 2,000 qualified as urban if 50 per cent or more of the labor force was devoted to secondary or tertiary activities; in which case, the services criterion was excluded. Consequently, the rural definition also changed given that, basically, it had been taken as the residue. The landscape criterion (places where the natural landscapes prevail) of the 1982 census is worth mentioning, if only because it was the unique case in the region.

The importance of the definition modification introduced in 1992 is, not surprisingly, reflected in the data on Chile's urbanization. For one thing, the officially-defined level rose only modestly between 1982 and 1992, contrasting with the important increase registered in the previous intercensal period. Even more dramatically, the figures falsely suggested that, for the first time for 40 years, the population of rural areas had increased.

Ecuador Since 1950 the definition of Ecuador's urban population has followed a political-administrative criterion, although its application has varied over time. In 1950, the urban population was that inhabiting the urban and suburban zones of each canton, the former being 'the area within the capital city or within the perimeter of the canton', and the suburban zone 'the area that, outside the urban perimeter, pertains to the jurisdictional territory of the urban parishes'. In 1962 the urban population was registered in a census of the cities – provincial capitals or heads of the cantons, distinguishing urban from peripheral population, the latter being that contained within the cities' boundaries but living in nonurbanized conglomerates. In the following censuses (1974, 1982 and 1990), the urban population is considered 'as living in the provincial capitals and the heads of the cantons previously defined as urban areas for census purposes', thus excluding the 'peripheral' population living elsewhere within the legal limits of provinces and capitals of the cantons.

For Ecuador, therefore, the existence of an urban area relies solely on the existence of a canton, being its political capital. The changes introduced in the intermediate (and inferior) ranges of the political-administrative division of the country resulted in the emergence of new urban centers. From 1950 to 1992 the number of cantons rose from 86 to 173, indicating that 87 entities that were rural in 1950 had been deemed urban by 1992. As a result, reclassification has constituted a key component of the country's urban population growth. Moreover, with this political-administrative approach, it has been the case that a nucleus of less than a thousand inhabitants is included as urban, whereas cities that surpass 20,000 are excluded. Such 'distortions' make the study of urbanization extremely problematic.

Conclusions and Suggestions

Our understanding of recent trends in the territorial redistribution of population in Latin America is evidently only partial and insufficient. On the one hand, we have an acceptable knowledge about the levels and trends of the urbanization process in Latin America in the second half of the twentieth century and about its context of very strong demographic growth. We also know how Latin America behaved in relation with all regions of the world and we can recognize the variety among the countries of the region. We also have at our disposal a fairly clear image that this process has involved very large cities concentrating a high proportion of the urban population for the region as a whole, and a very high proportion for some countries. Yet, when we try to penetrate the dynamics of the metropolitan areas and cities of different sizes and ask, for example, about the role played by the different demographic components, or when we inquire about the changes in the population composition of those areas, the new forms of social segregation, the rise of poverty, etc., the answers are rougher, partial and conditioned. In other words, in these aspects of the urbanization process, our knowledge is weaker and provisional. In many instances we only manage to gather a few useful indicators for the formulation of some middle-range hypothesis.

This chapter has also made clear that many of the deficiencies in our current knowledge are highly related to the use of inadequate concepts and to limitations in the population data collection systems. We have problems with the basic elements of our classifications; namely, points of concentration (Eldridge, 1942), nodes, cities, towns, metropolitan areas and so on – in other words, recognizing how human groups are territorially localized. We also have problems with the delimitation of the territorial units and with data collection for these, as shown above. Therefore, any attempt at suggesting how to improve data production for the study of settlement processes should begin by promoting the production of good-quality data in a timely fashion and accessible to every user.

Having said this, the region can boast examples of innovative policies that are providing more and better information for the development of local communities. One promising example is from Chile, where great progress has been made as a result of the use of new technologies combined with new political decisions. REDATAM[6] software has made census microdata available to municipal officers, researchers, businessmen, neighbors, etc., and permitted processing of very small units. Such access and usage was made possible by the political decisions to distribute the communal databases to municipalities, free of charge, and train municipal officers in the use and application of the software. Though this facility has not been used extensively yet, by eliminating a restrictive and centralized handling it is an important precedent for the future use of census information.

---

[6] REDATAM are the initials denoting a computer package aimed at the REtrieval of DATa for small Microcomputer Areas. The last available version is the fourth generation: Redatam+G4 (or R+G4), which is distributed on a free basis through INTERNET (www.eclac.cl). It can be used – either in Spanish, English or Portuguese – with Microsoft Windows 95, 98, NT4 or 2000 in any IBM compatible microcomputer. The computer package employs a compressed hierarchical database, containing micro-data (or aggregated information) with millions of records of people, housing, blocks of cities or any administrative division of a given country. It is also feasible to process a database in association with external information in common formats such as dBASE. With the program and from any database, any geographical area of interest may be defined (from blocks of a city), or make combinations of these areas, or create new variables, or very rapidly obtain several types of tabulations and export to other software. Moreover, the data of different geographical levels may be hierarchically combined in order to create aggregated variables and display the results in maps from REDATAM itself or by means of a Geographic Information System.

Chapter 6

Urbanization Trends in Asia: The Conceptual and Definitional Challenges

Gavin W. Jones

This chapter discusses levels and trends of urbanization in Asia, and the social and economic changes that are intimately bound up with them. It will go on to discuss the extent to which differences in definitions between Asian countries affect the comparability of levels of urbanization recorded. This provides the background for discussing conceptual issues relating to the urban-rural divide. A case study on developing an urban-rural classification for Cambodia provides an illustration of some of the issues involved. Finally, the paper discusses the definition of urban populations in the mega-urban regions that are playing an increasingly prominent role in the urban structures of Asian countries.

Levels of Urbanization and Social and Economic Change in Asia

As elsewhere in the world, the level of urbanization in Asia has been steadily increasing (UN, 2003). In 1950, 17.4 per cent of Asians were living in urban areas, by 2000 this proportion had climbed to 37.5 per cent, and by 2030 it is expected to increase further to 54.1 per cent. Urbanization in Asia is the lowest of any continent apart from Africa, which is only slightly below it. Even so, the enormous differences between Asian countries render an all-Asia figure almost meaningless.

Japan and the Republic of Korea, at around 80 per cent urban, and some of the oil states of the Middle East, represent the upper bound of Asian urbanization. At the other extreme are Bhutan at 7 per cent, East Timor at 8, Nepal at 12 and Cambodia at 17. These extremes do reflect very real differences. In between, however, serious doubts must be raised about some of the differences in levels of urbanization indicated by figures using national definitions of urbanization.

Urbanization levels and trends in Asia basically reflect the structure of the economies of Asian countries. A high level of urbanization can only be expected when an economy has experienced a major shift in its industrial structure, as witnessed by countries such as Japan and the Republic of Korea. A study of Asian urbanization over the 1960-1980 period argued that urbanization was yet to come, particularly in densely populated South Asia and in the still unurbanized countries of Southeast Asia (Lo and Salih, 1987). A review of the 1980-2000 period shows that, even by 2000, the same conclusion could be reached for much of Asia (UN,

2003). Urbanization over the 1980-2000 period was not spectacular in South Asia, or in some of the countries of Southeast Asia. This partly reflected the steady, rather than rapid, economic growth in these countries, as well as the inability of the official statistics in many countries to capture urbanization accurately.

Although economic growth in Asian countries has not always been as rapid as anticipated, the 'Asian tigers' (Republic of Korea, Taiwan, Hong Kong and Singapore) achieved spectacular economic growth over the last three decades of the twentieth century,[1] and some other countries (especially Thailand, Malaysia and Indonesia) also grew very strongly. This growth was sharply interrupted by the Asian economic crisis beginning in 1997, although China has continued to be a stellar economic performer since then, and India's economic performance has improved considerably (Asian Development Bank, 2002). Even in the Asian countries that have not grown so rapidly, however, social changes and transformation in living conditions have been far-reaching. For example, the percentage of Indian villages with access to electricity grew from less than 5 per cent in the 1940s to 84 per cent in 1992 (Jones and Visaria, 1997). School enrolment ratios have increased in virtually all countries and other social indicators have shown considerable improvement (see World Development Indicators published in World Bank, various years, and the Human Development Index published by UNDP, various years).

Despite the transformation of living conditions in the countryside in many Asian countries, the image of rural folk prevalent among middle-class city dwellers (especially government officials) tends to be quite negative (reflected by the pejorative Indonesian term '*kampungan*', suggesting that somebody behaves crudely, like a country bumpkin). In most Asian countries, because of the recency of the time when the proportion of the population living in urban areas came to exceed even one fifth, a substantial urban proletariat, divorced from any rural roots, has yet to develop. There is a great deal of movement back and forth between cities and rural areas at times of major festivals and holidays, not to mention the patterns of circular migration that often have their roots in dry-season underemployment in key rural areas.

The Main Issues

A transformation of the settlement pattern has been occurring in most Asian countries, reflecting the rapid and substantial social, economic, demographic and structural change taking place. It is important to understand these changes in people's residence, and how they relate to their work and other aspects of their lives. Yet for a number of reasons, this is rarely possible. In brief, the issues that will be addressed in the following pages include the following:

---

[1] 'East Asia achieved its high-growth "miracle" by boosting and sustaining investment rates while absorbing excess agricultural labour into industry' (Asian Development Bank, 2002, Box 3.1).

- Settlement classifications are mainly simple urban-rural dichotomies which were developed before the era of rapid change in the 1980s and 1990s. A dichotomous urban-rural classification fails to pick up the fact that many people live, as it were, in some intermediate state between what we usually envisage as 'urban' or 'rural'.
- Even if we accept the continued use of a dichotomous classification, inappropriate criteria are often used to designate an area as urban or rural.
- 'Building block' areas (i.e. the areas for which the designation of urban/rural status is made) are often too large and heterogeneous for satisfactory urban-rural determination.
- There is a need to develop new settlement classifications to incorporate extended metropolitan regions. This new settlement type, which may now be home to over 10 per cent of Asians, has specific characteristics that are not effectively defined using urban-rural categories.

Do Inter-Country Differences in Level of Urbanization Reflect Reality?

A good example of this issue is provided by comparing the Philippines and Thailand. It is generally assumed that, at a crude level of generalization, the level of urbanization in a country will be correlated with its level of economic development as reflected in per capita income levels, proportion of nonagricultural employment, and similar indicators. The Philippines-Thailand comparison challenges this assumption, as the figures in Table 6.1 show.

Table 6.1   Comparison of the Philippines and Thailand: development indicators and level of urbanization

	1960	1970	1980	1990	2000
Per capita income					
Philippines	295	410	690	730	1040
Thailand	200	380	670	1570	2010
% male employment in agriculture					
Philippines	59	57	62	53	47
Thailand	78	75	72	64	56*
% urban					
Philippines	30.3	33.0	37.5	48.8	58.6
Thailand	12.5	13.3	17.0	18.7	31.1
Difference in % urban	17.8	19.7	20.5	30.1	27.3

Per capita income for 1970 is actually 1976, and for 1990 actually 1991.
* Both males and females

Source: Per capita income: World Bank, World Development Reports. Male employment: Thailand: population censuses; Philippines, from Herrin and Pernia (2000, Table 12). Urban: UN (2003); Thailand figure for 2000 from National Statistics Office (2001).

Not only is the excess in percentage urban in the Philippines compared with Thailand widening over time, but this has been happening at a time when Thailand's economic development was running rapidly ahead of that in the Philippines. In other words, the widening in the differential in percentage urban is precisely the opposite of what might have been expected on the basis of the usual correlation between economic development and urbanization.

To explain this we need to examine the definitions of urban areas adopted in the two countries:

- *The Philippines*: All cities and municipalities with a density of at least 1,000 persons per square kilometer; administrative centers, *barrios* of at least 2,000 inhabitants, and those *barrios* of at least 1,000 inhabitants which are contiguous to the administrative center, in all cities and municipalities with a density of at least 500 persons per square kilometer; and all other administrative centers with at least 1,000 inhabitants which have predominantly nonagricultural occupations and possess certain minimal urban facilities.[2]
- *Thailand*: Municipalities.

The key to the difference is that the Philippines has a much more inclusive categorization of urban areas than does Thailand. Municipalities in Thailand have special administrative status, and not all towns are so defined. Nonmunicipal towns are designated 'sanitary districts' (subdivided as urban and rural sanitary districts). Some of these are quite large, with populations exceeding 20,000. Yet until 1999, according to the Thai definition, such towns were not urban! The designation of a growing sanitary district as a municipality was a rare event in Thailand; indeed, between 1947 and 1989, the number of municipalities rose only from 117 to 131, despite the substantial growth in population and urban functions of many small and medium-sized towns during the intervening period. At the same time, the number of sanitary districts almost doubled – from 439 to 862 – between 1960 and 1989 (Boonpratuang *et al.*, 1996). By contrast, in the Philippines, small villages with populations of only 1,000 were considered urban. Therefore, the comparison of Thailand's and the Philippines' level of urbanization using the official definitions was becoming increasingly meaningless.

There were political and administrative reasons why so few of Thailand's sanitary districts had been upgraded to municipalities. Sanitary districts were under tighter control of the provincial governor than were municipalities, which had popularly elected governments (Archavanitkul, 1989, pp.24-26). The costs of services which had to be provided to a municipality were also higher.

Clearly, a more realistic figure for Thailand's urban population was needed. A strong case was made for considering urban sanitary districts with a population exceeding 5,000 as urban, based on evidence that there is a continuum of urban characteristics between municipalities, urban sanitary districts, rural sanitary districts and rural areas, with urban sanitary districts being distinctly more 'urban'

---

[2] For more detail, see Mejia-Raymundo (1983, p.64).

than 'rural' according to a range of measures (Goldstein and Goldstein, 1978; Archavanitkul, 1989; Kirananda and Surasiengsuk, 1985). Such an adjustment would have raised Thailand's level of urbanization from 14.7 per cent to 22 per cent in 1970, from 18 per cent to 26 per cent in 1980 and from 19 per cent to 27 per cent in 1990. This level would still have been on the low side, especially for 1990, but it would no longer have been so strikingly inconsistent with levels in other comparable countries.

An alternative adjustment procedure, which used geographic rather than administrative area information to add to urbanized areas all residential and/or industrial areas with a total population of over 5,000 persons per square kilometer, measured the urban population in 1990 at 32 per cent of the total population (NESDB/UNDP/TDRI, no date: Study Area 2, p.9 and Table 4.7).

Reflecting growing dissatisfaction with the official designation of urban places in Thailand, in 1999 a major administrative reorganization took place whereby all sanitary districts were upgraded to municipal status, raising the total number of municipalities to 1,081, and the resulting proportion urban in the 2000 Population Census to 31 per cent. The figures in Table 1 show that even after this adjustment, the discrepancy between Thailand and the Philippines remains wide. There are two main reasons for this. One is that in Thailand, the *tambon* (i.e. sub-district) administrative organizations continue to be designated rural, although some of them, especially those just outside municipalities, have distinctly urban characteristics. The other is that the Philippines, compared not only with Thailand but also with other countries with higher or roughly equal levels of economic development, uses very generous criteria in allocating urban status to particular localities.

The important general point to be made here is that some Asian countries adopt a definition of urban that is administratively rigid, whereas others adopt a more flexible approach that is well fitted to adding to the areas designated urban as changes occur in the economy and in settlement patterns. This flexible approach, however, requires frequent and detailed investigations to track the need to reclassify localities from rural to urban and (occasionally) *vice versa*.

Urbanization: Conceptual Issues

It is not just the practicalities of urban definition that need to be considered, but the very meaning of our concepts of urban, and whether they continue to have relevance. As noted by Hugo (1992), the blurring of the distinction between urban and rural populations resulting from mobility of people, goods, services, capital and ideas raises issues about the utility of simple urban/rural dichotomous classifications. This is apparent, he argues, in Asian countries, where a convergence can be observed in the characteristics of the urban and rural populations of many countries in respect of indicators such as fertility and mortality, educational attainment, and economic activity. A quite widespread trend is the increasing proportion of the rural population who are engaged in nonagricultural activities.

I agree with this assessment, and indeed have argued that the UN estimates of urbanization in Southeast Asia, showing that urbanization remains fairly low in the region, are misleading (Jones, 1997). In 1995, 55 per cent of employment in areas defined as rural in West Java (Indonesia) was in nonagricultural occupations. In the 1980s, 50 per cent of farm family incomes in the heavily rural state of Kelantan in Malaysia were derived from off-farm employment (Shand, 1986). Throughout most of Southeast Asia the extent to which 'urban' facilities have permeated rural areas over the past 30 years has been astonishing, as the forces of modernization 'impinge on formerly isolated, inward-looking, self-sufficient and agriculturally-based communities' (Rigg, 1997, p.157). Somehow, the recorded statistical increase in urbanization fails to capture what has really been going on.

On the other hand, the continuing strength of certain differentials between urban and rural populations should not be understated. For example, although educational differentials may be narrowing when measured by school enrolment ratios or the proportion of adults with certain educational attainment levels, I have no doubt that the perpetuation of urban-rural difference would appear more marked if a suitable measure of the quality of the education they have received could be found. The contrast in educational opportunity between city populations and those in the remoter and poorer rural areas remains stark.[3]

Perhaps one way to express what has been happening in much of Asia is that the urban has been 'invading' the rural, in the sense that rural areas have been dramatically opened up to new ideas through transportation and communications developments, greater population mobility, and the spread of education. In Thailand, in Java, and in many parts of the Philippines, 40 years ago there were many isolated villages where the impact of the outside world was minimal. Such villages are now extremely difficult to find. Nevertheless, the extent of 'urban invasion' is not uniform, and in terms of a number of indicators, rural areas closer to large cities tend to be more comprehensively affected than are more isolated villages. This is certainly true when the indicator is the structure of employment, or access by reasonable road to a large town or city. It is less true of access to ideas through TV and radio, the presence of which is ubiquitous. On structure of employment, one study in Java showed that rural dwellers in districts closer to the larger cities tended to have a higher proportion engaged in nonagricultural activities than did those in districts further from those cities.

> There may be a number of reasons. One is that some cities had spilled over their boundaries, so that some of what was being recorded as rural population was really suburban. Another is that factories and service sector establishments built in rural areas were more likely to be located in areas close to the large cities than in areas further away. Another is that rural dwellers in areas close to the large cities could avail themselves of the opportunities to commute to urban jobs. (Jones, 1984, p.126)

---

[3] This is evident in fieldwork I have been conducting in recent years with colleagues in various parts of Indonesia. Both quality of buildings, level of training of teachers, level of absenteeism of teachers, and availability of teaching materials tends to be worse in the rural schools, particularly those serving very poor and isolated communities.

Our concepts of rural-urban distinctions are tightly bound up with our understanding of what leads to a rise in the urban share of the population. Typically, we distinguish between natural increase of the urban population, migration from rural to urban areas, and 'reclassification' of areas previously considered rural. Methodologically, many studies distinguish between the effect of differences in rates of natural increase between rural and urban areas on the urban share of the population, on the one hand; and the 'balance' of the difference between urban and rural rates of population growth, which is attributed to the combined effect of net migration and reclassification. The lazy researcher by this means conflates two entirely different processes. The natural increase and the migration relate to areas already by definition either rural or urban. Reclassification is a different matter altogether. It is frequently done in recognition – at some point in the process – that a locality is changing in very significant ways from an earlier 'rural' state to an 'urban' one. The process has been well documented for China by Yu Zhu's work on *in situ* urbanization (see Chapter 11). Another form of incipient *in situ* urbanization has been recognized in Vietnam by Douglass and DiGregorio (2002). They argue that extremely densely settled villages and communes, mainly in the Red River delta, have the potential to develop into proto-urban agglomerations, through further development of their rural craft and industrial base, a base that has grown in self-reliance and sophistication partly because of government policies to inhibit rural migration to cities.

Urban Definitions

Despite the many reservations about the very concept of an urban-rural divide, I believe that an urban-rural distinction still has something to offer to those seeking to understand the world in demographic, sociological and economic terms. Ideally, there should be nuanced definitions of a nonbinary kind, so that gradations of urban characteristics can somehow be captured. However, prevailing definitions are in many cases so crude that even in terms of conventional binary classifications, there is great room for improvement. Perhaps this issue can best be considered by comparing urban definitions (from UN, 2003) for a few more Asian countries besides the cases of Philippines and Thailand, already covered above:

- *India*: Towns (places with municipal corporation, municipal area committee, town committee, notified area committee, or cantonment board); and all places having 5,000 inhabitants or more, a density of not fewer than 1,000 persons per square mile or 390 per square kilometer, pronounced urban characteristics and at least three fourths of the adult male population employed in pursuits other than agriculture.
- *Indonesia*: Municipalities (*kotamadya*), regency capitals (*kabupaten*) and other places with urban characteristics.
- *Malaysia*: Gazetted areas with their adjoining builtup areas and with a combined population of 10,000 persons or more.
- *Nepal*: Localities with 9,000 inhabitants or more (*panchayats*).

From these examples, it is evident that some countries adopt a purely administrative approach to defining urban areas, whereas others use more functional criteria. The administrative approach is exemplified by Thailand (discussed earlier), Malaysia and Nepal. The key question for countries such as these is what criteria are adopted for designating a locality as a municipality or (in the Malaysian case) a gazetted area. As we have seen, in Thailand there are substantial barriers to a town graduating to municipal status, and without such graduation, the town will not be considered as urban. There is no such problem in Malaysia, where it is unlikely that there is any urban area of 10,000 or more that has not been gazetted. However, the higher population cut-off in Malaysia for inclusion in the urban population explains why Malaysia, with a much more advanced industrial economy than the Philippines, had an almost identical level of urbanization in 1990 (49.8 per cent and 48.8 per cent respectively).

The cut-off point for size of place to be included as urban is indeed a real issue. In Malaysia a town is actually a 'gazetted administrative area with a population exceeding 1,000', so the choice of a cut-off of 10,000 could easily be replaced, for some purposes, with a cut-off of 5,000, or even 2,000. A study by the author in the 1960s found that, using the criterion of the percentage of male employment in agriculture, towns in the 2,000 to 10,000 range varied widely (Jones, 1965). The 'artificial' creation of 'New Villages' into which mainly Chinese agriculturalists were forcibly resettled, in the context of the Communist emergency prevailing in Malaya in the late 1940s and early 1950s, was largely responsible for some sizeable settlements in which the great majority of those employed were in agriculture. Thus in order to ensure that the urban population was not exaggerated, the choice of a 10,000 population cut-off was reasonable, but this did serve to exclude many smaller towns that met virtually all criteria that might be used to define urban, including a high proportion of nonagricultural population.

The Philippines (discussed earlier), India and Indonesia are examples of countries that adopt mixed administrative-functional criteria to designate places as urban. In Indonesia, as previously shown in Chapter 1, every village is classified as either urban or rural according to its weighted score based on the three criteria of population density, proportion of the workforce in agriculture and possession of certain designated urban facilities.

The most populous country in the world provides an extreme illustration of the effect of changes in urban definition on the estimated urban population. China's percentage urban ostensibly increased from 23.5 per cent to 46.6 per cent between 1983 and 1987. While urbanization was undoubtedly quite rapid over this period, as a result of rapid economic development, it was clear that most of this increase resulted from definitional changes as discussed in Chapter 11. Basically, the increase was because of three factors; namely, a decision to include the nonagricultural population of cities and towns in the urban population, a rapid increase in the number of cities and designated towns, and the enlargement of many administrative urban areas. China is such a large population that the arbitrary changes in urban definitions there had a large impact on the estimated level of urbanization in Asia as a whole.

In the previous section, the point was made that *in situ* urbanization is occurring to a greater or lesser extent in most countries, and this heightens the importance of reclassification of urban-rural status in tracking the urbanization process. The point to stress about reclassification is that, if (1) a system is adopted whereby urban-rural status is determined for small geographic areas, such as villages, rather than larger areas such as subdistricts, and (2) regular reassessments of the status of these areas are conducted (e.g. before each decennial Census), then reclassification of areas from rural to urban is likely to capture some meaningful change in the characteristics of the locality. However, if reclassification is more a matter of political decisions about the boundary of a city, or a 'catch-up' decision recognizing an urban status long apparent in a particular subdistrict, then the rise in the urban share of the population resulting from such decisions is unlikely to give a timely and accurate reflection of the real changes in characteristics of the various localities that go to make up the national population.

Cambodia Case Study[4]

Cambodia provides an interesting case study of some of the problems in finding an urban definition that is suitable both conceptually and for planning purposes. At present, there is no functional classification of urban areas. The designation of places as urban or rural is based only on administrative criteria, which are unsatisfactory for planning for the needs of actual urban populations. At the time of the 1998 Population Census, the following areas were treated as urban:

(i)   All province capitals (which are whole districts);
(ii) Four of the seven districts of Phnom Penh municipality (the other three were considered rural);
(iii) The entire provinces of Sihanouk Ville, Krong Kaeb and Pailin, which are called Krongs or municipalities.

There are a number of problems with this designation:

• Districts are quite large in area, and a number of communes in the districts in which the province capitals are located are very rural in character. By considering these rural communes as urban, the populations of most province capitals are exaggerated.
• In some cases, adjoining communes in another district are part of the builtup area of a province capital, but are not included because of the restriction of the town population to the district in which it is located.
• Parts of Phnom Penh municipality are rural, but also some areas of Kandal Province immediately adjoining the Phnom Penh municipality are builtup areas, which for planning purposes should be included as part of the Phnom Penh urban agglomeration.

---

[4] This section draws heavily on Jones and Rao (2001).

- Large areas of the provinces of Sihanouk Ville, Krong Kaeb and Pailin are rural, and by considering the entire provinces as urban, their urban population is greatly exaggerated.
- A number of small towns that are not province capitals are missed by the current urban definitions, and thus are included in the rural population although they have distinct urban characteristics and in some cases their populations exceed 20,000.

The key problem with the current classification of urban places is that it makes decisions at the level of province or district. Many of the provinces or districts that are designated as urban encompass vast areas of agriculture and wastelands and uninhabited areas including mountainous terrain. A more fine-grained definition of urban places requires designations of urban or rural at a lower administrative level – the commune or, ideally, the village. While there are only 183 districts in Cambodia, there are 1,609 communes and 13,406 villages.

The 1998 Population Census data provided a reasonable basis for determining whether localities should be considered urban or rural. Ideally, the designation of localities as urban or rural should be focused on the village level, and contiguous urban villages combined to form towns. However, the available data did not permit the classification of villages as urban or rural, because their density is not known precisely, and there is no inventory of their urban facilities. An exercise was therefore conducted, focusing on the commune. Although this was not entirely satisfactory, the focus on the commune, rather than the province and district level, did represent a considerable refinement of the existing urban classifications.

Using the data from the 1998 Population Census, appropriate cut-off points were sought for three criteria for designating communes as urban or rural, these being population density, percentage of male employment in agriculture, and a minimum population size. The first – population density – is useful in indicating whether there is a sufficient concentration of population to be consistent with urban status. The second – percentage of male employment in agriculture – is important in distinguishing between densely populated agricultural areas and densely populated areas where the focus of economic activity is nonagricultural, a typical distinguishing feature of urban areas.[5] The third – a minimum population size – is typically used for defining urban areas, in order to avoid the designation of small groupings of households as urban areas.

After preparation of detailed compilation sheets from the Census data, field investigation of localities based on data provided in these sheets, and consultation with district officials, the cut-off points decided on for each of these criteria were:

---

[5] In the Cambodian context, where most women are recorded as in the labour force, a case might be made for focusing on the proportion of the workforce as a whole, rather than the male workforce only. However, because in Cambodia as well as in most countries, the range of male economic activities is wider than female, the restriction of the indicator to male workers is justified as a more sensitive indicator of the importance of non-agricultural activities in the particular locality.

- Population density exceeding 200 per square kilometer.
- Percentage of male employment in agriculture below 50 per cent.
- Total population of the commune exceeding 2,000.

In actual fact, it made little difference whether a minimum size of 1,000 or 2,000 was used for the population of communes, because a frequency distribution of communes by population showed that relatively few communes (10 per cent) had populations below 2,000.

As for population density, it made relatively little difference whether a population density figure of 100 or 200 per square kilometer was used, as most communes in the densely-settled central area of Cambodia, whether urban or rural, had densities well above 200, and some of those urban areas that did not, largely in isolated provinces, had densities well below 100.

As for the cut-off of 50 per cent for the agricultural share of male employment, the data sheets revealed that by raising the cut-off to 55 per cent, the urban population would rise by 172,800, or from 17.6 per cent to 19.1 per cent of the population. Although field visits indicated that a case could sometimes be made for raising the cut-off to 55 per cent, it was decided not to do so. For one thing, international comparisons of urban criteria showed that a cut-off of 55 per cent would be well above the normal, and also it was felt that as the Cambodian economy gradually develops, the proportion of males working in agriculture will fall, thus bringing more and more areas within the urban classification, even where a cut-off of 50 per cent is used. Another alternative, that of lowering the cut-off to 40 per cent, was considered, but rejected, because a proportion as low as this would be too rigid a criterion for designating a commune as urban in Cambodian conditions.

Application of the cut-off points decided on for this study, while it proved appropriate in most areas, ran into difficulties in four provinces with a small population and/or isolated location. Here not even one commune would be designated as urban by application of these criteria. Yet in each of these provinces, the provincial headquarters was indeed a small town. Because of the administrative and political need to ensure that every province contained at least one town, in these provinces the criteria were relaxed so that the provincial capital could be considered to be an urban area.

The application of the criteria noted above raised the estimated urban population of Cambodia in 1998 by 211,350, from 1,795,575 to 2,006,925. This resulted in a rise in the estimated urban percentage of the population from 15.7 per cent to 17.6 per cent. In other words, there is a small rise but not a drastic change in the estimated urban proportion of the population for the country as a whole.

On the other hand, at the province level, there are some striking changes in the estimated urban population: large increases in the urban population of three provinces, substantial increases in the urban population of four other provinces, and declines – some of them quite large – in the urban population of all other provinces.

The reason why the urban population declined in so many provinces using the new classification was the exaggeration of their urban populations by the prevailing procedure of designating as urban the entire district where the provincial capital is located and, in three cases, designation of the entire province as urban. The reason why the urban population increased substantially in some provinces is that towns that are not the provincial capital, some of which are of reasonable size, are not considered urban according to the prevailing procedures.

Unfortunately, because of using the commune rather than the village as the unit of analysis, certain areas which exhibited urban characteristics in the field did not qualify according to the criteria used. Therefore, work is needed on a further urban classification based on data at the village level. Communes in Cambodia contain an average of eight villages. When decisions about urban or rural status are made at the commune rather than the village level, some villages which are actually urban have undoubtedly failed to be categorized as such, because the commune as a whole failed to meet the criteria for urban designation. In such cases, the village itself would have easily met the urban criteria of population density exceeding 200 per square kilometer and low proportion of employment in agriculture, but these figures were 'diluted', as it were, by the figures for the more rural villages within the same commune. On the other hand, some rural villages will have been classified as urban because the commune in which they are located meets the criteria.

One interesting case study demonstrates the problems of using the commune rather than the village level in designating urban places. It refers to three adjoining communes in Bat Dambang district: Kouk Khmum, Ta Pung and Ta Meun. The relevant commune statistics are: Kouk Khmum (73 per cent males in agriculture, density 230, population 12,000), Ta Pung (67 per cent males in agriculture, density 129, population 13,705), and Ta Meun (74 per cent males in agriculture, density 127, population 14,959). Therefore, the agricultural percentage in each of these three communes fails to meet the 50 per cent criterion, and two of them are below the density cut-off as well.

Although none of these communes meets the criteria for classification as urban, they contain adjoining villages which form quite a distinct town (see Figure 6.1). Two of the communes contain one or more villages which satisfy the criterion of less that 50 per cent male employment in agriculture, and the other has one village which almost satisfies the criterion. The combined population of the contiguous villages which satisfy, or almost satisfy, the criterion is 6,953.

This may be a fairly unusual example, but important nonetheless, because it shows how a quite substantial town can miss out on urban designation if it is composed of villages straddling two or three communes, most of whose villages are rural. This town formed of adjoining urban villages, if located in only one commune, would probably have led to that commune meeting the criteria for designation as urban. But because each of these villages forms part of a largely-rural commune, none of the communes concerned qualify for urban status. It would certainly not be appropriate to declare the three adjoining communes as an urban agglomeration, because although their combined population is 40,000, the town at their intersection has a population of less than 7,000. But equally, it is unfortunate not to recognize this town at all.

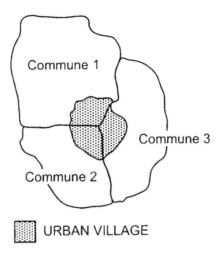

URBAN VILLAGE

Figure 6.1   Urban villages in three communes in Bat Dambang Province, Cambodia

It is impossible to tell whether a village-based designation of urban areas would increase or decrease the total urban population of Cambodia as measured in the study just discussed. What *is* clear, however, is that (1) it would increase the total number of urban places, especially those with populations below 5,000 and from 5,000 to 10,000; (2) it would lower the population of many towns as defined in the present study, because some villages which form part of communes which meet the urban criteria would be excluded from the urban population by a village-based study.

Extended Metropolitan Regions (EMRs)

One of the most interesting issues in Asian urbanization is defining the status of localities in the vicinity of large cities – the outer parts of so-called mega-urban regions. As originally highlighted by McGee (1991), there are densely settled agricultural areas surrounding major Southeast Asian cities including Jakarta, Bangkok and Manila, where complex physical and employment patterns are emerging. As cities grow, these surrounding areas change in character, both because of the location of many 'urban' facilities, housing estates and productive enterprises there, in the midst of continuing intensive agriculture, and because of the changing employment patterns of their populations, which are swollen through migration both from far away and from the city proper. In many cases, the *in situ* changes affecting the long-term population are just as important – or even more important – than those wrought by inmigration to these areas. Debate has focused on whether the special characteristics noted by McGee and others – for example, the intense mixture of land use, with agriculture, cottage industry, industrial estates, suburban developments and other uses existing side-by-side – distinguishes such EMRs in Southeast Asia as much as McGee contends from those removed from current Southeast Asian megacities in time and space (see Dick and Rimmer, 1998).

The definition of urban populations in mega-urban regions poses special issues. If building block areas (e.g. villages) can be categorized as urban and rural, then it is possible to map a patchwork of urban and rural areas in these regions, just as it is anywhere else. On the other hand, it is also useful for some purposes to define zones surrounding the metropolitan area, and in doing this, it is necessary to take contiguity into account. If some villages defined as rural are contiguous to urban areas within the mega-urban region, and indeed help link such urban areas to each other, then in defining zones in the mega-urban region, such villages would normally be considered a constituent part of the mega-urban region.

The change in settlement patterns in the zones of rapid change surrounding major metropolitan areas in Asia is often very 'messy', with a haphazard mix of land uses: cultivated fields, roads, houses, workshops, factories, small shops, unused land cleared for development, and waste land strewn with wrecked cars or used for temporary storage. In Bangkok, for example,

> The bulk of the decentralization has been occurring in the form of 'ribbon development' along the three major transport corridors leading out of the urban core to the southwest, southeast, and north ... It has involved conversion of paddy lands; encroachment on agricultural land; and leapfrogging types of development in the fringe areas, leaving large tracts of unused land in between ... Access to individual lots is typically provided by long and disjointed dirt roads built by the private developers along the boundaries of elongated agricultural lots. Subsequently, individual houses and all types of smaller and larger private housing estates emerge, mixed with commercial and industrial land uses ... Private developers have largely failed to provide an adequate secondary road system. The result is a series of 'superblocks', which in large areas are bounded by main roads and arterials with little or no internal road network to distribute traffic efficiently or provide access within the blocks. This leaves much of the land within the superblock unusable and tends to concentrate the traffic on the relatively few, already overcrowded main thoroughfares. (Robinson, 1995, pp.97-8)

Recent studies on Southeast Asian megacities have delimited an inner and outer zone outside the metropolitan core (which is itself quite large in some of these cities: 1565 square kilometers for Bangkok, and around 650 square kilometers for Manila and Jakarta). The inner zone (the area of greatest change in population, economic activity and transport and communications) tends to extend from 30 to 50 kilometers from the city center, and the outer zone even further (Jones *et al.*, 2000; Mamas *et al.*, 2001). Parts of the outer zone remain strongly rural in terms of some indicators: they are included in the mega-urban region more because of their potential for future change than for change that has already occurred. By the same token, the indicators of 'urbanness' available are very limited, and meaningful change toward a more 'urban' pattern of life, however that may be defined, may actually have proceeded further in outer zone areas than is obvious from these indicators. Even in terms of employment, census or survey data listing one occupation per respondent can be very misleading, because even in areas removed from large cities, studies show that off-farm employment can be half or more of farm-family incomes. This process is likely to have proceeded further in outer zone areas surrounding the very large cities, but more research is needed to determine whether this is in fact the case.

In most of the Southeast Asian megacity studies, some combination of the following indicators is used to define urban areas: population density, percentage of male employment in agriculture, rate of population growth, and availability of urban facilities. An alternative approach being used in studies of Hanoi and Ho Chi Minh City in Vietnam is to use commune-level data to identify high-density, populous urban-like settlements by subtracting land under cultivation from total land area. These studies show that the potential for *in situ* urbanization in the Red River delta is much higher than is recognized by usual treatments. Around Ho Chi Minh City, also, the leap in urban population to the south-west to the southern Mekong delta shows a substantial proto-urbanization process under way that, due to the flood plains, remains separated from Ho Chi Minh City but can, through bridges and highways, be expected to eventually become part of the expanding metropolitan region.[6]

Has Megacity Growth Been Slowing?

It is frequently argued that the growth of megacities is slowing, partly because of the slowing of overall population growth, and partly because as cities become very large, the rate of net inmigration tends to decline. However, we should be cautious about this claim. The pool of population remaining in rural and regional areas (including urban population in the regions) is still large enough to generate high rates of megacity growth through migration, if the factors inducing regional dwellers to move to the megacities are favorable.

What actually happened to megacity populations over the 1980s varied widely between cities. An assessment of trends is complicated by one factor that leads to a systematic tendency to understate the growth of megacity populations, namely that the spread of urban activities disregards existing urban boundaries. As already noted, growth of many megacities has spread well beyond the metropolitan boundaries normally used to define these cities. The population growth rate in areas outside the metropolitan boundaries is frequently much higher than that inside the boundaries. When these megacities were smaller, their core areas frequently had relatively slow growth, and their outer areas faster growth, but at that time both core and outer areas were contained within the metropolitan boundaries. Hence the growth of the metropolitan population took account of both the slower core growth and the faster peripheral growth. But with the further expansion of these cities, the peripheral areas with faster population growth are in

---

[6] Personal communication from Mike Douglass.

many cases located almost entirely outside the metropolitan boundaries. Therefore the growth of the metropolitan population may be quite slow, but it would be a grave mistake to interpret this to mean that megacity growth is slowing. Such an interpretation should only be made when the growth rates of the extended metropolitan region as a whole have been carefully studied, and found to be declining.

A proper understanding of the growth of the megacities is impossible unless it is based on careful analysis of the component zones within the extended metropolitan region. As shown in Table 6.2, the population growth rate in the different zones varies considerably, and in most cases, the inner zone (which lies outside the official metropolitan boundary) is growing considerably faster than the metropolitan area. In the Indonesian urban agglomerations, natural increase accounted for more of the population growth over the 1990-1995 period than net migration. But the role of migration differed between zones. The inner zones were the key areas of change – migrants came there from both the metro area and elsewhere in the country. Net migration in many cases contributes as much as two thirds of the population growth in the inner zones, whereas in the official metropolitan areas net migration contributes little to growth, and is in fact negative in many of the central-city areas. The age structure of migrants to the inner zone reflects the more prevalent 'family' migration, whereas migration to the metro area is frequently dominated by single young people.

Table 6.2   Annual average population growth rates of some extended metropolitan regions (EMR) in Southeast Asia, by zone

Zone	Bangkok 1980-1990	Manila 1980-1990	Taipei 1980-1990	Jakarta 1990-1995	Surabaya 1990-1995	Bandung 1990-1995	Medan 1990-1995
Metro area	2.3	3.0	2.0	2.1	1.7	2.8	1.9
Inner zone	3.3	3.8	2.8	6.0	2.5	2.9	5.5
Outer zone	2.0	2.9	0.7	1.8	0.9	2.8	0.5
EMR	2.4	3.2	2.2	3.4	1.8	2.8	2.9

Metro area refers to the official metropolitan area. Inner and outer zones lie outside this.

Source: Studies reported in Jones et al. (2000); Mamas et al. (2001).

Although detailed analysis of trends to the end of the 1990s will have to wait for studies using the results of the 2000 round of population censuses, preliminary findings from Jakarta and Bangkok suggest that the 1990s witnessed some important modification of earlier trends. The population growth rates of the official metropolitan areas of these cities have fallen distinctly – indeed, in the case of Jakarta, an actual *decline* in population was recorded between 1995 and 2000. But the regions immediately surrounding the metropolitan areas have continued to grow quite rapidly. To complicate the picture, however, the trends for Manila over the same period have been very different.

Table 6.3 gives some indications of recent trends in these three cities. Populations in the metropolitan areas of both Bangkok and Jakarta grew very slowly over the 1990s,[7] representing a sharp deceleration of their growth over the 1980s, and this slow growth was characteristic also of the other largest cities of Indonesia. Does this slowing of growth represent a key turning point in the history of these major megacities?

Table 6.3   Growth of population in the extended metropolitan regions (EMR) of Bangkok, Jakarta and Manila, 1990-2000

	1990 (000s)	2000 (000s)	Average annual increase (%)
Metro Bangkok	5,882	6,320	0.7
Rest of EMR	2,707	3,760	3.3
Bangkok EMR	8,590	10,080	1.6
Thailand	54,549	60,607	1.1
Metro Jakarta	8,259	8,385	0.2
Rest of EMR	8,876	12,749	3.7
Jakarta EMR	17,135	21,134	2.1
Indonesia	179,379	202,000	1.2
Metro Manila	7,945	9,933	2.3
Rest of EMR	6,481	9,855	4.3
Manila EMR	14,426	19,788	3.2
The Philippines	60,703	72,345	1.8

The answer is not clear, for a number of reasons. First, there is controversy over the extent of underenumeration in these large cities in the year 2000. Censuses tend to undercount most seriously in megacities, and this tendency may have increased in recent years as the size of cities has increased and their populations become increasingly mobile. Crowded cities with mobile populations provide the strongest challenge to census takers, and undercounts have been suspected in many of these cities over a long period – cities as widely divergent as Bangkok, Karachi and New York. In the 2000 round of censuses, a large undercount is suspected in Jakarta. The undercount in Jakarta and other cities may have been worse in 2000 than in 1990.

---

[7] It is unlikely that this growth rate was constant over the whole decade. It is likely that the growth rate in both cases was slowing towards the end of the decade, both because of falling rates of natural increase and a lessened attraction of the big city to migrants since the economic crisis began in 1997. Evidence for Indonesia based on comparison of the 2000 census results with the 1995 Inter-Censal Survey certainly indicates that this was the case for Indonesian cities.

Before concluding that this was the case, however, it is necessary to be aware that fertility rates have sunk very low in these large cities, thus reducing rates of natural increase of the population. Also, movement of people from the city to suburban areas outside the metropolitan area is certainly taking place. The data in Table 6.3 show that the zone surrounding the metropolitan area maintained considerable population growth over the intercensal period. The growth rate of population for the EMRs of both Jakarta and Bangkok turns out to have exceeded that of the nation's total population, and probably of the total urban population.

In the case of Manila, fertility rates are higher, no doubt leading both to more rapid natural increase in the city and to more migration from still rapidly-growing rural and regional populations. Within Metro Manila, population growth rates in the 1990-2000 period were very slow in some older, inner areas, and much more rapid in areas further from the center. The overall growth of the extended Manila region appears to have been very rapid indeed.

The Potential of Small Area Data Collection

Some Asian censuses have moved from sample censuses to complete counts in the 2000 round (Indonesia is an important example). This provides new opportunities for detailed small area analysis, and opens up the possibility of adopting smaller building block areas for identifying classification types than was previously possible when data were available only for larger regions. It should therefore enable more sophisticated settlement classification to be used, in particular when utilizing new developments in geographic information systems. Technological developments in satellite imagery also have great potential to advance the study of the spatial dynamics of urban change in major megacity regions by overcoming the usual need to rely on administratively defined areas. They have already been used effectively in studies of the Bangkok region and of Vietnamese metropolitan areas.

Conclusion

From Asia's low but rising level of urbanization in a context of lower urban than rural fertility rates, we can infer that rural-urban migration of the traditional kind is still important in mobility patterns. Although the differing extent of reclassification of areas from rural to urban complicates the interpretation of the contribution of migration to urban growth, migration has frequently contributed much to the population growth and changing patterns of urban amenities on which this reclassification was based. In any case, unlike most of the OECD countries, where there is now a very high level of urbanization, in Asia the rural-urban distinction is still meaningful, albeit not as sharp as it once was. Socioeconomic differences between urban and rural areas may be narrowing, but they remain quite marked.

Though there is great variation in levels of urbanization between Asian countries, many of them, including the giants of China, India, Indonesia, Pakistan and Bangladesh, are at a stage in their urbanization process at which urbanization normally proceeds quite rapidly. For this reason alone, it is very important to be able to follow these urbanization trends closely. But at the same time, the nature of the urbanization process, and of urban-rural interactions, is changing. Unfortunately, urbanization trends and differences within and between Asian countries, and the changing nature of urban-rural interactions, cannot be tracked very effectively using the currently available data on urbanization. Indeed, the problem goes even deeper than this. There is a need for considerable rethinking about the traditional binary categorization of urban and rural areas, and whether more suitable classification systems can be found. There are new and complex dimensions to the spatial patterning of population, perhaps the most notable being in the zones within the parts of the EMRs that lie outside the official metropolitan area. Even if the binary classification is continued, systems relying on administrative decisions about the urban status of rather large geographic entities are not well designed to track the changes in urbanization that are taking place.

Each country has its own needs, of course, and should not be pushed into the adoption of one system of classification in the interests of international comparability or with the aim of capturing the gradations in the rural-urban continuum. However, it would be quite conceivable for each country to adopt one system of urban classification for its own administrative needs, plus (where necessary) another system for purposes both of international comparability and of providing planning data for domestic needs that reflect contemporary changes in the urban-rural continuum and the specific characteristics of mega-urban regions. What is needed is for convincing arguments for such a procedure to be put to politicians, national development planners and national statistical agencies.

Chapter 7

Analyzing Urbanization in Sub-Saharan Africa

Philippe Bocquier

This chapter analyzes the patterns and processes of urbanization in Sub-Saharan Africa (SSA) and reviews the adequacy of existing practices of measuring urban growth there. It begins with some historical context on the importance of settlement systems in SSA's demography, highlighting the slow and sporadic evolution of urbanization up to the nineteenth century and the great acceleration after the Second World War. It goes on to survey the contemporary dynamics, pointing notably to a slowdown in urban growth from the 1980s onwards and showing a projection indicating that on recent trends the overall urbanization level for the region would stabilize at not much over 50 per cent. There follows a critical analysis of official definitions of urban areas, leading on to an assessment of the consequences for the measurement of population change in the settlement system. Lastly, some suggestions are provided on ways of improving the collection and handling of data on urbanization and migration, with particular attention being given to the revision of Census procedures and their fuller integration with intelligence from remote sensing. In attempting to end on a positive note, it is suggested that the current dearth of reliable statistical systems could be a blessing, if this means that there are fewer barriers to a great leap forward.

Importance of Settlement Systems in Sub-Saharan Africa's Demography

Historical studies have shown that lack of labor has always been a problem in SSA's societies and economies. In a continent where the natural environment has nurtured diseases, the population was prevented from growing significantly for several centuries. The rise of precolonial empires in SSA led to the concentration of power, natural and human resources in cities,[1] particularly through the influence of Islam, as in Sahel (Timbuktu, Gao, Djenne, Kano, Katsina) and along the East African coast (Kilwa, Lamu, Bagamoyo, Shanga). Historical concentrations of people are more difficult to establish deeper in the hinterland of SSA because of

---

[1] The following comments on African urban history are mainly based on the comprehensive and up-to-date introduction by Anderson and Rathborne (2000). Another good synthesis of African urban history can be found in Coquery-Vidrovitch (1991).

the degraded archaeological sites, but are confirmed in West Africa (Ibadan, Ife, Basoko), in the Great Lakes region (Bigo, Ntusi, the 'moving' capital of Buganda), in Abyssinia (Aksum) and in Southern Africa (Great Zimbabwe). Many of those civilizations were built on a traditional and controlled type of slavery. However, this never reached the point where these cities could compete on a global scale.

Apart from slavery, most societies in SSA found a solution to the lack of labor by scattering their members across large pieces of land, eventually founding new communities where the resources could be better exploited with less labor. The weak concentration of human population in SSA did not help the development of new agricultural techniques, although where it occurred, SSA societies were able to create quite effective techniques. But generally, migration of labor (be it through slavery or the splitting of existing communities) rather than its intensification has been the most commonly used solution to adjust to shortage of resources across the continent. As a consequence, urbanization had been sporadic and remained at a constantly low level up to the nineteenth century.

This pattern was considerably disrupted by colonization. Some urban centers were developed on the basis of existing precolonial centers (e.g. Mombasa, Kampala, Zanzibar, Ibadan, Kano, Kumasi, etc.). However, most commercial and administrative centers were created out of nowhere, often on the coast for ease of access to modern transport (as with most of the current capital cities of SSA states) and, more often than not, in total ignorance by colonialists of previous commercial centers which were thereafter marginalized. From the end of the nineteenth century up to the Second World War, the new centers that became cities were meant (by the colonialists) to facilitate the extraction of raw materials and goods. Industrialization was marginal and the processing techniques basic. Urbanization was stimulated by external needs, not by endogenous development.

However, urbanization did bring changes in SSA societies and those changes, particularly in the migration system, contributed to the urbanization process. The newly-born cities, at the same time as appealing to many Africans, created an enormous fear: they meant both exploitation and abuse, alongside access to new resources and goods. It took a while, from the Second World War onwards, before cities acquired a significant share of the population. However, when they did, their growth seemed limitless, raising new concerns among both the population and planners.

Contemporary Dynamics of Settlement Systems in Sub-Saharan Africa

Migration was instrumental in the development of the colonial city. As ordinary traditional agriculture was not deemed sufficient to provide the raw materials that Western countries wanted for their industries, forced labor was implemented by the colonialists (Cordell *et al.*, 1996). It also served the deconstruction of the local economies and societies to the benefit of the colonial power. Plantations and mines were first targeted, but forced labor in conjunction with, or followed by, other migrant workers (pushed by tax, land scarcity, etc.) also fuelled commercial and administrative centers' growth. Factory and railway workers, office clerks, domestics, etc., migrated towards the main centers (Coquery-Vidrovitch, 1992).

Forced labor was gradually replaced by waged employment, but the colonial economy had made a deep impression on the settlement system. The same migration routes were taken by workers, as they became more involved in a monetized economy. National independence did not stop the migration trend nor did it contradict an economy based essentially on cash-crops and raw material export. The urban growth rate has never been as high as in the early post independence years of the 1960s and 1970s. Some capital cities grew by more than 10 per cent a year (as for Abidjan around 1970). This facilitated the exploitation of the continent to the benefit of the importers of the Western world. Limited industrial development (mostly secondary industry) occurred and was in many cases stimulated further after independence by import substitution policies. The main driver of urban growth, however, was urban consumption and labor needs in the tertiary sector (commerce, public services and administration) rather than by industrialization. The informal economy took the main share of employment to satisfy the consumption needs of urban dwellers (Snrech, 1994; Naudet, 1996; Charmes, 1996; Arnaud, 1998).

From the 1980s onwards, as African exports suffered from the ups and downs of world markets and as economic policy inconsistencies, mismanagement and corruption came to an intolerable level, the pace of urbanization slowed down considerably. The urban economy, no longer fuelled by export revenues, became less attractive: living conditions deteriorated, unemployment levels rose, and Structural Adjustment Programs (SAPs) did not help or even accelerated the process. The informal sector, offering cheaper products, took some advantage of the situation but this proved insufficient to sustain the overall economy. So export-oriented were the SSA economies that they failed to create the internal market that could sustain development. Dependence on external aid, debt relief and expatriate remittances increased.

It should therefore not be surprising that, at the turn of the century, the pace of urban growth appears to have considerably decreased in SSA: its level of urbanization remains the lowest in the world, estimated by the UN (2003) at 37.2 per cent in 2000 (Chapter 3). However, this is an overestimate. Using a consistent definition of urban areas (10,000-inhabitants threshold), the projected rate of urbanization of SSA would be only 30 per cent in 2000 (Moriconi-Ébrard, 1994).

Figure 7.1 uses UN data to show the percentage-point difference between urban population growth rate and total population growth rate from 1950-55 to 1995-2000 (from left to right on each line), and we compare this differential to the level of urbanization reached at the end of each period. The thick curved arrow represents a global model of evolution of the urbanization process. It does not mean that all countries should follow the same path in history, but this line represents more an ideal type of evolution than an observed standard evolution. The actual observed pattern will depend on the time when a country (or a region) started its transition into a modern economy and level of involvement in the global economy so that the speed of the transition and the urbanization level attained at each point in time will not be the same.

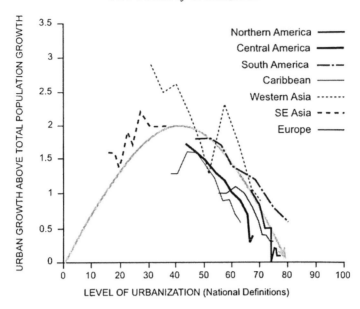

Figure 7.1   **Percentage-point difference between urban and total population growth rates in relation to level of urbanization, 1950-1955 to 1995-2000, for selected parts of the world**

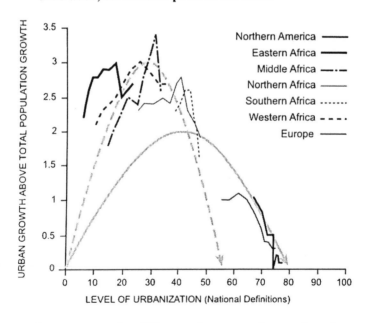

Figure 7.2   **Percentage-point difference between urban and total population growth rates in relation to level of urbanization, 1950-1955 to 1995-2000, for Europe, Northern America and regions of Africa**

When we compare the path of each developing region over the second half of the twentieth century, it is however clear that most regions followed the same pattern, except for Africa (Figure 7.2). Urbanization in Africa followed a very peculiar evolution: its urban growth was very high and well above the population growth average in the 1950s when the level of urbanization was still below 20 per cent. It reaches a peak (3 percentage points higher than population growth), and then drops when the level of urbanization (e.g. in Southern and Northern Africa) is above 40 per cent. The rapid evolution (only 50 years) represented by a thick pecked curve on Figure 7.2, is different for each part of Africa: Eastern, Middle and Western Africa are clearly lagging behind Southern and Northern Africa as far as urbanization is concerned.

Though the comparability of UN data is questionable due to a lack of consistent definitions of urban areas across countries and over time, the rapid slowdown of urbanization in Africa has been confirmed since the 1980s by other sources, albeit pertaining to a shorter historical period (Moriconi-Ébrard, 1994; Bocquier and Traoré, 2000). This slowdown does not seem to be temporary.

Several demographic factors explain why this situation will persist. In the first place, two-thirds of the rapid growth from the 1950s to the 1970s were due to reclassification and migration, whereas at the turn of the century two-thirds are due to natural growth (Arnaud, 1998; Chen *et al.*, 1998). Secondly, the young urban age structure that is still prevalent in African cities is a consequence of rapid growth in the 1950s-1970s. As urban dwellers grow older and fertility decreases in urban areas, the natural growth in the cities will be lower. Thirdly, the share of young women in rural-to-urban migration is increasing (Bocquier and Traoré, 2000). Whereas it has for a long time been thought that female migration would increase urban natural growth, this has been proved wrong by studies showing that female migrants' fertility is actually lower in the first years of migration and that it gradually increases to attain the level of other urban (nonmigrant) dwellers (Brockerhoff, 1998).

A further factor is an increase in urban-to-rural migration. This shows the importance of return migration flows (after retirement, redundancy, or simply failure to settle down or earn a living in the city). But a more disturbing phenomena is the emergence of substantial numbers of city-born persons in urban-rural flows, as in the case of the once-booming economies of Côte d'Ivoire, Cameroon and Zimbabwe where this has been studied (Beauchemin, 2000; Gubry *et al.*, 1996; Potts, 1997). Most urban-to-rural moves seem to be motivated by the high living costs and unemployment of urban areas and by the low living costs (food, school, housing) found in rural areas.

Using retrospective data from a survey in Côte d'Ivoire, Beauchemin (2000) has shown that urban outmigration outnumbered rural outmigration in the early 1980s (at the time when the country was experiencing an economic crisis). Whereas the relative risk of migrating from urban to rural areas increased gradually from as early as the 1960s, the risk of migrating from rural to urban areas decreased significantly and constantly from the early 1980s. In proportion to the population at risk, the reduction of rural-to-urban migration seems more dependent on the economic situation than the increase in urban-to-rural migration which appears to be a long-term phenomenon. Our hypothesis is that the slowdown in the urbanization process would have occurred even if the economic crisis had not struck as hard as it did in Côte d'Ivoire. This hypothesis is difficult to validate for the rest of SSA in the absence of relevant data and analysis.

A final hypothesis for the 'deurbanization' trend is that, in the outset of the AIDS epidemic, the disease strikes the urban population more than the rural, partly because the number of sexual partners is supposed to be higher in urban settings, notably through prostitution. Then, when it reaches a certain level, it should spread into rural areas. The effect on the reduction of the rural-urban growth differential would be higher in the first phase of the AIDS epidemic than in its second phase. However, data do not support this hypothesis: first, the percentage of young males and females having multiple partners (Table 7.1) or of males having commercial sex (Table 7.2) is only slightly higher in the urban areas than in the rural areas and, second, the median prevalence rate of women in antenatal care clinics (Table 7.3) is lower in major urban areas for countries with the lowest prevalence (those with less than 10 per cent). Therefore it seems that the effect of the AIDS epidemic on the rural-urban growth differential should be negligible. The effect of HIV/AIDS in reducing urban growth levels more than those in rural areas, if it occurred at all, would have been occurring only in the earliest stages of the epidemic when the disease was more strongly concentrated in urban areas from where it quickly diffused to rural areas.

Table 7.1 Percentage of young males and females having multiple partners in the past year in urban and rural Sub-Saharan Africa

	Urban		Rural		Total	
	Males	Females	Males	Females	Males	Females
Minimum	11	2	5	1	6	1
Median	44	9	38	6	39	7
Maximum	64	21	57	12	60	15

Source: Computed from DHS data of 12 countries: Burkina Faso, Cameroon, Côte d'Ivoire, Ethiopia, Kenya, Malawi, Rwanda, Tanzania, Togo, Uganda, Zambia, Zimbabwe. 1998-2000, except Zambia 1996. www.measuredhs.com/hivdata

Table 7.2 Percentage of males having commercial sex in the past year in urban and rural Sub-Saharan Africa

	Urban	Rural	Total
Minimum	2	0	1
Median	8	6	6
Maximum	19	22	21

Source: as for Table 7.1

Table 7.3   HIV/AIDS prevalence rates of women in antenatal care clinics in major urban areas and other areas in Sub-Saharan Africa

Indicator	All countries		Countries with lower prevalence (<10%)		Countries with higher prevalence (>=10%)	
	Major urban areas	Other areas	Major urban areas	Other areas	Major urban areas	Other areas
Minimum	5.2	4.4	2.7	2.0	14.4	8.2
Median	6.2	8.1	3.1	4.0	19.0	18.2
Maximum	7.3	12.6	3.4	6.6	28.0	27.1
No. of countries	39	38	22	23	17	15

Source: Computed from Epidemiological Fact Sheets by Country (UNAIDS) database. 1993-1999, centered on 1998. www.unaids.org/epidemic_update/report

On a more general scale, the macroeconomic conditions are not in favor of African urban growth. One has to consider the role of the city in the national and international economies. The concentration of population and activities depends on the relative importance of the city in the national economy and also (especially for the capital city) on the share of the national economy in the international economy. However, African economies, to say the least, rank low in the global economic system.

This international economic discrepancy has also direct consequences on the migration of labor. Just as rural-to-urban migrants' average level of education lies between the level of the nonmigrants in rural areas and the level of nonmigrants in urban areas, international migrants' education usually lies between those of the origin and destination countries. Underdevelopment encourages migration of skilled labor toward more developed countries: it can take the form of the infamous brain-drain, but it extends also to more modest parts of the population. The international migration from Africa to other parts of the world is not numerically very large: the net loss for Francophone West Africa was evaluated at only -0.09 per cent a year for the population aged 15 and more, and no more than -0.05 per cent a year for the total population (Bocquier and Traoré, 2000). However, this international migration can be qualitatively significant, as the selection-by-education effect is higher in the case of international, especially intercontinental, migration.

In Table 7.4, we computed from Network of Surveys on Migration and Urbanization in West Africa (NESMUWA)[2] data on six West African countries the

---

[2] The NESMUWA is the latest available data collection (1993) on both internal and international migrations. National surveys based on the same methodology were simultaneously carried out in 1993. Although Mauritania and Nigeria were part of the network, they have been excluded from our analysis of international migration because the surveys were not carried out on the same basis as the others.

proportion of migrants to Europe and North America (where the most educated are presumably migrating, due to higher selection effects) who originated in urban areas. In the six countries involved, the majority (51.6 per cent) of migrants come from urban areas but in Mali they are only 29.6 per cent and in Senegal 49.7 per cent. Therefore, migration is depriving not only the urban areas of their most educated labor force, but also the rural areas. Conversely, a higher proportion of immigrants (most of them return-migrants) are coming back to the urban areas of their countries of origin: 78.6 per cent for all six countries, 67.2 per cent for Mali and 69.4 per cent for Senegal. This shows not only that the rural areas are losing their skilled labor force but also that return migration, when it occurs, benefits urban areas more than rural areas.

Table 7.4   **Percentage of urban origin or destination among African international migrants aged 15 and over from six countries in West Africa**

	Percent coming from an urban area among emigrants to:				Percent migrating to an urban area among immigrants from:			
	Europe	North America	Total western countries	N migs	Europe	North America	Total western countries	N migs
Burkina Faso	71.9	n.s.	74.3	1,700	87.3	n.s.	87.3	1,000
Côte d'Ivoire	70.7	100.0	76.3	13,300	90.5	100.0	91.4	8,200
Guinea	84.4	58.7	77.6	8,500	94.3	n.s.	94.6	3,100
Mali	26.5	89.9	29.6	23,300	65.6	n.s.	67.2	4,600
Niger	100.0	n.s.	100.0	700	100.0	n.s.	100.0	1,000
Senegal	48.7	70.8	49.7	71,300	68.3	94.4	69.4	15,300
Six countries	49.2	79.0	51.6	119,200	77.4	98.2	78.6	33,400
N migs	109,900	9,300	119,200		31,500	1,900	33,400	

N migs Number of migrants extrapolated, n.s. not significant (*n*<200).

Source: Computed from NESMUWA data on migrations during the years 1988-1992 in 6 countries (Burkina Faso, Côte d'Ivoire, Guinea, Mali, Niger, Senegal).

Whereas south-to-north international migration can have positive consequences for the country of origin (in terms of investment or remittances to the family), it has not proved effective in terms of production or productivity enhancement. International migrants, qualified or not, are more often a net loss to the African economies, as they are selected from the most active and the better educated among their population of origin, be it urban or rural. The most educated among them will be integrated (often through acquisition of citizenship) in the host country, to its benefit.

Considering all the intertwining factors listed above, and contrary to the conventional wisdom, the rural way of life could stay predominant in SSA. Except for Northern Africa and the Republic of South Africa, the urbanization rate, if its current downward trend continues, could stabilize at a much lower level than previously expected. Drawing a new linear curve on Figure 7.2 that averaged the trends just for Eastern, Middle and Western Africa (rather than the line shown there for all of Africa), then it would lead to the level of urbanization stabilizing at below 50 per cent using UN data. With a more appropriate definition of urban areas (e.g. using the 10,000-inhabitants threshold), this projected rate would be only 35 per cent. By contrast, Northern Africa seems likely to join other Arab countries and Asian countries leading to a majority of the population living in urban areas. As for Southern Africa as a whole, it is not clear whether this group will join countries of intermediate level or stay behind. It all depends on the role of the Republic of South Africa as the leading country in the region.

Urban dynamics is not a cause of development, but it certainly reflects the underlying development process. It is by no means by chance that parts of the world like SSA, and possibly other parts of the least developed world, like Indo-China and the Indian subcontinent, should lag in urbanization evolution when one considers their relative rank in the global economy.

Capturing Migration and Urbanization Dynamics

To compare the different definitions of urban areas used across countries and time in SSA, we used the national definitions collected in the UN Demographic Yearbooks for the years 1967, 1975, 1985 and 1995. However, the definitions stated in those yearbooks are not consistent with those of some national reports. Therefore, operational definitions were added for the French-speaking West African countries where a more thorough investigation has been conducted in the context of the Network of Surveys on Migration and Urbanization in West Africa (NESMUWA, see footnote 2 above).

Deciphering Official Definitions

According to our compilation of urban definitions used in 38 countries in Africa (not presented in detail here), definitions provided in the UN Demographic Yearbooks are often incomplete or inaccurate. In some cases (e.g. for Benin, Burkina Faso, Guinea, Kenya), the Yearbooks list the definition as 'not stated', even though the information is available in national reports. More disturbing is the fact that the definition in the Yearbooks sometimes contradicts national reports (e.g. for Côte d'Ivoire, Mauritania, Niger). In fact for seven countries for which UN and national sources can be compared, only one (Senegal) had a consistent definition of urban areas in both sources.

There is also at least one case where both UN and national reports contradict actual practice. In Côte d'Ivoire, the 1975 definition (population and activity criteria) is said to be officially used, though in practice a list of cities is established with no consideration of the 1975 definition. This practice is probably more widespread than official (UN or national) reports indicate. When a list of cities is used, it probably means that no consistent and objective criteria are actually used at all. How the list is made remains difficult to say, as one cannot rely on the national official reports. The list method also tends to be conservative, in the sense that an urban area will rarely be removed from the list (declassification is rare) whereas others will be added.

In the case of Côte d'Ivoire, two localities which were considered as cities when they changed administrative status (to become *chefs-lieux de préfecture*) were added during the 1988 census to the 1975 list. This updated list did not take into account the different components of urban growth:

• The population growth is only considered within the existing administrative limits of the agglomeration. The analysis would then lay more stress on the increased density of urban areas than on their lateral extension. Suburban extensions will be taken into account only if the administrative limits are extended or if those extensions are listed as separate urban areas.

• The reclassification of areas from rural to urban as a consequence of population growth cannot be considered in a fixed list, even if one considers the same administrative boundaries. Urban growth will then be underestimated.

If one considers the total list of administrative localities of Côte d'Ivoire in 1988, out of the 68 officially considered as urban, only 57 had a population exceeding 10,000 inhabitants whereas 73 other (non listed) localities were above the 10,000 inhabitant threshold (Beauchemin, 2000).

Regarding demographic criteria, the highest threshold is 10,000 inhabitants (Senegal) but the 2,000 or 5,000 thresholds are more common. When the level of urbanization is as low as it is in Africa, the administration, the bureau of statistics, and researchers also tend to use low thresholds. This has led to an overestimation of African urbanization levels in international comparisons (such as in the UN's), where the 10,000 threshold is usually preferred.

When definitions do not explicitly include geographical criteria (most of them), it implies that administrative limits are used as boundaries ('implicit administrative limits'). The counting unit, i.e. the spatial unit used by the State to exercise its administrative power (be it 'localities', 'counties', 'communes', etc.) is never specifically defined in reports. The agglomeration component of the demographic definition of urbanization (concentration of population) requires a more precise definition of what are the limits of the urban space and what is the level of building block density from which a locality is considered urban.

The changes in the definitions over time, generally on the occasion of a new Census, reflect the complex interactions between, on the one hand, changes in urbanization theory and its implementation by administrators and the central bureau of statistics, and on the other hand, changes in political priorities and their implementation in administration (Dureau, 1993). Among the 38 countries we have studied, 14 had a constant definition over time. Most changes reflect the introduction of a functionalist approach to urbanization (mainly introduction of economic criteria in addition to the demographic and strictly administrative criteria), but this leads to a bias towards the analysis of economic and administrative urban centers.

The changes towards a functionalist definition of urban centers (as proposed in the 1960s and 1970s) which were applied in Africa with some delay in the 1980s and 1990s contrast with the demographic approach to urbanization stressed in most analyses of the late 1980s and 1990s (Moriconi-Ébrard, 1993; Dureau, 1993). The demographic approach appears to be less restrictive in theoretical terms (functionalist criteria can always be used after data collection, on a nominative and hierarchical list of urban centers), more practical in the field, and allows international and chronological comparisons. But this approach still leaves open the problem of the geographical criteria: which geographical units should one take into account? Administrative units or others? Satellite images have not so far been combined with demographic data to produce uniform geographical criteria in any SSA country.

Consequences of Urban Definition on Measurement of Urban Growth

Direct collection of migration data is limited in Africa as it is elsewhere in the world. In the case of Africa, this is exacerbated by poor statistical systems that prevent analyzing migration through secondary sources. However, two surveys have successfully collected information on migration in Africa: the pioneer survey was carried out in 1974-75 in Burkina Faso (former High Volta), and the second – the NESMUWA, largely inspired by this (see above) – covered eight West African countries simultaneously in 1993 (see Table 7.5). Both operations were able to measure migration flows during the five-year period preceding the surveys. Details on the data collection procedures and analysis of results can be found elsewhere (Cordell *et al.*, 1996; Bocquier and Traoré, 2000). Here we will focus on the implementation of urban definitions and their implications for the measurement of urban growth.

Table 7.5   Strata used in the NESMUWA surveys and enumeration areas update

Country	Date last census (sampling base)	Number of EAs	Strata (urban/total)	Update since census
Burkina Faso	1985	333	10 / 20 = 50%	1993 (2 main cities)
Côte d'Ivoire	1988	297	20 / 30 = 67%	None
Guinea	1983	368	5 / 9 = 56%	1992 (DHS)
Mali	1987	547	15 / 22 = 68%	None
Mauritania	1988	317	13 / 28 = 46%	1992
Niger	1988	310	16 / 30 = 53%	None
Nigeria	1991	n/a	30 / 83 = 36%	None
Senegal	1988	307	19 / 29 = 66%	1992/93 (EDS)
Average	5¾ years before the survey	354 (Nigeria excluded)	58% (Nigeria excluded)	4 years before the survey

Source: Bocquier and Traoré (1998).

It must first be stressed that all surveys used the urban definition from the last census. There was no attempt to use a different or better definition, nor was there an attempt to update the number or the extent of urban areas. Generally the stratification led to sampling a larger number of urban areas than rural areas (58 per cent on average, to be compared with a 30 per cent urbanization rate for the seven NESMUWA countries apart from Nigeria). This was to get a better evaluation of the migration flows towards different categories of cities in the country (capital city, main other cities and other cities) and also to evaluate the flows between those urban categories.

The conservative definition of urban areas probably led to an underestimation of urban growth. New towns emerging since the last Census (i.e. during the four years before the survey after consideration of the update) because of migration or because of reclassification will not be taken into account. Also the extension of existing cities outside their boundaries (as defined in the last census) will not be considered. Migration from the city center towards the outskirts of the city will then be considered as urban-to-rural migration if the destination area is still considered rural. In short, the urban growth as measured by the surveys essentially reflects the increasing (or decreasing) density of urban centers within outdated limits.

It is very difficult, using published data, to evaluate the impact of urban definitions on the measurement of the level of urbanization in SSA. A more thorough investigation has been done by Beauchemin (2000) who evaluated the impact of different urban definitions in Côte d'Ivoire through the EIMU 1993 survey (the Ivorian part of the NESMUWA project). To do so, he classified all of the 297 sampled enumeration areas according to the population size of the agglomeration they belonged to at the time of the census (1988), along a hierarchical scale (less than 5,000 inhabitants, 5 to 10,000, 10 to 20,000, 20 to 40,000, 40 to 80,000, more than 80,000 and Abidjan). This classification considers, but does not question, the validity of the official boundaries of localities at the time of the Census. Nevertheless, the comparison of the results obtained with the official urban definition and with the demographic definition is still interesting.

The change to a demographic definition of urban areas (10,000 threshold) leads to the urbanization rate being 40.6 per cent instead of the official 39.0 per cent in 1988 using a list of urban localities (Beauchemin, 2000). According to the EIMU 1993 survey, the urbanization rate increased to 40.7 per cent using the demographic definition and to 40.1 per cent using the official definition (Table 7.6). The difference is not large (0.66), but 1.23 per cent of the population would not be classified in the same category depending on the definition: according to the official definition 0.28 per cent are considered urban though they live in a locality of 5 to 10,000 inhabitants, and conversely 0.95 per cent are considered rural though they live in a locality of more than 10,000 inhabitants (0.20 per cent in 10 to 20,000 and 0.75 per cent in 20 to 40,000 inhabitants). The misclassification is even worse (7.24 per cent in total, i.e. 0.95 per cent plus 6.29 per cent are considered rural though living in a locality of 5 to 10,000 inhabitants) if one considers a lower urban threshold of 5,000 inhabitants.

Table 7.6 **Urban and rural areas in Côte d'Ivoire in 1993, according to the official and demographic definitions**

Demographic definition Class of locality in 1988 (in 000s)	Official definition					
	Urban areas		Rural areas		Total	
	000s	%	000s	%	000s	%
0 to 5	0	0.00	7202.2	52.69	7202.2	52.69
5 to 10	38.6	0.28	860.5	6.29	899.1	6.58
10 to 20	131.7	0.96	27.1	0.20	156.8	1.16
20 to 40	1079.5	7.90	102.5	0.75	1182.0	8.65
40 to 80	587.4	4.30	0	0.00	587.4	4.30
80 and more	1162.2	8.50	0	0.00	1162.2	8.50
Abidjan	2478.3	18.13	0	0.00	1162.3	18.13
Total	5477.8	40.07	8192.2	59.93	13670.1	100.00
Urban (5000 threshold)	5477.8	40.07	990.0	7.24	6467.9	47.31
Urban (10000 threshold)	5439.2	39.79	129.5	0.95	5568.8	40.73

% is percentage of total population

Source: Enquête Ivoirienne sur les Migrations et l'Urbanization 1993; Cris Beauchemin (2000), for allocation of localities according to population size.

Table 7.7 **Annual net migration rate in Côte d'Ivoire, 1988-1992, for the population aged 15 and over**

	Including international migration	Without international migration
From official definition (68 localities)		
Rural areas	+1.98	+0.99
Urban areas	-0.06	-1.22
From demographic definition (>10,000 people)		
Rural areas	+2.02	+1.15
Urban areas	-0.21	-1.57

Source: Enquête Ivoirienne sur les Migrations et l'Urbanization 1993; Cris Beauchemin (2000), and personal communication.

One notices immediately from Table 7.7 that, contrary to the conventional wisdom, the rural population increased due to net migration during the 1988-1992 period. This surprising effect is not due to a conservative definition of urban areas. On the contrary, a change in the demographic definition (population of 10,000 inhabitants or more) increases the growth differential between the urban and the rural areas. This differential is 2.04, i.e. 1.98 - (-0.06), with the official definition (when including international migration), and it reaches 2.23 with the demographic definition, an increase of more than 9 per cent (23 per cent without international migration).

Of course, the change in definition would not have the same consequence for all countries and at all times. The definition effect will depend on the type of official definition used and on the urbanization trend. There remains to see the effect of revising the official boundaries of the localities at the time of the Census: this would lead to extending the geographical limits of the agglomerations. But this extension would probably not change the estimation of the overall urbanization rate or of the net migration rate, as it would lead to an inclusion of small urban centers within bigger ones.

The example of Côte d'Ivoire not only shows the potential bias in using a list of urban areas, but it shows the merits of using the population size criteria, i.e. an exhaustive list of localities classified by size. Even though the localities' geographical limits are defined along administrative criteria, the list of localities and their population size allows us to group them according to the needs of the classification. This can be done at two levels: first, the list enables us to choose the appropriate threshold (be it 5,000, or the more internationally accepted 10,000 inhabitants) for comparison purposes, and second, the list enables us to group localities into agglomerations to take account of the spatial extension of cities. Of course, it would be best to use a smaller and more neutral unit than the administrative locality but a list of localities ranked by size is generally more available than a list of cities classified by economic and administrative criteria.

From an historical standpoint, the urban areas existed first as political space (where the city limits were mainly defined by an administrative authority) and then some evolved to the rank of agglomerations with boundaries defined by city planners using building-density criteria. In a later stage, the morphology criterion gives way to connectivity criteria: the metropolitan areas of the developed world are defined by flows around one or several centers (Moriconi-Ébrard, 2001) and migration is no longer the engine of urban growth which occurs largely through an extension of the network based on a city.

With the possible exception of Johannesburg-Pretoria (with about 12 million inhabitants) in South Africa, most urban areas in Africa – even the largest among them (like Lagos or Ibadan in Nigeria) – follow the classical scheme of an urban agglomeration. The use of modern technology (from public transport to information technology), and the predominance of services in the employment structure, are not the main characteristic of African cities, where a large bulk of the work force is still occupied in commerce and crafts, notably in the so-called informal sector. In some circumstances, mainly civil wars, some cities in Africa revert to the political criteria. Freetown, Kampala, Monrovia, Mogadishu, Bujumbura, etc., during their wars, became fortified cities (a refuge for the persecuted population) held by one political force.

The historical evolutions outlined in Figures 7.1 and 7.2 do not mean that each country or region in the world is bound to pass through the same steps towards a fully urbanized (and metropolitan) society. The international division of the economy leads to increasing disparities between the poorest countries where the raw materials are produced (as in Africa) and the richest countries where human capital prevails. Therefore, if countries are ranked along an urban scale that reflects the level of economic capacity, SSA lags behind other areas.

In this context, simple population density criteria are able to depict most of the settlement schemes encountered in SSA. The distinction between urban and rural areas, however imperfectly measured using the present official definitions, is still valid for depicting differences in housing as well as employment characteristics. Published census data being at an aggregate level, it is difficult to confirm the above hypothesis. For that we would need small-scale analyses of the continuum between urban centers and rural areas, which are not currently available.

Nevertheless, profession and housing conditions strongly depend on the size of agglomeration, however imperfectly the latter is measured. Again using the Côte-d'Ivoire example, Tables 7.8 and 7.9 show for both these a clear-cut distinction between rural and urban areas, and then a more gradual change of characteristics along the urban scale. Concerning profession (Table 7.8), an overwhelming proportion of the active population in rural areas (about 85 per cent for both sexes) is engaged in farming, which is to be expected. Then this proportion falls abruptly in small cities for males (31 per cent) but even more so for females (only 17 per cent). This shows that, even for agglomerations that are close to rural areas, the employment profile is already much urbanized. Perhaps even more significant is the proportion of females engaged in trades: from less than 10 per cent in rural areas, their proportion varies between 60 per cent and 70 per cent in urban areas.

Table 7.8   Main profession by sex in different areas of Côte d'Ivoire for the active population aged 6 and over, 1993

Sex	Main profession	Abidjan	Main cities	Other cities	Rural	Total
Males	Intellectuals	28.96	17.11	16.54	2.66	9.63
	Factory workers	22.09	21.51	19.69	7.66	12.52
	Craftsmen	25.97	30.54	20.37	3.09	11.13
	Traders	20.61	20.84	12.12	1.98	7.77
	Farmers	2.37	10.00	31.28	84.60	58.95
	Total	100.00	100.00	100.00	100.00	100.00
Females	Intellectuals	13.62	7.94	5.52	0.81	4.10
	Factory workers	14.33	8.93	10.55	1.23	5.01
	Craftswomen	10.71	9.28	6.41	0.80	3.75
	Traders	61.11	69.91	60.12	9.75	28.18
	Farmers	0.23	3.95	17.41	87.42	58.95
	Total	100.00	100.00	100.00	100.00	100.00

Source: Computed from the Enquête Ivoirienne sur les Migrations et l'Urbanization 1993.

As for household characteristics (Table 7.9), only 7 per cent of households live in a rented house in rural areas as compared to almost 49 per cent of households in small cities, rising to 73 per cent in Abidjan. Clearly, renting of housing is associated with urbanization. Walls constructed of hard materials (cement, bricks, concrete) are also a function of the type of area, rising from 31 per cent in rural areas to 78 per cent to 92 per cent in urban areas. One notices a typical form of habitat in Abidjan, makeshift houses where walls are mainly made out of wood. Sewerage mains equip less than 43 per cent of houses in rural areas compared to 84 per cent in small cities. The use of modern types of energy (electricity, gas, paraffin) for cooking is expected in urban areas, but what is even more significant is the use of charcoal (from 32 per cent in small cities up to 64 per cent in Abidjan, but less than 3 per cent in rural areas) compared to the use of wood (96 per cent in rural areas, 61 per cent in small cities down to 11 per cent in Abidjan). Also different is the access to electricity for lighting and to taps or vendor supplied water, and also possession of a television: here again, the data show a clear cut difference between rural and urban areas, and in urban areas, a hierarchy from the smaller to the bigger cities. Though the situation depicted here for Côte-d'Ivoire cannot be generalized to all SSA, it certainly suggests further investigation on the (dis-) continuities of the urban network elsewhere in the continent.

Table 7.9   Proportion of households with selected housing characteristics and equipment in different areas of Côte d'Ivoire, 1993

		Abidjan	Main cities	Other cities	Rural	Total
Status of occupation	Landowner	19.95	30.43	37.72	79.74	58.30
	Rental	73.39	61.99	48.76	6.99	30.30
	Housed	6.65	7.58	13.51	13.27	11.39
	Total	100.00	100.00	100.00	100.00	100.00
Type of walls	Cement, bricks or concrete	78.23	92.34	80.94	31.11	52.24
	Wood	20.33	1.41	5.50	1.14	5.34
	Mud and other	1.44	6.25	13.57	67.75	42.42
	Total	100.00	100.00	100.00	100.00	100.00
Type of sanitation	Main sewer	30.53	13.06	10.01	2.02	9.59
	Latrine	61.82	86.09	74.07	40.82	53.38
	Other	7.65	0.85	15.92	57.16	37.03
	Total	100.00	100.00	100.00	100.00	100.00
Type of cooking	Electricity, gas, paraffin	24.74	6.85	7.07	1.18	6.99
	Charcoal	64.35	51.18	32.29	2.71	23.09
	Wood	10.91	41.97	60.64	96.10	69.93
	Total	100.00	100.00	100.00	100.00	100.00
Access to	Electricity	74.63	67.14	52.76	12.22	34.62
	Commercialized water	81.43	38.94	30.19	8.57	28.28
	Television	44.70	32.79	28.04	5.33	18.37

Source: Computed from the Enquête Ivoirienne sur les Migrations et l'Urbanization 1993.

Conclusions: Alternatives to Existing Classification Schemes

The low population threshold used in most African official statistics that leads to the overestimation of urbanization levels probably counterbalances the underestimation of cities' extension outside administrative boundaries and (unknown) reclassification from rural to urban areas. Though this is difficult to prove, it has certainly led to an overestimation of urban dispersion and underestimation of the urban primacy of the capital city. However, those distortions, which can appear important at the national level, are not so great for the evaluation of urbanization trends at the continental level. Using the national definition (as in the UN data) or a shared and consistent definition (e.g. using a 10,000 inhabitants threshold) leads to the same result: the pace of urbanization in Africa slowed down rapidly in the 1980s and 1990s, and the primacy indices (share of the capital city among urban areas) remained relatively constant over time.

Being underdeveloped, most African cities have not exhibited many new forms of settlement seen in the Western world ('rurbanization', transition zones, etc.). The morphology of African cities is rather classical and reflects the simple structure of the African economies, most 'upper order functions' being fulfilled in other countries by way of new technologies. What is less classical is the way Africans (as is also the case in other poor parts of the Developing World) are maintaining links between rural and urban areas, and not only in terms of remittances. Migration is a dominant demographic phenomenon which is not fully understood (partly because of the lack of data) and on which the future pattern of urbanization depends. Whether migration has the effect of extending the urban way of life (even in rural areas), or restraining it, is not clear. Neglected movements from urban to rural areas need to be explained better as they can be part of the explanation of the recent slowdown in urbanization trends. In that regard, future migration surveys should pay particular attention to the identification of the different residences the migrants occupied during their lives. A rural-urban dichotomy or the simple administrative identification is not enough, but that can only be improved if a proper codification scheme (with a list of localities) is available and used.

The lack of reliable and continuous statistical systems in most African administrations may prove paradoxically an advantage to improve the monitoring of urbanization. Because it is not possible to expect much from existing systems, in a context whereby the administration has less and less the means and power to improve them, it is not unreasonable to think of a great technological leap forward. In the absence of a permanent statistical system at the local level in most African countries, satellite images could be instrumental in a time-evolving definition of urban agglomerations (focusing on their morphology rather than on their political dimension) that could readily be implemented in African censuses using enumeration areas (EAs) as the smallest geographical units. This would not contradict the use of administrative limits for other purposes so the new definitions will encounter less (political) opposition than, say, a redefinition of current administrative boundaries, be it in town or in rural areas.

Satellite images could be used to delimit agglomerations independently of administrative boundaries, which have been the main impediment to a better definition of urban areas. The main criteria would then be the density of building blocks per square kilometers, which would replace the demographic criteria (be it 5,000 or 10,000 inhabitants within administrative boundaries). In more practical terms, satellite images could be used to delimit EAs for the census or for other surveys. The EAs would then become the smallest spatial unit to agglomerate population in towns. In most African censuses they include between 500 and 1,000 individuals. A threshold for building blocks density would have to be chosen after examining the correlation with other variables typical of urban areas (infrastructure, activities, etc.) at a global level, so as to allow for international comparison. A building blocks density scale (along 4 or 5 thresholds) could also be used and combined with population thresholds (on a quasi-exponential scale in thousands of inhabitants, such as 5, 10, 50, 100, 500, 1,000, etc.) to offer a subtler categorization of settlements. This morphologic approach, as opposed to a functionalist approach, is probably flexible and operational enough to be implemented in all African countries.

This revision of urban definition should also be linked to the revision of Census procedures. Censuses are costly for public budgets in Africa, and the lapse between two censuses is usually deemed too long, particularly to provide for reliable and up-to-date sampling base for surveys. A permanent system of partial censuses would probably be better to provide for recent economic, social and demographic data. A census using satellite images and providing simple lists of building blocks and households for about a quarter of the EAs in the country every three years would serve as a master sampling base for more thorough surveys. The urban areas could be updated every nine years but the 3-years extrapolation could be used to evaluate urbanization growth and characteristics (INSEE, 2000). The new morphological definition detached from functionalist criteria could always be attached to the basic density and population criteria at a later stage of the analysis. To be implemented and useful for cross-countries comparisons, the new method would have to be agreed on internationally through a body like the UN.

PART III
CASE STUDIES

Chapter 8

The Transformation of the Urban System in Mexico

Gustavo Garza[1]

In the first of four country-based case studies of recent urban transformation, this chapter examines the situation in Mexico. This is an obvious choice for case-study treatment for several reasons, most notably because of switching from being a predominantly rural country to a mainly urban one within a lifetime. The pace of change was particularly rapid between 1940 and 1980 when the proportion of the national population living in localities of at least 15,000 inhabitants leapt from 20 to 55 per cent. Subsequently, while the pace of urbanization has slowed somewhat due mainly to economic factors, Mexico has received considerable attention as one of the first countries in Latin America to have seen the growth rate of its intermediate-sized cities apparently overtake that of its largest urban centers. This is, however, partly statistical artifact, in that Mexico City – hemmed in by mountains – has emerged as the driving force behind the emergence of a set of interlinked metropolitan areas that together constitute one of the world's largest concentrations of urban population.

The objectives of this chapter are twofold. First, the chapter describes the principal dimensions of urbanization and settlement change, concentrating on the period since 1970 and giving particular attention to departures from established trends. This examination draws extensively on the results of previous studies, notably Unikel *et al.* (1976), Graizbord (1988), Consejo Nacional de Población (1994), Ruiz (1994), Aguilar and Hernández (1995), Garza and Rivera (1995), Aguilar *et al.* (1996) and Garza (2001). Necessarily, this account portrays developments largely through the 'spectacles' afforded by contemporary conceptual and definitional frameworks, starting with trends in level of urbanization and distribution of city sizes and also examining the main features of large-city changes. The second objective, therefore, is to assess the adequacy of the picture provided by these conventional measures and discuss what changes are needed in order to provide a sound basis for monitoring and analyzing the evolution of Mexico's urban system over the next few decades. Within the context of a projected urbanization level of 92.4 per cent in 2050 (Garza, 2000), this part of

[1] I am very grateful to Rosalía Chávez for her important assistance in processing the statistical information and the bibliographical reviews. I would also like to express my thanks to Jaime Ramírez, Emelina Nava and Raúl Lemus, for their cartographic support.

the chapter begins by questioning the continued use of an urban-rural dichotomy in Mexico. The chapter goes on to discuss the definitional challenge posed by the growth of metropolitan areas arising from accelerating urban decentralization, and finishes by considering the implications of the emergence of a polynuclear 'megalopolis' in the central region dominated by Mexico City.

Urbanization and Uneven Population Development in Mexico

Defining Urban Localities in Mexico

The analysis of urban development requires an empirical city definition in order to classify the population into urban and rural components, if adopting a dichotomous approach, or into groups of places, normally by population size, if following the *continuum* theory. In Mexico the first censal definition was made in 1910, when localities with 4,000 or more inhabitants were considered urban. In the 1921 census, the threshold was reduced to 2,000 people, and from the census of 1930 to that of 1960 it was 2,500 inhabitants. From the 1970 census onwards, however, following some international recommendations to allow the use of alternative cut-off points (see Unikel *et al.*, 1976), there has not been any official definition of urban population. Instead, since then the census data has been tabulated by size of localities. The census of 2000, for instance, provides national-level information for eight population-size groups as follows: 1-2,499; 2,500-4,999; 5,000-9,999; 10,000-14,999; 15,000-49,999; 50,000-99,999; 100,000-499,999; and 500,000 and over. In addition, it is possible to obtain information for every single municipality and locality from the census volumes for individual Mexican states.[2]

The choice of urban threshold by both academics and the public sector continues to be greatly influenced by a major research project carried out in Mexico in the 1970s (Unikel *et al.*, 1976). In that study, selected characteristics were examined for localities aggregated into five groups spanning where the threshold was expected to be found, starting at 2,500 and with breaks at 5,000, 10,000, 15,000 and 20,000. In all, five variables were considered: labor force in nonagricultural activities; population literate; population with primary schooling; salaried employed population; and Spanish-speaking population and people using footwear.

Analyzing each variable for these intervals (for instance, 94 per cent of the localities of 15,000-19,999 had more than 70 per cent of nonagriculture labor force, compared to 59 per cent of localities in the 10,000-14,999 class and only 8 per cent for the 5,000-9,999 size range), the study recommended, in a kind of *continuum* approach, the following classification of the localities: (i) rural: less than 5,000

---

[2] Localities are places with at least one permanent dwelling and are delimited on the basis of continuous built-up area. In 2000 there were 199,369 localities distributed between 2,428 municipalities, which are the lowest political division in the 31 states. The biggest locality is the urban area of Mexico City with 18 million inhabitants, while there were 182,335 localities with less than 500 inhabitants in 2000.

inhabitants; (ii) mixed-rural: 5,000-9,999; (iii) mixed-urban: 10,000-14,999; (iv) urban: 15,000 and over (Unikel *et al.*, 1976, p.347). Following a macroanalysis of the urban process in Mexico between 1900 and 1970, it was concluded that only localities of more that 15,000 people could be considered clearly urban. Given that practically every Mexican urbanization study since then has adopted this criterion, this is what is done here.

Urbanization Stages

Mexico has undergone a continuous process of urbanization since the beginning of the twentieth century (Table 8.1). In 1900 the total population of Mexico was 13.6 million people, of which only 1.4 million lived in urban localities, giving a 10.6 per cent level of urbanization. The Mexican Revolution interrupted overall population growth between 1910 and 1921, but by 1940 the total had grown to 19.6 million, of whom 3.9 million, or 20 per cent, were urban dwellers. The next 40 years saw, alongside very strong economic growth, a doubling of the national population and a tenfold increase in the number of people living in localities of more than 15,000 people, resulting in urbanites being in the majority before the end of the 1970s. By the year 2000 the proportion had risen to two out of every three people.

Table 8.1   Mexico: Population by rural and urban sectors, level and rate of urbanization, 1900-2000

Year	Population (millions)			Urbanization		Annual change in urban population (thousands)
	Total	Rural	Urban	Level (%)	% change in level per year	
1900	13.61	12.17	1.44	10.6	n.a.	n.a.
1910	15.16	13.38	1.78	11.8	1.1	35
1921	14.34	12.24	2.10	14.6	2.0	32
1930	16.55	13.66	2.89	17.5	2.1	79
1940	19.65	15.72	3.93	20.0	1.3	104
1950	25.78	18.57	7.21	28.0	3.3	328
1960	34.92	21.41	13.51	38.7	3.2	631
1970	48.23	25.50	22.73	47.1	2.0	922
1980	66.85	30.11	36.74	55.0	1.5	1,401
1990	81.25	29.76	51.49	63.4	1.5	1,475
2000	97.48	31.83	65.65	67.3	0.6	1,416

Source: Garza (2001, Table 2).

The pace of this process, however, has not been consistent over time, as is also clear from Table 8.1. Broadly, three different periods can be identified in Mexico's urban experience over the past century: 1900-1940, 1940-1970 and 1970-2000.

Their precise characterization, however, depends on whether the main emphasis is on the rate of urbanization (as is useful for linking these trends to economic and social change) or the absolute increase in the urban population (as required for studying the needs of urban services, infrastructure, etc.). Using the former, defined as the relative change in the level of urbanization, the three broad stages can be labeled as: (i) *moderate*, at 1.5 per cent per year between 1900 and 1940; (ii) *rapid*, at 2.7 per cent from 1940 to 1970; (iii) *slow*, at 1.2 per cent from 1970 to 2000. In terms of the absolute increase in the total urban population, the same periods translate into: (i) *low growth* with a yearly average increase of 62 thousand; (ii) *medium growth* with an annual average of 627 thousand; (iii) *rapid growth* with 1.4 million extra a year. The focus of the remainder of this chapter is on the last of these stages, associated with the largest absolute volume of growth in Mexico's urban population of the past century but at the same time with a slowdown in the pace of urbanization.

Change in Mexico's Urban Hierarchy, 1970-2000

The last three decades have witnessed impressive changes in the distribution of population across Mexico's urban hierarchy. At the same time as the level of urbanization rose from 47 to 67 per cent, the share of the urban population living in large cities – localities of at least a half a million people – increased from 52.7 to 69.4 per cent. Yet, while this indicates an increasing degree of concentration in large cities, it also represents a spreading of population in this part of the urban hierarchy. As is shown by Table 8.2, the number of these large cities shot up from only 4 in 1970 to 28 in 2000, with three-quarters of the increase being accounted for by localities of between a half and one million inhabitants. Even the 50 per cent of urban people living in the million-plus cities was spread across 9 localities in 2000 as opposed to just 3 in 1970. Moreover, it was only in the final decade that the main surge occurred in the number of these very large cities: up to 1990, their contribution to Mexico's urban population had been slipping back according to the analysis in Table 8.2. It is to this feature of urban-system change – the proliferation of large cities – that the following decade-based account therefore gives most attention.

Continuation of a Highly Primate Urban Hierarchy, 1970-80

At the top of the urban hierarchy in 1970 were only four large cities. Measured in terms of their metropolitan zones (see Garza, 2001, and the relevant section later in this chapter), Mexico City – with 8.6 million inhabitants – was by far the largest, 5.8 times the 1.5 million of second-placed Guadalajara, a level of primacy far in excess of the figure of 2 expected of a normal rank-size relationship. Monterrey came next, with 1.2 million, followed by Puebla with 623 thousand. Clearly at this time, the Mexican urban hierarchy was highly primate, with one city as its core.

Table 8.2   Mexico: Population distribution by city size, 1970-2000

	Urban total	Small 15,000 to 19,999	20,000 to 49,999	Sub-total	50,000 to 99,999	Midsize 100,000 to 499,999	Sub-total	500,000 to 999,999	Large 1,000,000 and more	Sub-total
1970										
Urban pop.	22,730	740	2,123	2,863	1,750	6,142	7,892	629	11,346	11,975
%	100.0	3.3	9.3	12.6	7.7	27.0	34.8	2.8	50.0	52.7
Cities	174	43	72	115	25	30	55	1	3	4
1980										
Urban pop.	36,739	947	2,947	3,894	1,633	10,275	11,908	2 553	18,384	20,937
%	100.0	2.5	7.8	10.3	4.3	27.3	31.7	6.8	48.9	55.7
Cities	227	55	96	151	24	44	68	4	4	8
1990										
Urban pop.	51,491	1,396	3,755	5,151	2,800	10,990	13,790	10 076	22,474	32,550
%	100.0	2.7	7.3	10.0	5.4	21.3	26.8	19.6	43.6	63.2
Cities	304	80	124	204	39	42	81	15	4	19
2000										
Urban pop.	65,653	1,205	4,810	6,015	3,259	10,815	14,074	12 590	32,974	45,564
%	100.0	1.8	7.3	9.2	5.0	16.5	21.4	19.2	50.2	69.4
Cities	350	70	164	234	46	42	88	19	9	28

Source: Garza (2001, Table 2).

By 1980, at the end of Mexico's four decades of high economic growth, the overall situation was not far different. Mexico City's population had now reached 13 million, making it the third largest city of the world according to the UN (2003). Even though Guadalajara had grown to 2.3 million, the index of primacy was still as high as 5.7. Puebla, now with 1.1 million, had joined the rank of cities with more than one million inhabitants, while four other cities had reached the 500 thousand mark, these being León, Torreón, Toluca and Ciudad Juárez (Garza, 2001). Along with the growth of the three largest cities, this pushed the large-city share of Mexico's urban population up from 52.7 to 55.7 per cent (Table 8.2). Clearly, the high degree of primacy at the very top of the national urban hierarchy continued, though there were initial signs of a more diffuse pattern of large-city growth.

Urban Growth Patterns in the 'Lost Decade', 1980-1990

While annual GDP growth slumped from 6.6 per cent in the 1970s to only 1.6 in the 1980s and indeed was stationary between 1982 and 1988 (Garza, 2001), Mexico's urban system underwent two apparently contradictory trends. On the one hand, the overall pace of urbanization during this 'lost decade' continued at much the same level as in the previous ten years. On the other hand, growth at the top of the urban hierarchy was checked somewhat.

Within this broad picture, several features are worthy of mention. In the first place, while the economic conditions were bad enough at national level, they impacted especially severely on the more rural parts of the country. Between 1980 and 1988 the number of waged jobs in agriculture dropped by 9 per cent, while those in manufacturing rose by 27 per cent and those in commerce and services by 73 per cent (Rendón and Salas, 2000). Along with real wages collapsing to barely two-fifths of their 1982 level and a significant rise in the proportion of nonwaged workers (Garcia, 1994; Hernández, 2000), such a dramatic change in the labor market not surprisingly pushed more rural people to the cities and helped the proportion of national population living in localities of at least 15,000 inhabitants rise from 55 to 63 per cent (Table 8.1).

The economic crisis's clearest impact on the population distribution was, however, the check on the growth of cities at the very top of the urban hierarchy. No new cities entered the million-plus category, while its four cities did not grow as fast as the urban population as a whole, seeing their share of the latter fall quite steeply to under 44 per cent (Table 8.2). On the other hand, this did not prevent a substantial increase in the urban share of all large cities which was swelled by the incorporation of eleven new urban areas with between a half and one million inhabitants (Tijuana, San Luis Potosí, Mérida, Mexicali, Culiacán, Acapulco, Coatzacoalcos-Minatitlán, Tampico, Querétaro, Aguascalientes and Chihuahua).

An indication of the way in which Mexico's urban system was evolving at this time can be obtained by highlighting the types of cities that matched or exceeded the 3.5 per cent annual growth in urban population over the decade (Garza, 2001). Five main groups can be identified:

- Manufacturing cities around Mexico City that had grown very fast since the 1960s: Puebla (with an annual growth rate of 4.1 per cent between 1980 and 1990), Querétaro (10.5), Pachuca (6.6), Tlaxcala (6.6) and San Juan del Rio (8.7). Also, the city of Toluca had a growth rate of 3.4 per cent, only slightly below the average, and at this time its metropolitan expansion began to overlap with Mexico City's.
- Cities located in the regions along the northern border with the United States, principally Tijuana (5.7 per cent growth a year), Ciudad Juárez (4.0), Mexicali (6.0), Matamoros (5.0), Nogales (4.3) and Piedras Negras (3.7). The main determinant of their dynamics at that time was the high growth of 'maquiladora' manufacturing firms (inbond plants), up in number from 620 in 1980 to 1,703 in 1990 and in total workforce from 119 to 446 thousands (Bendesky *et al.*, 2001).
- Fast growing tourist and port cities, the outstanding example being Cancún with annual growth of 18.6 per cent. Other important cases are Acapulco (7.2 per cent), Puerto Vallarta (9.5), and some colonial tourist cities as Oaxaca (6.2), Guanajuato (4.2), and San Miguel Allende (5.1).
- Some interior manufacturing cities such as Saltillo (5.6), Aguascalientes (6.6) and San Luis Potosí (3.5).
- Some central places in capitalist agriculture regions, including Culiacán (5.9), Hermosillo (5.2), Celaya (8.4), Irapuato (8.1) and Los Mochis (9.7).

In sum, therefore, the 1980s economic crisis did not stop urban development in Mexico, but just halted the growth of the four main metropoles. Perhaps the most impressive new features are the extent to which cities of over 0.5 million were added to the urban hierarchy and the strong growth of a range of medium-sized cities. These served to strengthen the network of urban centers across the country, notably along the US border but in other regions too, gradually helping to reduce Mexico City's dominance in the urban system. This effect, however, was partly offset by the fact that many of the largest population gains were recorded by cities situated relatively close to Mexico City, which together were beginning to evolve into a single, more integrated population concentration.

Towards a 'Polycentric' Urban Hierarchy, 1990-2000

Despite the Mexican economy recovering to moderate growth in the 1990s, with GDP up by 3.5 per cent a year, the rate of urbanization slowed. This can be explained, in part, by the relatively high degree of urbanization reached by this stage – 67.3 per cent in 2000, not far behind the average for the developed world especially allowing for the urban size threshold being higher than most other countries. At least two other factors also contributed. With agricultural production bouncing back after its collapse in the 1980s, rural population switched from decline to a 2.1 million increase (Table 8.1). Also important was the acceleration of Mexican emigration to the United States, totaling 2.2 million for the 1990s, three times more than in the previous decade (Corona, 2002; Morales, 1989).

Nevertheless, the proliferation of large cities that had characterized the 1980s now proceeded further. On the one hand, the four largest cities continued to grow more slowly than the 2.5 per cent a year of Mexico's urban population as a whole, though the four largest cities – by virtue of their large base – were still absorbing nearly one-third of the total urban population increase. Meanwhile, the cities with above-average growth rates were very largely the same as in the 1980s, namely:

- Cities in the hinterland of Mexico City, including Toluca, Cuernavaca, Pachuca, Tlaxcala and San Juan del Río.
- Manufacturing cities out of the immediate influence of Mexico City and the border, such as Hermosillo (in the state of Sonora), Saltillo (80 kilometers southwest of Monterrey), San Luis Potosí (located midway along the main NAFTA road corridor) and Aguascalientes (two hours by road north of Guadalajara).
- Tourist port cities such as Cancún, Puerto Vallarta, Zihuatanejo and Cozumel.
- Northern border cities headed by Tijuana and Ciudad Juárez, plus others in the northwest such as Nogales and San Luis Río Colorado as well as Reynosa, Matamoros and Nuevo Laredo in the northeast.

The overall result was a filling out and deepening of the Mexican urban system. The dominance of Mexico City was further counterbalanced by the appearance of five new million-plus cities as well as the growth of the three other existing ones. Meanwhile, transport infrastructure improvements, not least along the main NAFTA road corridor, helped to increase the degree of system integration. This is particularly the case for the fast-growing northeast border cities, being part of Monterrey's subsystem of cities and having close contact with the highly urbanized central area of Mexico via the NAFTA corridor (Garza, 1999). At the other extreme, however, the cities of northwest Mexico remain very largely separate from the rest of the national urban system, as in the case of Tijuana that is located 3,000 kilometers away from Mexico City and '... clearly falls under the orbit of the San Diego cititstate ...' (Peirce, 1993, p.5).

Mexico thus began the new millennium with a system of cities that was considerably transformed from that of 30 years earlier. In particular, the trebling of the number of million-plus cities has led to a much broader basis for urban population growth than was provided by just Mexico City, Guadalajara and Monterrey in 1970. As such, the Mexican urban system can be said to have moved from having a highly primate city rank-size distribution towards what might be called a more 'polycentric' urban hierarchy. Even in 2000, though, Mexico City's population was still greater than that of the eight other million-plus cities combined, at 18 million compared to their 15 million. Moreover, two of these – Puebla and Toluca – are within the hinterland of Mexico City and, along with other nearby cities like Querétaro and Cuernavaca (with 787 and 660 thousand people in 2000), are helping to consolidate the weight of this central region within the national urban system. The only components of the Mexican urban system that would seem to stand some chance of competing with this evolving polycentric urban concentration are, firstly, the northeast subsystem based on Monterrey, some

800 km north of Mexico City and straddling the NAFTA corridor, and Guadalajara, the center of the Bajio high-density urban region some 300 km west of the capital (Figure 8.1).

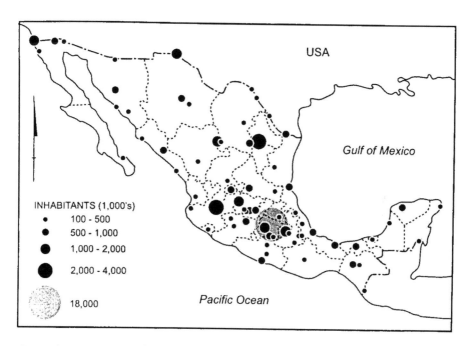

Figure 8.1   Mexico: Cities with more than 100,000 inhabitants, 2000

The Metropolitan Character of Recent Urbanization

Now that Mexico has reached such a high level of urbanization, it is important to consider other ways of measuring urban change that capture more satisfactorily the evolving distribution of population and economic activity in the national landscape. In this section, therefore, the focus is on the Mexican experience of seeking alternative representations of the new patterns of urban development. Primary attention is given to the efforts made so far towards identifying a metropolitan dimension to the country's settlement pattern, with a particularly close examination of the changes taking place in the region dominated by Mexico City.

Metropolitan Areas in Mexico

Increasingly over the last three decades, urbanization in Mexico has taken on a metropolitan character, in that the builtup areas of cities have spread beyond their administrative boundaries and physically separate localities nearby have fallen under their influence, as evidenced notably by commuting flows. Academic research took the lead in acknowledging this development, undertaking the first systematic delimitation of Mexico's metropolitan structures in the 1970s. By contrast, it is only very recently that the National Institute of Statistics, Geography and Informatics (INEGI in Spanish) has begun to publish statistical information on a metropolitan basis (e.g. INEGI, 2000).

The basic approach to metropolitan delimitation in Mexico has followed quite closely the example of the USA (Chapter 18). The starting point is provided by municipalities of at least 100 thousand people where the builtup area has spilt over into the territory of a neighboring one. Two different units are involved. One is more or less equivalent to the US concept of urbanized area, though rather confusingly here labeled 'metropolitan area' (MA). The other, known as the 'metropolitan zone' (MZ), surrounds the MA and comprises whole municipalities that may or may not be part of a continuous builtup area but should be urban in character (e.g. low agricultural labor force, significant amount of manufacturing and rapid rate of population growth) as well as be a relatively short distance from the central-city municipality.

The results of applying such an approach will, of course, vary according to the precise criteria used. The original research, keeping strictly to this approach, identified 12 MZs for 1970: Mexico City, Guadalajara, Monterrey, Puebla, Torreón, León, Tampico, San Luis Potosí, Chihuahua, Orizaba and Veracruz y Mérida (Unikel *et al.*, 1976). Using a similar methodology, 37 MZs were estimated for 1995, the biggest one (Mexico City) with 16.3 million and the smallest (Tlaxcala) with 104 thousand (Sobrino, 2000). In addition to these, however, it has been recognized that increasing numbers of other cities have similar characteristics, even though their main builtup areas do not overspill their administrative boundaries. On the criteria of at least 200 thousand inhabitants and 0.25 per cent of national product from manufacturing, commerce and services, Garza (2000) was able to add just one to the basic list of 12 in 1970 but as many as another 19 to the list in 2000.

Table 8.3   Mexico: Metropolitan and nonmetropolitan population, 1960-2000

	1970		1980		1990		2000	
	N	%	N	%	N	%	N	%
Urban pop	22.73	100.0	36.74	100.0	51.49	100.0	65.65	100.0
Metro	14.58	64.1	25.27	68.8	41.69	81.0	54.48	83.0
Nonmetro	8.15	35.9	11.47	31.2	9.80	19.0	11.18	17.0
Urban areas	174	100.0	227	100.0	304	100.0	350	100.0
Metro	13	7.5	26	11.5	51	16.8	56	16.0
Nonmetro	161	92.5	201	88.5	253	83.2	294	84.0

N Number, Urban pop Urban population in millions.

Source: Garza (2001, Tables A-2 and A-3).

On this basis, Table 8.3 summarizes the part played by metropolitan growth in Mexico's urbanization since 1970. The number of metropolitan centers rose fourfold over these three decades, up from 13 to 56, and their total population increased almost as rapidly. This type of settlement clearly dominates the Mexican urban system in demographic terms. Already in 1970, while making up less than one in 12 of Mexico's urban localities, the 13 metropolitan areas were accounting for almost two-thirds of the nation's urban population. By 2000, Mexico was even more of a 'metropolitan nation' with five-sixths of its much larger urban population living in this form of settlement.

The Metropolitanization of Mexico City

Mexico City was already taking on a metropolitan character in 1950 when it began to spill out of the Federal District, which had been set up at the time of national independence. This expansion started on the northern side, as the city's urban fabric extended into Tlalnepantla, one of 122 municipalities in the State of México, itself one of the 31 States which, along with the Federal District, make up the whole country. In the following 30 years to 1980, 20 further municipalities of the State of México were incorporated into this Metropolitan Zone of Mexico City (MZMC). By 2000 a further 19 had been added from the State of México, together with one (Tizayuca) from the State of Hidalgo, even further north. By then, MZMC comprised the 16 'delegations' of the Federal District together with 41 municipalities, all but one of which lay in the State of México (Table 8.4).

Table 8.4 Population of the Metropolitan Zone of Mexico City (MZMC), 1950-2000

Territory	1950		1980		2000	
	N	Pop(m)	N	Pop(m)	N	Pop(m)
MZMC total	12	2.95	37	12.99	57	17.97
Federal District	11	2.92	16	8.36	16	8.61
State of Mexico	1	0.03	21	4.63	40	9.32
State of Hidalgo	0	0	0	0	1	0.05

N Number of administrative units (delegations in Federal District, municipalities in States) included in MZMC; Pop(m) Population in millions.

Source: For 1950 to 1990, Garza (2000), 'Ámbitos de expansión territorial', Gustavo Garza (comp.), *La Ciudad de México en el Fin del Segundo Milenio*, Mexico: Gobierno del Distrito Federal, El Colegio de México; for 2000, INEGI (2000), *XII Censo General de Población y Vivienda*, Mexico: INEGI.

An impression of the geography of MZMC, defined in this way, can be obtained from Figure 8.2, which shows the extent of the builtup area against the background of the administrative boundaries and the physical environment. Seven 'delegations' in the south of the Federal District are only partially builtup, and Milpa Alta is not physically joined to the main fabric. In part, this happened because the Chichinautzin Mountain in the south is protected ecological land and, in part, due to the fact that the main thrust of the city's growth was northwards across the relatively flat plains. It is also apparent that several of the northernmost municipalities contain only relatively small urban areas that are physically separate. These localities were deemed metropolitan because more than 80 per cent of their labor force is engaged in manufacturing and services and because they are functionally linked to the central city, helped by being connected to the metropolitan public transportation system.

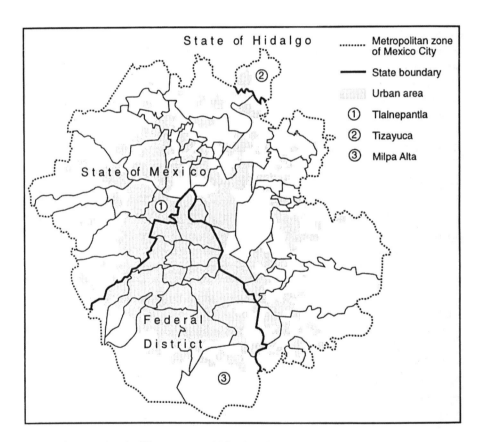

Figure 8.2　Metropolitan zone of Mexico City, 2000

It should be stressed, however, that there is no universal agreement on how MZMC should be delimited. For instance, INEGI's definition includes only 34 of the State of México's municipalities, thereby omitting seven that are not attached by continuous urban development (INEGI, 2000). Meanwhile, a National Council of Population study (Consejo Nacional de Población, 1998) defined the MZMC to include 39 of the State's municipalities, only one fewer than those shown in Figure 8.2. By contrast, a government program took a more expansive approach, including 59 municipalities outside the Federal District, 18 more than shown here (Ciudad de México *et al.*, n.d). In addition, problems of nomenclature have arisen because the term 'Mexico City' strictly refers to the Federal District only. Recognizing this, the first plan for MZMC was entitled 'Programa de ordenación de la Zona Metropolitana del Valle de México', referring to this unit of territory as the Valley of Mexico.

Mexico City: From Metropolis to Megalopolis

The problems of defining Mexico City have become greater in recent years, not just because of its lateral expansion but also due to the growth of other cities in the same region. It was in the 1980s that MZMC and the Metropolitan Zone of Toluca (MZT) came to abut each other, when the municipality of Lerma was incorporated into MZT, itself following the incorporation of Huixquilucan into the western edge of MZMC during the previous decade. This simple event could be seen as heralding the emergence of a much larger urban conglomerate possessing 'megalopolitan' features.

Not that this broader type of settlement form actually needs to have all its constituent parts immediately adjacent to each other, as was not the case with the original Megalopolis extending from Boston to Washington in the northeast of the United States (Gottmann, 1961). Instead, high levels of interaction are more vital in this context, and various studies provide ample evidence of the existence of such linkages in the Mexico City case. Unikel *et al.*'s (1976) analysis of vehicle flows identified an urban subsystem that, besides Mexico City, comprised Toluca, Puebla, Cuernavaca, Querétaro and Pachuca. This configuration was confirmed by a later study based on intensity of telephone calls (Consejo Nacional de Población, 1991). Defined in this way (Figure 8.3), Mexico City's urban subsystem also contains a number of smaller cities. Some of these contain important manufacturing plants relocated from Mexico City since the early 1980s. They include San Juan del Rio (a city of around 100 thousand inhabitants and a substantial industrial center in its own right), Cuautla (322 thousand inhabitants), Tulancingo (94 thousand), and San Martín Texmelucan (70 thousand).

Even if a contiguity constraint is imposed, the dimensions of this developing urban structure are impressive enough. In 2000 the megalopolis comprising MZMC and MZT was home to 19.4 million people. The most likely next step is its combining with the Metropolitan Zone of Cuernavaca (MZC), which will occur when the municipality of Huitzilac gets incorporated into MZC or possibly MZMC. If the same process were to draw the Metropolitan Zones of Pachuca, Puebla and Querétaro into the megalopolitan structure, then this entity could reach 36.7 million inhabitants in 2050 (Garza, 2000). This speculation about the future pattern of urban development, even more than the interpretation of events so far, raises fundamental questions about the way that population statistics are reported for complex polycentric urban regions like this.

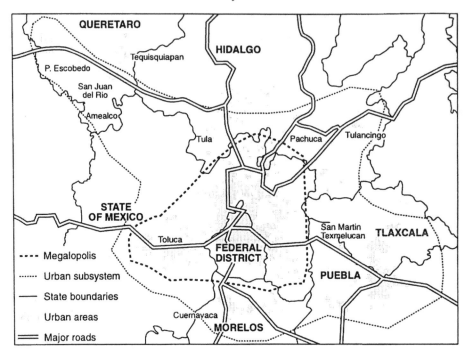

Figure 8.3 Urban subsystem and megalopolis of Mexico City

New Settlement Measures for Urban Mexico

In the above analysis of recent Mexican urbanization, several spatial units have been used: rural and urban sectors, cities or urban localities, metropolitan areas and zones, subsystem of cities, and megalopolis. These concepts have allowed the main characteristics of the country's urban development over the last three decades to be identified fairly successfully. Certainly clear is Mexico's transformation from a rural society to a mainly urban one, as also is the evolution of its pattern of cities from a highly primate urban hierarchy towards a polycentric one with Mexico City's megalopolis as its core. We now turn to the question of how suitable these spatial units are for studying Mexican urban development in the future.

Urban Mexico in 2050

In 2000 Mexico's total and urban populations were 97.5 and 65.7 million respectively, giving a 67.3 per cent level of urbanization. According to a demographic forecast made by Garza (2000), these figures will be 122.1 and 103.8 in 2020; 135.8 and 120.9 in 2030; and 141.5 and 130.2 million in 2050. By then the proportion of national population living in localities of at least 15,000 inhabitants will be 92.4 per cent and Mexico will be essentially a completely urban nation, at least from a demographic point of view. Yet how useful will this sort of information be by itself?

A definition of urban population based on locality size has, at least, two important limitations. First of all, equal urbanization levels between countries do not mean that they are at the same stage of economic development. The United States, for instance, had in 1998 a GDP per capita of 30 thousand dollars and 77 per cent of urban population, and Brazil 5 thousand and 80 per cent, respectively. Secondly, the mere percentage of urban population does not reveal the great differences in life style, education, culture, and income level between urban citizens across the world. Conventionally, the 'level of urbanism' is a relative concept and it should be possible to compare Mexico's urban-society achievements with the urban modernization of the most economically advanced countries. A measure that has a finite limit of 100 per cent is not able to do this.

To partially overcome these disadvantages, Garza *et al.* (2002) have proposed a mixed index to measure modern urban life styles or the 'degree of urbanism'. In this measure, called the 'index of urbanization-urbanism', the urban population percentage is weighted by data on the use of personal computers and cellular phones, with the scores of individual countries being set to a ratio of that of the lowest country. The indices derived on this basis for the USA and Brazil was 283.0 and 170.7 respectively, clearly demonstrating the qualitative differences in urban life style in both countries. It is therefore suggested that, when countries reach a high level of urbanization, as measured in demographic terms, a more composite measure such as this should be used to monitor further trends nationally, and be adopted for making international comparisons of urban development as well.

The Urban-Rural Dichotomy

The urban-rural delimitation used in this paper has, as has been mentioned, followed the practice recommended by research done 30 years ago, which identified the 15,000 population size cut-off that seemed then to best differentiate between urban and rural localities on the basis of five socioeconomic variables. It is surely time to review this approach in the light of subsequent developments. For one thing, one of those five variables – Spanish speaking population and people using footwear – is clearly out of date, and it would be advisable to consider new data. Mexico's 2000 census has 132 variables for every one of the country's 199,369 localities, providing the basis for establishing a new size cut-off for urban localities, if this is what is considered necessary.

There is, however, mounting evidence that size alone is no longer a reliable predictor of whether a Mexican locality is urban or rural. Consider, for example, the case of a small locality, Real Montecasino, which had a population of 678 in 2000 and is located in the countryside municipality of Huitzilac between Mexico City and Cuernavaca. This locality had only 1.8 per cent of its labor force in primary activities, and 83 per cent of its houses had electricity, piped water and sewage, 70 per cent a telephone, 66 per cent an automobile and 92 per cent a television set. By all conventional standards, this is clearly an urban place in spite of having less than one thousand inhabitants. How many other cases are there of localities with less than 15 thousand people that have urban characteristics?

Data from the 2000 census already allows us to have a tentative answer to this question. An analysis in process[3] has selected 15 variables as measures of urbanness including, among others, secondary and tertiary labor force, natural population increase, children aged 6-14 in primary school, people within the social security system, average income, and availability to houses of piped water, drainage and electricity, telephone and automobile. These variables have been studied for all localities with a population of between 1,000 and 14,999. It has been found that 463 localities with between 1,000 and 2,499 inhabitants had urban characteristics, along with 356 localities with 2,500-4,999, 287 with 5,000-9,999, and 135 with 10,000-14,999. The total population of these localities with urban characteristics but under 15,000 people sums to 5.9 million, almost one in five of the 31.8 million rural population shown in Table 8.1. On this basis, in 2000 the urban population comes to 71.6 million, giving a level of urbanization of 73.4 instead of 67.3 per cent and taking Mexico quite close to the 75.4 per cent level quoted by UN (2003) for developed countries.

What implications should we draw from this analysis? In the first place, it prompts recognition of the fact that the size threshold for urban localities may alter over time with socioeconomic development. Secondly, though, it cautions us against relying too heavily on any particular cut-off based on the population size of localities in Mexico. Thirdly, if we are really interested in distinguishing places with urban characteristics, then maybe these should be defined on the basis of these other types of variables regardless of size, albeit that suitable data may be available only in census years. Finally, whether the basis continues to be size or switches to an alternative measure based on one or more other variables, it might now be timely to question whether a simple division between urban and rural settlements is still essential for most requirements or whether some graduated scale would serve users better.

Metropolitan Delimitation and Beyond

Mention has already been made of the relative arbitrariness in delimitating the metropolitan zones and areas – an issue that is by no means unique to Mexico (see, for instance, Klove, 1959 and Dahmann, 1999, as well as other chapters in this book). This problem is due, in part, to the different purposes of these taxonomic exercises, and in part to the lack of suitable data. In relation to the former, one can recall the example, mentioned above, of the seven municipalities in the outer part of the Metropolitan Zone of Mexico City (MZMC), the builtup areas of which are physically separate from the main agglomeration (see Figure 8.2). On the one hand, they could be considered as municipalities in transition that do not yet belong to the metropolitan zone. The demographic bias of including them or not is negligible, as they together account for just 0.9 per cent of the total metropolitan

[3] I am very grateful to Luis Jaime Sobrino, of El Colegio de México, for allowing me to present the above data from research in progress.

population. However, in planning for future urban growth, they are very important because they contain most of the building land over which the city can grow.

The data restriction is encountered when one considers alternatives to using whole administrative units as building blocks for delimiting metropolitan areas. For instance, in the 1990 and 2000 Mexican censuses all localities of at least 2,500 inhabitants are subdivided into Basic Geoeconomic Statistical Areas (AGEB in Spanish), each of which comprises up to 50 street blocks. In 1990 the MZMC, as defined with 27 municipalities plus the Federal District, was divided in 3,195 AGEB. In order to have a more precise metropolitan zone delimitation, it could be possible to analyze the 'urban' AGEB around the already-defined metropolis and choose an urban standard to decide which to incorporate into a more carefully delimitated metropolitan zone. The problem with this suggestion, however, is that commuting data, so commonly used for delimiting metropolitan areas, are not available at the AGEB level.

The conclusions to be drawn on this aspect of Mexico's settlement definitions are as follows. In the first place, it would be worth considering the use of the AGEB as building blocks for defining metropolitan areas and zones. Secondly, in order to make this work well, it would be advisable to make commuting data available at this building-block scale from the next census in 2010. Moreover, the use of these building blocks would help to sidestep a problem that was mentioned earlier in the chapter about the current official criteria for considering localities as metropolitan, namely that this designation should apply only when the central municipality's urban fabric goes beyond its boundaries. Until these building blocks can be used, it is strongly recommended that, as in the approach used to identify metropolitan areas for the analysis reported in Table 8.3, the definition should include single municipalities of similar size and/or character to those entities that have outgrown their administrative limits.

More generally, it may well be helpful to consider developing different definitions for different purposes. This is not just a matter of having available the major alternative of using a physically-defined agglomeration versus a functionally-defined metropolitan zone, but also of being able to choose between different versions of the latter. In this connection, if the overall path of urban population distribution in twenty-first century Mexico is to be portrayed accurately, it is important to consider a higher-order definition that can handle polycentric urban regions like that developing in Mexico's central region. This chapter has suggested that already Mexico City's urban subsystem is beginning to take on some of the features associated with certain large and heavily-urbanized regions around the world that are now treated as 'megalopolitan'. What rules should be used to decide which metropolitan areas and other cities are part of this type of settlement? Should physical contiguity of metropolitan zones be required, or should more emphasis be given to the intensity of interactions between these potential component parts? How useful is it likely to be in the future to conceive of a single urban complex extending over 400 km from Puebla in the east to Guadalajara in the west and with Mexico City at its central pole – an area covering seven States that together already contained 44 million residents in 2000?

Concluding Comment

In a rapidly evolving planetary-scale market, the cities, metropoles, megalopolises and polycentric urban regions of all nations are increasingly competing with each other in order to attract the most dynamic and high-tech industries. The overriding conclusion of this chapter is that, in all developed and developing countries which have reached high levels of urbanization, it is becoming ever more necessary to construct a suite of new concepts and measures that will successfully capture and analyze this dynamic urban phenomenon. Arising from the Mexican experience reviewed in this chapter, such tools could include an urbanization-urbanism index, the polycentric macroregion, and the identification of urban populations regardless of locality size. These, and other measures that surely will be developed, will allow us to clarify the new kind of economic and demographic spatial organizations observed and to provide technical and theoretical support to those developing more innovative and effective urban and regional policies.

Chapter 9

Urban Development and Population Redistribution in Delhi: Implications for Categorizing Population

Véronique Dupont[1]

Metropolitan areas in India are undergoing major transformations. This chapter focuses on the case of Delhi and highlights recent developments in urban forms and processes with a view to assessing the adequacy of local definitions and categorization of human settlements. The dynamics of the metropolitan area of Delhi are examined from two interrelated perspectives. One is the evolving urban form, with emphasis on the processes of 'periurbanization' and 'rurbanization', including expansion of suburbs, formation of new residential quarters in surrounding rural areas and the creation of satellite towns. The other is population redistribution within the metropolitan area.

The chapter is arranged in three main sections. Firstly, the demographic and spatial dimensions of Delhi's metropolitan dynamics in terms of population growth, distribution (and redistribution) and spatial expansion are analyzed. Then the factors contributing to urban deconcentration and outward expansion are examined. Finally, some implications are drawn regarding categorizing population for further demographic analysis (reflecting upon the inadequacy of a simple rural-urban dichotomy), defining relevant limits for measuring urban growth and delimiting zones for the purpose of town and country planning.

These analyses are based on two main sources of data: decennial population censuses including the most recent, conducted in 2001 (see Annex 1 for details of the reports used), and a survey on population mobility conducted in 1995 complemented by indepth interviews and field visits (Dupont and Prakash, 1999). That survey included five peripheral zones that illustrate the dynamics of urban

[1] The study presented here is part of a larger research program on spatial mobility and residential practices of Delhi's population, and its effect on the dynamics of the metropolis. This program has been financed by the *Institut de Recherche pour le Développement* (ex-ORSTOM) with additional funding from the CNRS within the framework of *Action Concertée en Sciences Sociales* (Concerted Action in the Social Sciences) ORSTOM-CNRS and of *PIR-Villes*. In India, the program was conducted with the collaboration and support of the *Centre de Sciences Humaines* based in Delhi (French Ministry of External Affairs) and the Institute of Economic Growth (Delhi).

expansion, and covered a sample of 1,249 households containing 5,981 usual residents.

Delhi's Metropolitan Dynamics: Rapid Population Growth and Outward Expansion

Demographic and Spatial Growth of Delhi Urban Agglomeration

The development of Delhi and its metropolitan area bears witness to a major tendency in the urbanization process in India: an increasing concentration of the urban population in metropolises of a million or more inhabitants.[2] Yet, the domination of the Indian urban scene by the bigger cities takes place within the context of a country which is predominantly rural and is likely to remain so in the medium term (in 1991 only 26 per cent of the population lived in urban areas, and 28 per cent in 2001).[3]

The demographic evolution of Delhi during the 20th century is deeply marked by the country's turbulent history. Following the promotion of Delhi as the capital of the British Indian Empire in 1911, the population of the city expanded from 238,000 in 1911 to 696,000 in 1947, while quadrupling in area extent (Table 9.1). After independence in 1947 Delhi became the capital of the newly formed Indian Union and had to face a massive transfer of population following the partition into India and Pakistan. The 1941-51 period thus recorded the most rapid population growth in the history of the capital city, from almost 700,000 inhabitants in 1941 to 1.4 million in 1951, corresponding to an annual growth rate of 7.5 per cent. Nevertheless, in the post-independence period, the population growth of Delhi has been remarkably rapid for an urban agglomeration of this size, oscillating between

---

[2] In 1951, there were only five cities or urban agglomerations (see next note) with one million or more inhabitants, accounting for 19 per cent of the total urban population of the country; in 2001 there were 35, accounting for 38 per cent of the total urban population.

[3] The definition of an 'urban unit' or town that has been applied since the 1961 Census of India is as follows:

a) All places which answer to certain administrative criteria, such as the presence of a municipality, a corporation, a cantonment board, a notified town area committee, etc. These are called the statutory towns.

b) All other places which satisfy the following three criteria: i) a minimum population of 5,000 inhabitants; ii) at least 75 per cent of the male working population engaged in non-agricultural pursuits; iii) and a population density of at least 400 persons per square kilometers. These are called the census towns.

In addition, the concept of urban agglomeration was introduced at the time of the 1971 Census and remained unchanged in the 1981 and 1991 Censuses: 'An urban agglomeration is a continuous urban spread constituting a town and its adjoining urban outgrowths, or two or more contiguous towns together and any adjoining urban outgrowths of such towns.' For the census of 2001, two other conditions were added: 'the core town or at least one of the constituent towns of an urban agglomeration should necessary be a statutory town and the total population of all constituents should not be less than 20,000 (as per 1991 Census)'.

4 per cent and 5 per cent per year, to reach 12.8 millions in 2001. Since 1961 Delhi has been the third largest Indian urban agglomeration, overshadowed only by Mumbai and Kolkata.

Table 9.1 Population, area and density of Delhi Urban Agglomeration, 1901-2001

| Year | Population | | | Area | | Density |
	Number	Decennial growth rate %	Annual growth rate %	Square kilometer	Decennial growth rate %	Persons/ hectare
1901	214,115	-	-	n.a.	-	-
1911	237,944	11.3	1.06	43.3	-	55
1921	304,420	27.9	2.49	168.1	288.6	18
1931	447,442	47.0	3.93	169.4	0.8	26
1941	695,686	55.5	4.51	174.3	2.9	40
1951	1,437,134	106.6	7.52	201.4	15.5	71
1961	2,359,408	64.2	5.08	326.6	62.1	72
1971	3,647,023	54.6	4.45	446.3	36.8	82
1981	5,729,283	57.1	4.62	540.8	21.2	106
1991	8,419,084	46.9	3.92	624.3	15.4	135
2001*	12,791,458	51.9	4.27	791.9	26.9	162

The Delhi urban agglomeration comprises the urban area circumscribed within the statutory boundaries of the city (the three statutory towns corresponding to the Municipal Corporation of Delhi, the New Delhi Municipal Council and the Cantonment Board), together with contiguous urban entities and extensions falling beyond these statutory boundaries. Its limits are redefined at each census in order to take into account the most recent urban extensions.

* Provisional results for population figures; area was estimated on the basis of the published census maps.

Source: Census of India, Delhi, 1951, 1961, 1971, 1981, 1991 and 2001.

The population growth was concurrent with a spatial expansion in all directions, including to the east of the Yamuna river. The official area of the urban agglomeration was almost multiplied by four between 1951 and 2001 (Table 9.1), and its share in the total area of the National Capital Territory of Delhi[4] (covering 1483 square kilometers) increased from 14 per cent to 53 per cent. Delhi's geographic situation, in the Gangetic plain, and more particularly the absence of any real physical barrier to urban progression (the Aravalli Hills – the Delhi Ridge

[4] The National Capital Territory of Delhi is an administrative and political entity: a Territory of the federal Union of India, identified by the Constitution of 1949; its boundaries are fixed (Figure 9.1) and correspond to the ancient Province of Delhi under the British rule in India.

– to the west and south do not constitute an effective obstacle), have favored the multidirectional spreading of the urbanized area (Figure 9.1).

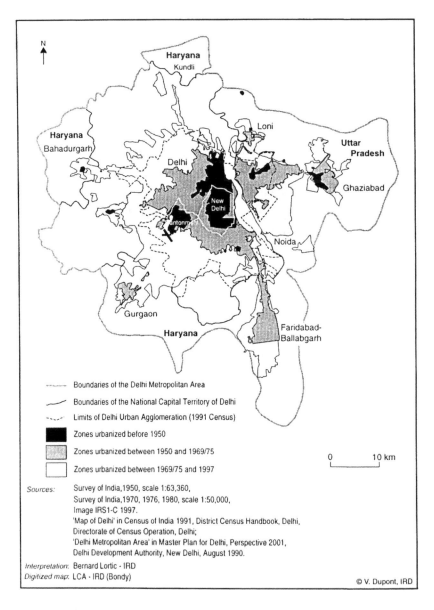

Figure 9.1   Spatial expansion of urbanized zones in the Delhi metropolitan area, 1950-1997

Contribution of Migration to the Population Growth of Delhi

Migration has played a major role in the demographic evolution of the capital. Following the partition of the country, Delhi – with a population was about 900,000 in 1947 – had to receive 495,000 refugees from Pakistan, while 329,000 Muslims left the capital.[5] In the postindependence era, migration continued to make a significant contribution to urban growth although it slowed down in the eighties. Migrants with less than 10 years of residence accounted for 62 per cent of the population of the National Capital Territory of Delhi in 1971, 60 per cent in 1981, and it declined to 50 per cent in 1991.[6]

Although the majority of migrants in Delhi come from rural areas, as many as 44 per cent of the total migrants residing in the Territory of Delhi in 1991 were from urban areas. This underlines the specific pull effect of a big metropolis in a predominantly rural country. Over two thirds of all migrants living in Delhi in 1991 were from neighboring states in North India: Haryana, the Punjab, Rajasthan and Uttar Pradesh (Dupont, 2000b).

An estimation of the respective share of the three components of urban growth (natural increase, net inmigration and reclassification of the urban-rural population due to changes in the spatial delimitation of the urban area) in the Territory of Delhi was attempted for the 1971-1981 intercensal period by the National Institute of Urban Affairs (NIUA, 1988). It was estimated that natural increase contributed 35 per cent to the total urban population growth, net inmigration 41 per cent and reclassification of population 25 per cent. However, as the data are not available to estimate the three components of growth in recent years estimates are made based on the following assumptions. Since the population of the National Capital Territory is mostly concentrated in the urban agglomeration of Delhi (90 per cent in 1971 and 93 per cent in 2001), and the area of the Territory is constant, an estimation of the two components of population growth (natural increase and net migration) for the entire Territory provides a good approximation of the population dynamics of the urban agglomeration.

We attempted this exercise for the last three decades. The average annual rate of natural growth for each intercensal period was computed on the basis of estimated rates provided by the Sample Registration System, and the contribution of net migration deducted as the residual from the total growth rate. The estimated results are presented in Table 9.2. They not only confirm the crucial contribution of migration to the population growth of the National Capital Territory, but also suggest that this contribution did not slow down during the last decade (1991-2001).

---

[5] Source: Ministry of Rehabilitation, Annual Report on Evacuation, Relief and Rehabilitation of Refugees, 1954-55 (quoted in Datta, 1986).

[6] Since the 1971 Census, migrants are those who had resided in a place outside the place of enumeration.

Table 9.2   Contribution of natural growth and net migration to the total population growth of the National Capital Territory of Delhi, 1971-2001

Intercensal period	Average annual rate of growth (%)	Average annual rate of natural growth (%)	Average annual rate of net migration (%)
1971-81	4.34	2.11	2.23
1981-91	4.24	2.13	2.11
1991-2001*	3.88	1.70	2.18

* Provisional results

Source: Census of India and Sample Registration System.

Differentials in Population Growth and Densities within the Territory of Delhi

Differentials within the urban agglomeration The overall demographic change in Delhi urban agglomeration conceals differences within the urban area. Between 1981 and 1991, the pattern of growth in Delhi was 'clearly centrifugal' (Dupont and Mitra, 1995), continuing the trend highlighted by Brush (1986) for the 1961-1971 decade. Absolute decrease in population, indicating important net outwards moves, has occurred in the historical city core known as Old Delhi and the population has also declined in some parts of New Delhi (the area corresponding to the new capital built by the British). On the other hand, the highest growth rates above 10 per cent were recorded in neighborhoods of the outskirts (Figure 9.2).

During the 1991-2001 decade, these trends persisted. The depopulation of the old city area continued (-1.9 per cent in ten years). Population growth has also been very low in New Delhi district (only 2.5 per cent in ten years), whereas the districts including the peripheral zones of the urban agglomeration have recorded higher decadal growth (for example: 62.5 per cent in the North East district, 61.3 per cent in the South-West district, and 60.1 per cent in the North-West district).

In 1991, the highest population densities were registered in the historical city core: 616 persons per hectare on an average (740 in 1961) in the Walled City of Shajahanabad, established by the Mughals in the seventeenth century and covering an area of almost 600 hectares. The old city also has a high concentration of commercial and small-scale industrial activities with a mixed land-use pattern typical of traditional Indian cities. On the other hand, New Delhi, the area planned in the 1910s and 1920s according to a garden city model, had an average density of only 70 persons per hectare. The Delhi Cantonment, which includes military land and the international airport, recorded an even lower density of 22 persons per hectare. The average population density in the urban agglomeration was 135 persons per hectare. The classical model of population density gradients, characterized by high densities in the urban core and a sharp decline towards the periphery, and whose 'original causes ... can be summed up in three words:

protection, prestige, and proximity' (Brush, 1962, p.65), had largely survived in Delhi until 1991.

Notable changes in the distribution of population densities have taken place over the 1991-2001 period. For the first time, the highest residential densities are not recorded in the old city core, but in two northeastern *teshils* (administrative divisions below the district level): Shahdara (422 persons per hectare) and Seemapuri (402 persons per hectare), while in Old Delhi and its adjoining neighborhoods densities are now lower than 350 persons per hectare. The lowest density is recorded in a *teshil* of New Delhi district (32 persons per hectare in Chanakyapuri, a high status residential area where many embassies' quarters are also located).

A more refined analysis of the pattern of population growth and changes in density during the last decade was not permitted by the data available at the time of writing. Furthermore, 15 out of the 27 new *teshils* constituting the Territory of Delhi in 2001 include both rural and urban areas. It is thus not possible at present to test whether the official limits of the urban agglomeration are relevant in terms of rural-urban differentials in the sociodemographic and economic characteristics of the concerned populations. Even after the forthcoming publication of the results of the 2001 census at the ward level, the comparison between 1991 and 2001 at the level of a fine spatial division will not be possible for ordinary census data's users, due to some changes in boundaries of spatial divisions and the lack of published information about the correspondence between the former and the new classification. This difficulty that we already encountered for the previous censuses is compounded by the absence of published maps showing the basic spatial divisions. There is a lack of concordance allowing accurate intercensal comparisons of settlement classification.

Two distinct migration processes are contributing to the rapid population growth in peripheral areas of this megacity. One involves new inmigrants to the city, and the other involves natives of Delhi or migrants of longer standing who, having lived previously in inner zones of the urban agglomeration, moved to new residential sites. The 1995 survey of population mobility in the Delhi metropolitan area allows us to evaluate the respective contribution of the two types of moves. We will focus here on three peripheral neighborhoods, which illustrate the dynamics of settlement in zones that have recorded a rapid population growth and include various types of housing estates and different income groups. These are:

- *Tigri*, a working class neighborhood, with high residential density located in the southern periphery;
- *Badli-Rohini*, an extensive zone located in the west-northern periphery, including an industrial area, and housing low and middle income groups, with a population density in 1991 that was still low;
- *Mayur Vihar-Trilokpuri*, a residential zone located in the eastern periphery, including a large variety of housing estates, corresponding to a range of income groups, with population densities in 1991 that varied from middle to very high.

In these three peripheral areas, Table 9.3 shows that most household heads had been born outside Delhi. Most have moved from elsewhere in Delhi urban agglomeration to the periphery, although newcomers are a significant group.

Table 9.3   Delhi sample survey of peripheral neighborhoods: place of last dwelling of household heads and birthplace outside Delhi

	Tigri (South)	Badli-Rohini (North-West)	Mayur Vihar-Trilokpuri (East)
Place of last residence of household heads			
Same dwelling since birth	0.6	21.6	4.2
Other Delhi urban agglomeration	72.5	67.7	77.7
Outside Delhi urban agglomeration	26.9	10.8	18.1
Total	100.0	100.0	100.0
Number of cases	171	167	337
Born outside Delhi			
Per cent of household heads	90.8	68.3	82.8
Per cent of all household members	62.6	57.2	43.5

Source: Mobility survey, 1995 (see text).

Differentials between the urban agglomeration and the rural hinterland The centrifugal pattern of population dynamics extended beyond the city limits. As shown in Figure 9.2, population growth from 1981 to 1991 was faster in the 'rural' periurban fringe of the National Capital Territory of Delhi than in its urbanized area – 9.6 per cent per year as against 3.8 per cent respectively (in the urban/rural limits as defined by the 1991 census). These figures can be compared to the natural growth rates during the same period, that is 2.5 per cent annually in rural areas and 2.1 per cent in urban areas, thus underscoring the contribution of net inmigration. However, the rural zones accommodated only 10 per cent of the total population of the Territory of Delhi in 1991 and the population densities remain significantly lower in the rural zones than in the urban agglomeration (12 inhabitants per hectare, as against 135 in 1991). Nevertheless, these movements reflect the real attraction exerted by the rural hinterland of the capital on new migrants or residents of Delhi who have left the inner city in search of less congested and financially more affordable localities in which to settle.

This process of periurbanization around the capital is also expressed in economic terms, insofar as the composition of the working population residing in the rural zones of the Territory of Delhi is closer to that of the national urban population than the rural population. Thus, in 2001, only 11 per cent of the working population were employed in agriculture, as compared with 73 per cent in all India's rural areas and 8 per cent in its urban areas.

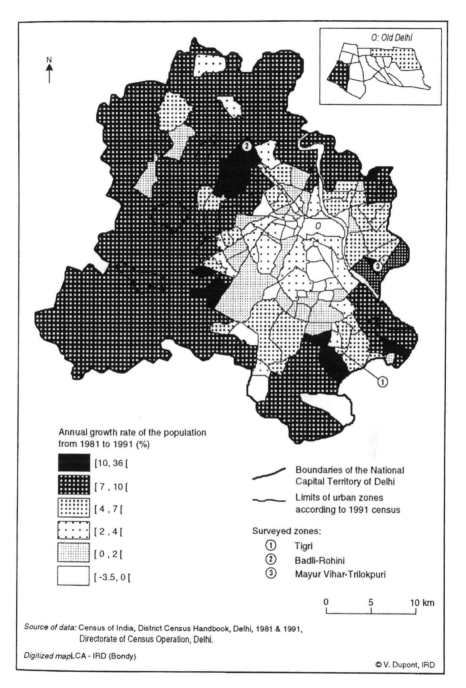

Figure 9.2   Annual population growth rate in the National Capital Territory of Delhi, 1981-1991, by census divisions

Although the administrative limits of the Delhi urban agglomeration have been extended several times, the rapid growth of the rural population in the National Capital Territory as well as changes in its economic characteristics underline the discrepancy between the administrative demarcation of urban Delhi and the real pattern of urban development. This points towards the development of a transitional periurban type of area around the Indian capital, as observed in other megacities of Asia (Ginsburg *et al.*, 1991). Yet, the dichotomous classification of human settlements in India does not recognize this development.

The Rapid Development of Peripheral Towns

The slowing down of the population growth rate in the urban agglomeration of Delhi during the eighties as compared to the previous decades was not the result of a decline in the rate of natural increase.[7] There was deliberate planning from the 1960s onward to develop towns on the periphery of Delhi to accommodate population growth; eventually these areas grew faster than the central agglomeration of Delhi (Table 9.4). The urban sprawl has followed the main roads and railway lines, hence connecting the builtup area of the core city – Delhi – with that of the peripheral towns, leading to the development of a multinodal urban area (Figure 9.1). The inappropriateness of current definitions of urban agglomeration is shown by the fact that the continuous urban spread of Delhi overlaps State borders. The presently contiguous ring towns of Delhi are located in other states (Uttar Pradesh and Haryana) and are not considered as being part of the Delhi urban agglomeration whose actual population size is thus underestimated by more than two millions (Table 9.4).

An example: the new town of Noida The case of Noida provides an illustration of the population dynamics in a new town of the metropolitan area. It is situated at the southeastern border of the Territory of Delhi, in the State of Uttar Pradesh, at fifteen kilometers from the center of the capital. Noida (the acronym for New Okhla Industrial Development Authority) is a new industrial center, founded in 1976, a product of the town and country planning policy. It was created from the clustering together of around twenty villages. The population of Noida underwent very rapid growth during the 1980s and the 1990s, to reach 294,000 in 2001 (Table 9.4).

According to our 1995 mobility survey, in the town as a whole (including the original villages, the planned areas and the slums), 69 per cent of the households were headed by a migrant. This figure rises to 89 per cent in the new planned sectors and 99 per cent in the slums. A majority of all migrants (56 per cent) had arrived directly from outside the Delhi-Noida conurbation. Among the households not having always lived in the same dwelling, 33 per cent were previously living outside the Delhi-Noida conurbation, 39 per cent in Delhi itself (56 per cent among

---

[7] According to estimates from the Sample Registration System, the average natural rate of increase in the urban areas of Delhi was 2.0 per cent per year from 1971 to 1980, and 2.1 per cent from 1981 to 1990.

households in the new planned sectors) and 27 per cent in a different dwelling in Noida. The acquisition of a house or of an apartment was the reason for 44 per cent of the last changes of residence from Delhi or a different dwelling in Noida (66 per cent for the households of the planned sectors), and better or cheaper housing conditions for 41 per cent of them. These figures show to what extent the power of attraction of the new industrial center extends beyond the metropolitan area of Delhi, and they also testify to a manifest influx of population from the capital, arising out of considerations related to housing.

Table 9.4   Population growth of cities, towns and villages in Delhi metropolitan area, 1951-2001

Towns/zones	Population					
	1951	1961	1971	1981	1991	2001
Delhi NCT	1,744,072	2,658,612	4,065,698	6,220,406	9,420,644	13,782,976
i) Delhi UA	1,437,134	2,359,408	3,647,023	5,729,283	8,419,084	12,791,458
ii) other census towns	-	-	-	38,917	52,541	28,303
iii) rural Delhi	306,938	299,204	418,675	452,206	949,019	963,215
Ring towns	114,543	187,981	351,240	802,199	1,505,670	2,787,151
Ghaziabad UA	43,745	70,438	137,033	287,170	511,759	968,521
Loni	3,622	5,564	8,427	10,259	36,561	120,659
Noida	-	-	-	35,541	146,514	293,908
Faridabad CA	37,393	59,039	122,817	330,864	617,717	1 054,981
i) Faridabad	31,466	50,709	105,406	-	-	-
ii) Ballabgarh	5,927	8,330	17,411	-	-	-
Gurgaon UA	18,613	3,868	57,151	100,877	135,884	229,243
Bahadurgarh UA	11,170	14,982	25,812	37,488	57,235	119,839
Towns/zones	Annual growth rate (%)				
	1951-61	1961-71	1971-81	1981-91	1991-01
Delhi NCT	4.31	4.34	4.34	4.24	3.88
i) Delhi UA	5.08	4.45	4.62	3.92	4.27
ii) other census towns	-	-	-	3.05	-6.00
iii) rural Delhi	-0.25	3.42	0.77	7.69	0.15
Ring towns	5.07	6.46	8.6	6.5	6.35
Ghaziabad UA	4.88	6.88	7.68	5.95	6.59
Loni	4.39	4.24	1.99	13.55	12.68
Noida	-	-	-	13.31	7.21
Faridabad CA	4.67	7.60	10.42	6.44	5.50
i) Faridabad	4.89	7.59	-	-	-
ii) Ballabgarh	3.46	7.65	-	-	-
Gurgaon UA	7.36	4.20	5.85	3.02	5.37
Bahadurgarh UA	2.98	5.59	3.80	4.32	7.67

NCT: National Capital Territory; UA: Urban Agglomeration; CA: Complex Administration. The Delhi metropolitan area identified by the planners consists of the National Capital Territory of Delhi and the first ring of towns around the capital as listed in this table, plus their rural hinterland (Figure 9.1). In addition, the village of Kundli was proposed for enhancement and included in the metropolitan area.

Source: Census of India, 1951, 1961, 1971, 1981, 1991 and 2001 (provisional results).

Consequently, about one fourth of Noida's working inhabitants commute daily to work outside their town of residence, this proportion undergoing large variations according to the sector of housing (1995 mobility survey; Dupont, 2001a). Almost all workers living in the slums (98 per cent) work in Noida. It is the attraction of employment opportunities offered by this new industrial center that has made them migrate. But among those residing in the planned sectors, only 52 per cent work in Noida, the rest commuting to Delhi. Thus, for a notable section of its population, Noida is merely a satellite town of the capital, playing the role of a dormitory.

Processes of Outward Expansion: Contributing Factors and Variety of Urban Forms

The pattern of population distribution and growth is related to a number of factors: patterns of landuse, the availability and price of land or residential property, and the accessibility of employment opportunities and urban services. If this last factor helps explain the centripetal force of the past, the actual centrifugal tendency is certainly associated with the scarcity of land for new residential constructions and its consequent appreciating value in central areas. The less congested peripheral zones provide more affordable housing possibilities, as well as more accessible sites for squatting. The expansion of the urban periphery is the outcome of the interactions between planning attempts and private initiatives and responses.

The Planned Development of Peripheral Zones

The evolution of Delhi and its region have been strongly influenced by a town and country planning policy that was initiated in the late 1950s, prompted by the desire to control the growth of the capital and to curb inmigration flows by reorienting them towards other towns in the region. Within the capital itself, this interventionist policy was given concrete shape by means of a master plan, the first of its kind in India, implemented in 1962. Particularly restrictive land control measures were taken, housing programs were launched, while some old central quarters and slums were destroyed and their inhabitants resettled in peripheral areas. However, these measures did not prevent a high level of speculation in land and proliferation of informal – considered as illegal – quarters (the 'squatter settlements' and 'unauthorized colonies').

The Delhi Development Authority (DDA), the central administration created in 1957 and responsible for the elaboration and the execution of the Master Plan, has played a direct role in the urban spread of the capital. The DDA set aside large land reserves, primarily through the acquisition of agricultural lands geared towards the implementation of various land development and housing programs:

- the construction of blocks of flats for sale to private households of different income groups;
- the development of land and the allotment of plots on a 99 year leasehold basis to private households and cooperative group housing societies;
- the servicing and allotment of land for the resettlement of slum dwellers and squatters evicted from central areas of the city. This policy, which resorted to

coercive measures, was pursued most actively during the 'Emergency' (1975-77), during which time about 700,000 persons were forcibly displaced to 'resettlement colonies' located on the urban outskirts (Ali, 1990, 1995; Tarlo, 2000).

In some cases, these schemes were part of large-scale projects aimed at developing new peripheral zones and leading to the creation of satellite townships (Rohini, Dwarka-Papankala, and more recently Narela subcity) that were planned to receive up to one million inhabitants or more (Jain, 1990).

The 1995 mobility survey indicated that the housing and plot allotments schemes of the DDA benefited mostly those households which were already settled in Delhi. For instance, in Mayur Vihar, 85 per cent of the households surveyed in the DDA flats moved from another dwelling located in Delhi as did 97 per cent among those surveyed in blocks of flats built by cooperative group housing societies. In Rohini, 92 per cent of the households surveyed in the planned sector had followed a similar residential trajectory. Thus, these public urban development programs have contributed more to the redistribution of the population within the urban agglomeration than to the direct attraction of new migrants, in spite of the emergence of a significant private rental sector among this segment of the housing stock.[8]

At the regional level, planning policy laid emphasis on the promotion of peripheral towns through the strengthening of their economic base, including the creation of the new industrial town of Noida. The first Master Plan of Delhi (1962) introduced the concept of metropolitan area, that encompassed the Territory of Delhi and the towns located within a radius of 35 kilometers around the capital and whose demographic and economic development was interdependent with that of Delhi and involved large-scale commuting. The development of the metropolitan area was further integrated within the larger planning framework of the National Capital Region, a region covering around 30,000 square kilometers, and provided with a planning board since 1985 (NCR Planning Board, 1988). However, the initial stress put on the promotion of the first ring of towns eventually strengthened the attraction of the whole metropolitan area and intensified commuting within it (NIUA, 1988). Due to their proximity to the capital, these ring towns did not emerge as autonomous, alternative growth centers, and most of them can be considered satellite towns, alleviating housing problems in the capital but exerting a heightened pressure on its amenities.

In the late 1980s, however, a new strategy of regional planning aimed at promoting regional urban centers situated beyond the metropolitan area, at a distance large enough to discourage daily interactions with the capital. It is proposed to develop five regional metropolises beyond the borders of the National

---

[8] In the sample of dwellings surveyed in Mayur Vihar, 19 per cent of the DDA flats were rented (including accommodation provided by the employer) as were 49 per cent among the apartments of the co-operative sector; in Rohini, 16 per cent of the dwellings surveyed in the planned sector were occupied by renters.

Capital Region to act as countermagnets and intercept future migratory inflows towards the metropolitan region (NCR Planning Board, 1988, 1996).

Informal Urbanization of the Periphery

Public housing policies have failed to meet the needs of large sections of the urban population, in particular the lower-middle classes and the poor who have had to resort to the informal housing sector.

Unauthorized colonies on agricultural land The proliferation of unauthorized colonies has contributed in a decisive way to the urbanization of the rural fringes of Delhi. These estates involve agricultural land not meant for urbanization, bought from farmers by unscrupulous real estate developers who indulge in illegal subdivisioning and selling of unserviced plots. In 1983, 736 unauthorized colonies were enumerated, housing an estimated population of 1.2 million, that is almost 20 per cent of the population of the capital (Billand, 1990). In 1995, their official number had reached 1,300 (Government of NCT of Delhi, 1996), and their total population in 1998 was estimated at about 3 millions.[9]

These housing estates are not recognized by the municipality and therefore do not have the benefit of its services. Authorities have repeatedly introduced regularization procedures to legalize these unauthorized colonies. However, it seems that this policy has had the perverse effect of indirectly encouraging the development of new unauthorized colonies, since prospective buyers hope their settlement would obtain regular status in the future, thereby guaranteeing the long-term economic profitability of their investments.

Initially, these colonies appealed to lower to middle income groups, people whose limited resources meant that they could not rent or buy in the legal housing market and were prepared to accept limited utilities and resources. In order to make their investment profitable, the new house owners often rent out one or several rooms, or one story, in their house, hence contributing to the increasing residential densities of these colonies.

The unauthorized colonies surveyed in Mayur Vihar were almost all occupied by migrant households[10] (they represent 92 per cent of households according to our 1995 mobility survey). Among the migrants, 44 per cent arrived directly from a town or village situated outside the capital, and 56 per cent lived previously in another locality (or several) within the Delhi urban agglomeration. Three quarters of all the households had occupied another dwelling in Delhi before settling in the present one. In spite of a significant rental sector (37 per cent of the households surveyed in this type of quarters are tenants), the unauthorized colonies of Mayur Vihar are mainly a place of resettlement within the urban agglomeration, rather than a place of initial reception for new migrants.

---

[9] Estimation provided by *Common Cause*, a citizens' association that took the matter of unauthorized colonies to the Delhi High Court, against their regularization.

[10] By 'migrant household' we mean household whose head is a migrant.

Squatter settlements on vacant land The poorer sections of the urban population live in squatter settlements (locally called *jhuggi-jhonpri*), which have continued to proliferate despite the 'slum clearance' policy (Ali and Singh, 1998; Majumdar, 1983; Suri, 1994). In 1999, about 600,000 families lived in a thousand *jhuggi-jhonpri* clusters which varied in size from a dozen dwelling units to 12,000; these squatter settlements altogether housed about 3 million persons or 20 to 25 per cent of the total population of Delhi.[11] Though squatter settlements are found throughout the capital, insinuating themselves into all the interstices of the urban fabric wherever there is vacant land and where surveillance by the legal authorities is limited,[12] the two biggest clusters are located on the periphery, on what was still the urban-rural fringe at the time of their initial occupation. The population density in squatter camps can be very high owing to the cramming together of families in one-room huts and very narrow lanes. In many squatter settlements, the structures are reinforced and further extended by the frequent addition of a story to respond to families' expansion, but also for rental purposes. A process of increasing residential density is at work in quarters that are already crowded and lacking basic infrastructure and access to services.

Delhi's squatter settlements shelter mostly migrant households attracted by the employment opportunities provided by the city. Yet, all the migrants have not settled directly in their present squatter settlement upon their arrival in the capital. A significant proportion among them – that varies from one slum to another, depending on its specific history – have stayed previously in another place in Delhi (this was the case for 38 per cent of the migrants surveyed in the Tigri *jhuggi-jhonpri* camp, and 70 per cent of those we surveyed in the *jhuggi-jhonpri* clusters of Badli-Rohini). Often, the residential trajectory of the slum dwellers is marked by eviction from one place, then squatting in another, until they are evicted again and eventually sent to a resettlement colony. They may also move on their own initiative to a better location, in the vicinity of employment sources (adjoining, for instance, an industrial area like Badli-Rohini).

Deconcentration of the Rich to the Rural Fringe

Residential strategies aimed at gaining access to more space and a better environment outside the city proper have seen many move to Delhi's periphery and the processes of periurbanization and rurbanization have proceeded apace.[13] Given the lack of a mass transit system in the capital and its metropolitan area, it is the tremendous increase in private means of transportation that has allowed the emergence of residential estates in distant rural fringes suitable only for those who

[11] Slum and Jhuggi Jhonpri Department, Municipal Corporation of Delhi.

[12] Numerous evictions of squatter settlements in 2000-2001 are, however, likely to have altered this spatial pattern.

[13] Rurbanization is understood here as 'the fixation in peri-urban countryside of residences of city dwellers, the interweaving of rural and urban spaces', that is, 'one of the forms of periurbanization', without 'continuity between the town and the rurbanized countryside' (George, 1993, p.411).

can afford the price of commuting daily by car, or who compensate for the increased transport cost by the cheaper housing costs. Two resulting types of urban form have developed in the rural fringes: 'farm houses', and large-scale housing schemes.

Farm houses The deconcentration of upper-class families to the rural fringes has created competition for land use, in particular in the southern agricultural belt where numerous 'farm houses' have been built (Soni, 2000). As they were initially genuine farms within agricultural lands, such zones are governed by planning regulations applying to farmlands, seeking to limit the builtup area in relation to the natural green and cultivated spaces. The agricultural nature of such lands is, however, often distorted. Instead, luxurious sprawling villas, surrounded by large parks and protected by high walls have become the fashion. Usually, 'farm house' owners are people from the top income bracket who have been able to build havens of tranquility on the outskirts of one of the most polluted capitals of the world.

Large-scale housing schemes   The direct control exercised by the Delhi administration on land suitable for urbanization has induced some private real estate developers to implement large-scale housing schemes outside the limits of the National Capital Territory of Delhi, often well beyond the perimeter of its urban agglomeration. The informal urbanization of the fringes of the capital has also prompted planners in the bordering states to intervene according to a different strategy, allowing private building societies to acquire large tracts of land in the framework of their master plans. These residential projects are designed for well-to-do city dwellers, looking for a better quality of life. Thus, some property developers make use of the very outlying character of these new residential areas to emphasize the rustic 'green' nature of the fringe areas to attract high income settlers from central Delhi (Dupont, 2001b).

The 1995 mobility survey conducted in the largest residential complex of this type, DLF Qutab Enclave located 23 kilometers away from the center of Delhi near the southern ring town of Gurgaon, provides insight into the population dynamics associated with this mode of periurbanization. The peopling of this residential neighborhood began in the 1980s and resulted mainly from a deconcentration movement within the capital. Thus, 65 per cent of the inhabitants had lived previously in Delhi itself, only 9 per cent in the town of Gurgaon or its surroundings, and 26 per cent outside of the metropolitan area of Delhi (although some of them had already familial or professional links with the capital).[14] The high status of the area is reflected in the fact that nearly three quarters of the households own their house or flat. Financial considerations are also involved: the cost of plots or dwellings being more affordable here than in neighborhoods of comparable standing in the capital. Nevertheless, environmental considerations are

---

[14] To compute these percentages, we have excluded the persons who have lived in the same dwelling since their birth (i.e. the children born after their family moved to DLF Qutab Enclave), representing 2.5 per cent of residents (out of a total sample of 566 residents).

also important in the choice of residential location by settlers in the area (Dupont, 2001b).

The indispensable condition for having access to real estate outside the capital and to a better environment is the possession of a personal vehicle to make possible daily journeys to distant workplaces, to undertake certain types of shopping and to maintain one's social network through visiting. About half of the gainfully employed inhabitants surveyed in DLF Qutab Enclave worked in Delhi proper, while half of the students attended a school or university in Delhi; yet, at the same time, bus services, either public or chartered by the developers, were still limited.

The construction of business and commercial centers has supplemented the development of residential complexes in this decentralized area, and the spatial expansion of the builtup area over largely spread zones is now combined with clusters of high-rise buildings in a similar way to the edge cities (Garreau, 1991) of the United States. The scale of the development schemes and the rapidity of transformation of this peripheral zone of Delhi has seen its rural components quickly shrink (Dupont, 1997). In the early years of the development, the discontinuity of builtup area between the city and these residential quarters in the rural fringes was much more pronounced than today, and the countryside more present. The extension of the urban fabric and the increasing density of construction has altered the panorama, contracting the rural space while encircling the village cores, and in the years to come these housing estates will be progressively transformed into a continuous suburb. This illustrates the difficulty in 'demarcating urban and rural spaces' and in 'distinguishing what is continuous suburb and discontinuous periurban' in a context of rapid urban growth common to numerous metropolises in the developing countries (Steinberg, 1993, pp.10-11).

Increasing Density and Transformation of the 'Urban' Villages

The process of urban expansion involves the annexation of agricultural land and the absorption of the surrounding villages in the urban agglomeration. Over the 1901-91 period, 185 new villages were incorporated within the limits of Delhi urban agglomeration (Diwakar and Qureshi, 1993), and 17 more during the 1991-2001 decade (Census of India, 2001). Many of these urbanized villages (designated 'urban villages' by the planners) appear like spontaneously developed enclaves within highly planned areas. They are subject to very great pressures on land and important transformations of their economic functions, morphology and population (Sundaram, 1978; Lewis and Lewis, 1997; Tarlo, 1996; Bentick, 2000). The habitat is transformed in response to the housing needs of numerous migrants with low incomes who find in the urban villages rent levels which are less than in the other planned areas of the capital. These urban villages enjoy a special status and remain outside the purview of most town planning rules, the objective being to preserve the original identity of village life and its traditional values. There is thus no restriction on the type of construction erected nor on the type of activity conducted in these zones. Paradoxically, this special status has accelerated the transformation of the original village nucleii. It has encouraged their commercialization and the proliferation of small industrial workshops, by offering to entrepreneurs working space at rents lower than in the recognized commercial or industrial zones, while at the same time avoiding the controls of the municipality. The manifold increase of economic activities has also attracted a working class of laborers who live in the villages if possible, thus contributing to an increased density of population and of housing.

The case of Harola, a village enclosed in the new industrial town of Noida, is examined in detail elsewhere (Dupont, 2001a). This urbanized village exemplifies in a spectacular way the radical transformations that may occur in the context of disruptions in sources of livelihood and in the local labor market, combined with a high demand for rental lodging, in the absence of restriction on constructions. Although the town authorities have deployed an active housing policy in Noida, we see borne out there a situation that is classic to cities in developing countries. This is the lack of any central measure in the rental sector, which is left entirely to uncontrolled private initiative.

The population dynamics in urban villages are exemplified in the 1995 mobility survey in the peripheral zones of Mayur Vihar-Trilokpuri and Badli-Rohini, as well as in the new town of Noida. In these areas the urban villages are the only type of settlement where one can find household heads who have been living in the same dwelling since their birth. Yet, there are still significant percentages of migrant households in the urban villages (22 per cent in Badli-Rohini, 28 per cent in Noida, and 67 per cent in Mayur Vihar), indicating that newcomers contribute to the current dynamics at work.

Implications for Categorizing Population

The processes that underlie urban development in the metropolitan area of Delhi contribute to an interweaving of urbanized zones and countryside, as well as to a blurring of the distinction between rural and urban population categories. This is especially evident at the fringes. The continuous geographical expansion of the urban agglomeration of Delhi entails, first of all, a physical integration of urban and rural spaces through the incorporation of villages in the urbanized zone. The process of periurbanization and rurbanization around Delhi is also expressed by a functional integration of the metropolis and new residential neighborhoods established in the rural fringes, without (necessarily) continuity of builtup space (at least during the initial phase of emergence of these outlying clusters). The daily commuting of the new dwellers in the rural-urban fringe between their decentralized housing estates and the centers of employment in the capital reflects the link of economic dependency between the different spaces.

However, the functional integration of urban and rural spaces is also at work in the central urban agglomeration of Delhi due to the continuous settlement of considerable flows of inmigrants, mainly from rural origins. Although we have not elaborated on this aspect here, these migrants usually maintain relations of a diverse nature (economic, social, emotional, etc.) with their native place (Banerjee, 1986; Basu, Basu and Ray, 1987; Dupont, 2000a): their life space transcends the urban-rural borders, exceeding the limits of the city to incorporate their home villages. Thus, the integration of urban and rural spaces extends beyond the geographic continuum through circular movements of individuals (commuters as well as migrants) between the different places with which they have relations (Dupont and Dureau, 1994).

The integration of urban and rural spaces, physical and functional, also induces a crossing – a certain symbiosis – of urban and rural characters of populations and the emergence of composite identities. Many inhabitants of metropolitan areas (like that of Delhi) appear to be neither exclusively urban, nor exclusively rural, whether it be a matter of populations in the rural fringes in the process of urbanization, of commuters from rural hinterland, of city dwellers who have shifted their residences into the surrounding rural zones, or of migrants still linked to their native villages.[15]

The process of Delhi's metropolitanization must be viewed as a system of reciprocal influences. On the one hand, the urbanization of peripheral zones and of surrounding rural populations, as well as the introduction of urban goods, information, ideas, social and cultural values and behavioral patterns in faraway villages through circulating or returning migrants. On the other hand, a certain ruralization of the metropolis and of its inhabitants is occurring, through rural migrants importing their original values and behavioral patterns, and retaining them to some extent in the city. This influences demographic behavior, like patterns of nuptiality and fertility. Thus, constructing population categories that would be based on the criterion of a single place of residence at the time of observation proves to be too restrictive for apprehending the spatial distribution pattern of populations and for further demographic analysis.

In order to analyze pluri-polar residential spaces, some authors have introduced the notion of intensity or density of residence (Poulain, 1985; Dureau, 1987) and tried to understand the residential space as a system of residence, a spatiotemporal configuration defined in relation to the various places of stay and the density of residence in each one of them (Barbary and Dureau, 1993). Yet, in the context of populations exposed to intense commuting, the geographical and social environment of the workplace may exert an influence as significant as – or even more significant than – the place(s) of residence of the individuals to explain some demographic behaviors.

A longitudinal approach proves also to be necessary in order to understand a pattern of behavior at a given time. Rather than the place of residence, or even the system of residence, of an individual at the time of observation, what matters more for explaining his/her demographic behavior is the duration of stay in successive places and the successive modifications of his/her system of residence. Some of these issues have been tackled by life history surveys and event history analysis in demography (GRAB, 1999).

---

[15] The fact that many individuals, by virtue of their multipolar residential and work spaces, and the effect of circulation, are neither exclusively urban nor exclusively rural, has been acknowledged for a long time by several authors. In the Asian and Pacific context, see among others: Goldstein (1978); Hugo (1982); Chapman and Prothero (1983).

One should also be able not only to characterize the settlement pattern prevailing in the residential space of reference at each step of the individual life course, but also to take into account the transformations undergone by these spaces. In this perspective, one specific difficulty for demographic analysis in the context of many metropolises of developing countries, as in the case of Delhi, is the speed of urban spread and transformations, especially in the urban-rural fringes. This is also a difficulty for urban and regional planning: the evolution of the Indian capital and of the towns on its periphery shows some growing discrepancies between the objectives of the planners and the actual development of the metropolitan area.

The rapidity of the urbanization process in the 'rural' hinterland also invalidates the pertinence for demographic analysis of the administrative limits of the Delhi urban agglomeration, despite their periodic redefinition. The rise of a transitional periurban type of area around the Indian capital further underscores the inadequacy of the dichotomous classification of human settlements in India and the need for the recognition of an intermediary category between rural and urban. At the level of the National Capital Region, the distinction made between three planning zones (the Territory of Delhi, the ring towns in the metropolitan area, and the zone beyond the metropolitan area) runs the risk of becoming an obsolete theoretical distinction, overtaken by the rate at which the actual dynamics are evolving. In particular, the development of a multinodal quasi-continuous urban area calls for a revision of the limits of the Delhi urban agglomeration, in order to encompass the contiguous towns located beyond the National Capital Territory borders.

Annex 1 Sources of Population Statistics

Census of India, 1951, 1961, 1971, 1981, 1991, 2001.
Census of India 1951, Punjab Population Sub-zone, General Population, Age and Social Tables.
Census of India 1961, Volume XIX, Delhi, Migration Tables.
Census of India 1971, Series 27, Delhi, Migration Tables.
Census of India 1971, Series 27, Delhi, *District Census Handbook*. Delhi: Directorate of Census Operations, 1972.
Census of India 1981, Series 28, Delhi, *District Census Handbook*, Delhi: Directorate of Census Operations, 1983.
Census of India 1981, Series 28, Delhi, Migration Tables, Socio-cultural Tables.
Census of India 1991, Series 31, Delhi, *District Census Handbook*, Village and Townwise Primary Census Abstract, Delhi: Directorate of Census Operations, 1992.
Census of India 1991, Series 31, Delhi, Migration Tables, Social and Cultural Tables, Delhi: Directorate of Census Operations (*on floppies*).
Census of India 2001, Series-8 Delhi, *Provisional Population Totals*, Paper 1 of 2001, JINDGAR Bimla, Directorate of Census Operations, Delhi.
Census of India 2001, Series-8 Delhi, *Provisional Population Totals*, Paper 2 of 2001: *Rural-Urban Distribution of Population*, JINDGAR Bimla, Directorate of Census Operation, Delhi.
Sample Registration System, *Sample Registration Bulletin*, Office of the Registrar General, Vol. 6 (1972) to Vol. 35 (2001).

Chapter 10

Urbanization and Metropolitanization in Brazil: Trends and Methodological Challenges

José Marcos Pinto da Cunha[1]

As is true for almost all of Latin America, Brazil has gone through enormous economic, social and demographic transformations since the beginning of the 1980s, with major implications for the spatial redistribution of the country's population and for the nature and dimensions of its settlements. One of these changes has been a slowing – indeed, in some parts of the country, a cessation – of the trend towards concentration that, for decades, had characterized the country's demographic dynamics. In addition, the types of settlements have become more diversified and new regional spaces have emerged in Brazil's demographic dynamics. Moreover, as rural-urban migration has waned, other forms of mobility have taken on greater importance. The resulting changes in settlement patterns have not only affected the most urbanized regions of Brazil but can be found in less densely inhabited areas as well.

Within the context of the case studies in this part of the book, this chapter has two primary objectives. One is to document the main features of settlement change, giving particular attention to the aspects that seem to have altered most over the last 20 years according to the evidence of the official statistics. The second is to subject this picture, and the concepts and data on which it is based, to critical scrutiny and suggest ways in which data providers and researchers can help towards producing a more meaningful and accurate representation of recent and future developments. It is shown that reliance on administrative criteria for distinguishing urban from rural localities seriously distorts the measurement of urbanization levels, while similarly the designation of metropolitan areas appears to owe more to political considerations than functional realities. Amongst its several suggestions, the chapter describes a study of São Paulo State that involves a more detailed classification of settlement in the 'transitional' zones around metropolitan centers than has previously been available.

Before proceeding, however, it is important to acknowledge the challenge posed by any attempt at developing a single set of urban and metropolitan

[1] In preparing this chapter, the author received support from the Population Research Center (PRC) of the University of Texas, where he was carrying out postdoctoral studies.

standards for such a large and varied country as Brazil, let alone internationally. With an area of over 8.5 million square kilometers and a current population of approximately 170 million, Brazil displays great diversity in its cultural, economic, social and demographic aspects. As a simple example, the meaning, characteristics and way of life in the rural areas of the north are very different from those seen in the southeast. The same can be said in regard to the range of large cities across the nation, in terms of their size, complexity, configuration, etc. It is very hard to perceive a single pattern of human settlement, especially when what might be considered near or far, and large or small, varies so much between regions.

Demographic Dynamics and Urbanization Processes

This section briefly documents the main characteristics of Brazil's urbanization and metropolitanization processes. Based mainly on data from the population censuses of 1980, 1991 and 2000, it begins with an overview of national and regional population trends. It goes on to examine what the official statistics tell us about urban and rural population change, the pace of urbanization and the scale of metropolitan growth. Attention is drawn to certain anomalies that are addressed in more detail in the following section.

National and Regional Population Trends

Brazilian demographic dynamics have gone through a major transformation in recent decades, with a substantial decline in the rate of national population growth. As shown in Table 10.1, the rate almost halved between the 1950s and the 1990s, and indeed has fallen by one third since the 1970s. Basically, the effects of impressive increases in life expectancy over these decades have been hugely outbalanced by massive reductions in fertility.

By contrast, the regional picture, whilst overlain by this reduction in national rates, has generally seen less alteration in broad pattern. The decades before the 1980s principally featured, on the one hand, progressive demographic concentration in the Southeast region, especially in São Paulo, and, in lesser measure, the increased relative importance of the agricultural frontier regions. Since then, the latter has continued, with both North and Center-West regions recording growth rates well above the national level. The Southeast, however, has not been as dynamic as in the past, very largely owing to the State of São Paulo suffering a much larger cutback in growth rate than Brazil as a whole. Nevertheless, over the past two decades, the relative weight of São Paulo State continued to increase, and the Southeast remained Brazil's most populous region by far, with its share falling only marginally between 1970 and 2000, down from 43.5 to 42.6 per cent of the national total.

Table 10.1 Brazil's population growth by region and São Paulo State, 1950-2000

Region and state	Annual average growth rate (%)				
	1950-1960	1960-1970	1970-1980	1980-1991	1991-2000
Brazil	3.04	2.89	2.48	1.93	1.62
North	3.40	3.47	5.02	4.06	2.62
Northeast	2.12	2.40	2.16	1.82	1.31
Southeast	3.11	2.67	2.64	1.76	1.61
São Paulo	3.50	3.30	3.50	2.02	1.79
South	4.14	3.45	1.44	1.38	1.42
Center-West	5.45	5.60	3.99	2.99	2.38

Source: FIBGE, Demographic Censuses.

The Urbanization Process

As shown in Table 10.2, Brazil's urbanization has been proceeding swiftly since the middle of the twentieth century, notably forced on by the intense rural migration to the cities of the 1960s and 1970s. The proportion of people living in urban localities passed the 50 per cent mark during the 1950s and had exceeded two-thirds by 1980. The pace of the rural-urban shift slowed somewhat after this, due principally to the economic crisis of the 1980s and the widespread process of industrial restructuring (see IPEA/IBGE/NESUR, 1999, for a more detailed account). Nevertheless, by the year 2000 the country had over 81 per cent of its population living in cities, according to the official census statistics (Table 10.2).

Table 10.2 Brazil's level of urbanization by region, 1950-2000

Region	Level of urbanization (%)					
	1950	1960	1970	1980	1991	2000
Brazil	36.2	44.7	55.9	67.6	75.5	81.2
North	31.5	37.4	45.1	51.7	59.0	69.9
Northeast	26.4	33.9	41.8	50.5	60.7	69.1
Southeast	47.5	57.0	72.7	82.8	88.0	90.5
South	29.5	37.1	44.3	62.4	74.1	80.9
Center-West	24.4	34.2	48.0	67.8	81.3	86.9

Source: IBGE, Demographic Censuses.

Largely paralleling the unevenness in the distribution of total population across Brazil, there is considerable regional variation in level of urbanization. At one extreme, the North and Northeast are the least urbanized of the five regions shown in Table 10.2, but even here 7 out of every 10 people were urbanites by 2000 according to the census data. The intermediate regions of the South and Center-West, with a level of at least 8 out of 10, are the ones that have seen the sharpest rise over the past half-century. Meanwhile, at the top, the Southeast, having been the most heavily urbanized region throughout this period, had reached the position of its population being almost 83 per cent urban by 1980, with its level rising to just over 90 per cent in 2000.

Despite the high levels of urbanization across much of Brazil by 2000, it is important to note the continued existence of a large rural population. Officially, there are still nearly 32 million people living in rural settlements in Brazil. In fact, as will be shown later, some commentators believe that this figure is an underestimate. This is partly because of the way the census data were collected, but especially is seen as a consequence of Brazil's laws and regulations for defining urban areas. In particular, they allow any hamlet or village to be raised to the status of a city when it becomes the seat of a municipality, not considering any other criteria such as demographic and functional ones (Veiga, 2002).

Weaknesses in the measurement of urbanization are also evident from mapping the rate of growth in the rural populations of States for the latest intercensal period (Figure 10.1). It can be seen that the strongest growth during the 1990s occurred in two areas that are completely disparate in their economic, social and demographic characteristics: the Northern States such as Amazonas and Rondônia, and São Paulo and the Federal District. In the former, this performance can be explained by the production structure and form of occupation. In the latter, however, the data presented clearly reveal an anomaly that, as will be shown below, has to do with methodological problems involved in the definition of what is urban and what is rural.

The Metropolitan Phenomenon

At the same time that Brazil's population has urbanized, it has generally become more concentrated in the country's larger cities. As a result, just as in most Latin American countries including Mexico (see Chapter 8), the Brazilian urban system can, in no way, be characterized as 'balanced'. In the year 2000, almost three-quarters of municipalities (all those with less than 20,000 inhabitants) accounted for under 20 per cent of the country's population, whereas almost 30 per cent of all Brazilians were living in less than 0.6 per cent of the municipalities (those with over 500,000 inhabitants).

The municipality level, however, is not at all satisfactory for portraying what is happening in this process of population concentration at the upper end of the urban hierarchy. For this purpose, we need to turn to the 'metropolitan' concept that attempts to include not only all the often extensive builtup areas associated with these large centers but also any nearby settlements that are physically separate but functionally integrated with them. Altogether, 23 metropolitan areas had been recognized in Brazil before the 2000 census. Their importance in the country's demography is reflected in the fact that between 1991 and 2000 they accounted for almost half the country's population growth, adding approximately 11 million over this period and increasing their share of national population by 1.3 percentage points to 39.9 per cent.

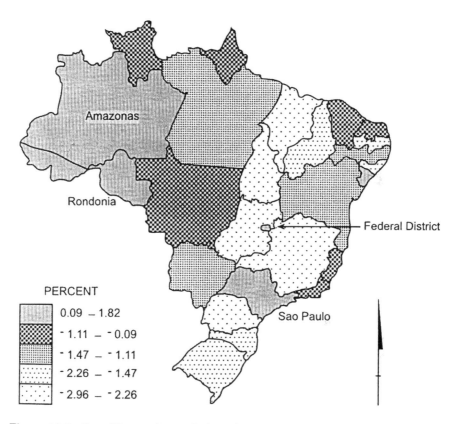

Figure 10.1   Brazil's rural population: Average annual population growth rate, 1991-2000

To provide further information on the metropolitan phenomenon in Brazil, Table 10.3 lists the ten largest of these 23 areas. These alone were home to 53 million people in 2000, 31.5 per cent of Brazil's total population, with half this made up of just the two largest, São Paulo (with 17.6 million) and Rio de Janeiro (with 10.9). In both 1980-1991 and 1991-2000 the annual population growth rate of the majority of these ten areas was higher than the national average of 1.9 and 1.6 per cent respectively. The only exceptions, bar Recife in the economically depressed Northeast, were the two largest. Moreover, São Paulo and Rio de Janeiro were the only cases where the metropolitan areas were not growing faster than the population of their State (Table 10.3). While these would appear to be experiencing population deconcentration within their regional contexts, albeit in relative terms rather than involving absolute decline, elsewhere in Brazil the process of metropolitanization was continuing apace.

Table 10.3   Brazil's ten largest metropolitan areas in 2000: Population change and share of national population, 1980-2000

Metropolitan area (State)	Population (millions)			Annual growth rate (%)				Share of national population (%)		
				1980-1991		1991-2000				
	1980	1991	2000	MA	State	MA	State	1980	1991	2000
São Paulo (SP)	12.59	15.44	17.63	1.88	2.12	1.49	1.78	10.57	10.51	10.41
Rio de Janeiro (RJ)	8.77	9.81	10.87	1.03	1.13	1.15	1.31	7.37	6.68	6.42
Belo Horizonte (MG)	2.62	3.45	4.33	2.53	1.48	2.60	1.99	2.20	2.35	2.56
Curitiba (PR)	1.49	2.05	2.73	2.95	0.98	3.24	1.39	1.25	1.40	1.61
Porto Alegre (RS)	2.31	3.05	3.72	2.58	1.48	2.23	1.21	1.94	2.08	2.19
Federal District (DF)	1.56	2.16	2.85	3.03	2.82	3.15	2.77	1.31	1.47	1.68
Belém (PA)	1.02	1.40	1.79	2.92	3.64	2.81	1.99	0.86	0.95	1.06
Fortaleza (CE)	1.59	2.33	2.97	3.50	1.70	2.80	1.72	1.34	1.58	1.76
Recife (PE)	2.39	2.92	3.33	1.85	1.35	1.49	1.17	2.00	1.99	1.97
Salvador (BA)	1.77	2.50	3.02	3.19	2.08	2.15	1.09	1.48	1.70	1.78
All 10 Metro Areas	36.10	45.11	53.24	2.05	-	1.88	-	30.32	30.71	31.45

Source: FIBGE, Demographic Censuses.

Even with this degree of metropolitan growth, however, centrifugal tendencies are much in evidence, at least internally within each metropolitan area. As early as the 1970s, the majority of these ten areas had seen higher growth rates in their peripheral municipalities than in their main city. In both subsequent decades, there were no exceptions to this generalization, according to the census-based evidence in Figure 10.2. While no central municipality recorded absolute population loss between 1980 and 2000, in every case the growth rate for the remainder of the metropolitan area was higher than for the center. According to studies of the population dynamics involved (e.g. Cunha, 2001b; Lago, 1998; Matos, 1995; and Rigotti, 1994), the most significant factor in this peripheral growth is the centrifugal movement of population resulting from intrametropolitan migration, usually comprising mostly persons in the poorer social strata.

Clearly, as found in many countries and commonly modeled, metropolitan growth in Brazil would seem to be following a predictable path. At least since 1980, metropolitan concentration has been proceeding more through lateral extension, or 'peripheralization', than by population densification within the administrative area of the central city. Then the stage can be reached whereby areas beyond the delimited boundary of the entire metropolitan area grow more rapidly than the latter, as appears to have been the case for São Paulo and Rio de Janeiro throughout this period. On the other hand, the precise dimensions of these types of changes, and indeed even their existence in some cases, can be highly sensitive to the way in which metropolitan areas are defined and the manner in which their boundaries, both internal and external, are delimited. We now turn, therefore, to an assessment of the methods used in Brazil to monitor and analyze urbanization patterns and trends.

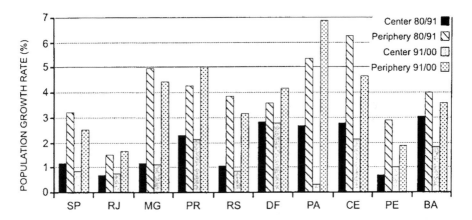

The actual growth of the periphery of the metropolitan area of Pará (PA) in 1991-2000 is 14.3 per cent. However, this fact was disregarded in order to make the scale of the graph more suitable for visualizing the remaining data.

Source: FIBGE, Demographic Censuses.

Figure 10.2   Brazil's ten largest metropolitan areas: Population growth rate for central and peripheral municipalities, 1980-2000

Theoretical and Practical Issues in Measuring Settlement Change in Brazil

As stated above, the rural sector still carries considerable weight in Brazil, yet its dimensions are not adequately reflected by census data, for at least two reasons. The first is theoretical in nature and relates to the growing interrelationships between town and country. The second is the practical issue of the way in which urban and rural areas are classified. In particular, definitions currently depend much more on administrative decisions than on any theoretical considerations as to the meaning and function of such areas. This section discusses these overlapping topics and presents data that indicate some of the problems arising from the type of classification used in Brazil to identify town and country populations.

The Challenge of Differentiating Urban and Rural

The problems posed by the increasing integration of town and country have been examined by several commentators. Faria (1978) considered that the diversification and expansion of the social division of labor in Brazil held true for 'the entire economy and society, and redid, or undid, the generic differences between country and city, and unified the urban and rural labor markets' (p.100). In the same vein, Silva (1997) stressed the increasing difficulty of defining what is rural and what is urban in Brazil, stating that, 'From the spatial standpoint, the rural (world) today can only be understood as a continuation of the urban context .... From the standpoint of the organization of economic activity, the cities can no longer be identified solely with industrial activity, nor can the countryside be identified just with agriculture and cattle raising' (p.43). According to Abramovay (1999, quoted by Patarra, 2000, p.39), 'The dichotomy between rural and urban ... was dismantled ... during the 1960s by the continuum between the rural and urban (spheres), meaning the nonexistence of basic differences in their respective ways of life and social organization and culture'.

Turning to more operational matters, Patarra (2000, p.34) warns of problems arising from definitions based on the census:

> On the one hand, they are definitions drawn up on the basis of the supposed dichotomy between rural and urban aspects of social reality; ... secondly, as such definitions are anchored in administrative criteria and defined at the local level of authority, they fall easy prey to political criteria.

Veiga (2002) goes into more detail about this, arguing that rural Brazil is in fact much greater than what the Census Office's data would seem to indicate. In fact, for that author, the country's rural population still surpasses its urban population numerically, since the current definition of 'urban' in Brazil – with only minor changes, basically the same since 1938 – 'transformed all seats of municipalities existing in the country into cities, regardless of their structural and functional characteristics'. According to Veiga (2002, p.3), 'Due to this policy, all such municipal seats were considered, even if they were in fact no more than tiny hamlets or villages.'

This practice has led to serious distortions in the delineation of Brazil's urban system. For example, it currently classifies as a city the seat of the municipality of União da Serra, in the State of Rio Grande do Sul, even though this boasts a mere 18 inhabitants (Veiga, 2002). Due to the minimalist nature of the requirements set down for a village to acquire the status of a city (or an urban area), a 'swelling' of the urban sector would seem to be occurring. On the basis of this evidence, Brazil's urban population must have a lower relative weight than that indicated by the census data used in the previous section, as is also argued by Egler (2001).

Metropolitanization, Urban Expansion and the Growth of 'New Rural Areas'

Also related to the problem of drawing a line between what is urban and what is rural, but with the opposite consequences for the statistics on urban population, is the situation found in the outskirts of Brazil's largest cities and particularly in the officially designated metropolitan areas. Here the issue is that there would appear to be zones that are in a process of transition from rural to urban worlds, which under the prevailing approach of the rural-urban dichotomy have to be classified one way or the other rather than be recognized for what they really are. Table 10.4 provides an illustration of this phenomenon, showing that in the majority of cases the parts of metropolitan areas that were still classified as rural in the 1990s were recording faster population growth than the urban parts and indeed than the urban areas in the rest of the State. This element of rural population growth is, however, intimately related to urban expansion, but is not treated as such by the national census.

Table 10.4   Rural and urban population growth according to metropolitan and nonmetropolitan location, selected states of Brazil, 1992-1999

State	Metropolitan area			Rest of the state		
	Urban	Rural	Total	Urban	Rural	Total
Ceará	2.25	0.50	2.20	2.56	-0.50	0.86
Pernambuco	1.00	2.17	1.07	1.25	-0.88	0.44
Bahia	1.37	5.01	1.48	1.39	0.39	0.90
Minas Gerais	1.57	6.99	2.01	1.39	-0.30	0.91
Rio de Janeiro	0.77	-0.09	0.76	1.51	1.27	1.47
São Paulo	1.41	1.57	1.42	1.66	1.86	1.69
Paraná	2.92	9.68	3.35	2.03	-2.58	0.71
Rio Grande do Sul	1.05	4.33	1.19	1.80	-0.84	0.99
Distrito Federal	2.13	9.84	2.67	na	na	na

This uses the rural-urban administrative division as defined in 1991; na not applicable.

Source: Special tabulations of 1992 and 1999 PNAD, provided by FIBGE.

This issue has been examined in some detail by Cunha and Rodrigues (2001) in an attempt to find a way of capturing this better. Based on data from the 1991 Census and 1996 Population Count reworked by the Census Bureau, their study demonstrated that this rural population growth was indeed closely related to metropolitan dynamics, strongly corroborating the hypothesis of urban sprawl as the key factor. The areas involved are seen as 'transition areas' – spaces where not only the landscape and form of settlement are much closer to urban patterns but especially where a large proportion of the resident population is dependent on the city, especially in terms of work. Most notably, the research on these transition areas showed that only five per cent of their economically active residents were engaged in primary-sector work, compared to 31 per cent in secondary activity and 53 per cent in the tertiary.

The key feature of this reworked data was the use of a more refined classification of settlement than the simple rural-urban dichotomy. Based on criteria that consider the form of space like streets and buildings along with distance and social infrastructure in rural areas, this classification comprises eight categories, as follows:

(1) Urbanized area.
(2) Nonurbanized area.
(3) Isolated urban area.
(4) Rural agglomerate of urban extension.
(5) Isolated rural agglomerate or village.
(6) Isolated rural agglomerate or nucleus.
(7) Other population agglomerates.
(8) Rural area (excluding rural agglomerate).

Together, as can be seen from the fuller set of definitions provided in Annex 1, the categories form a scale of broadly decreasing urbanness, ranging from large builtup areas through to isolated rural groups.

When it comes to the task of deciding how to portray the settlement system, the approach used by the Census Bureau is to class (1), (2) and (3) as urban and the remainder as rural. It is category (4) that Cunha and Rodrigues (2001) treat as the 'transition areas'. This is on the basis of this category being described as 'characterized by a complex of permanent and adjacent buildings, forming a continuously constructed area with streets' and '*located less than 1 km* from an urbanized area of a city or village or from a rural agglomerate already defined as urban extension and *contiguous to same*' (author's emphasis) (FIBGE, 2000). This type of locality is seen as being quite separate from (5), which is defined as 'a place ... located one kilometer or farther from the actually urbanized area of a town or village or from a rural agglomerate already defined as urban extension'. Certainly, while these areas retain some features of rural life, they also tend to offer attractive possibilities for real-estate development and, at least for a while, represent an alternative place to live for low-income households squeezed out of the high-cost areas closer to city centers.

This way of moving beyond the simple urban-rural dichotomy reflects the new reality of the relationships between the urban and the rural resulting from economic transformations, changes in the structure of personal or family preferences, new forms of use and occupation of the soil, etc. The 'transitional area' category helps to capture the growth of areas that are increasingly distant from the central municipality and are affected by the intensification of urban-to-rural migration. Although with limited share in the total population, the 'Rural agglomerate of urban extension' category registered a relative increase in almost all the Brazilian metropolitan areas between 1991 and 1996 (Table 10.5). This fact, and the reclassifications that undoubtedly occurred between these two surveys, confirm the importance that the expansion of the urban areas has today on the dynamics of the rural areas.

Table 10.5 **Brazil's metropolitan areas: population distribution according to place of residence, 1991 and 1996**

Metropolitan area	Place of residence, 1991				Place of residence, 1996			
	Urban	IUA	RAUE	Rural	Urban	IUA	RAUE	Rural
Belém	66.48	0.00	27.71	5.81	60.89	0.00	33.12	5.99
Grande São Luís	33.34	0.00	52.78	13.87	83.91	0.00	6.77	9.32
Fortaleza	95.96	0.02	0.00	4.03	96.28	0.05	0.00	3.68
Natal	87.58	0.40	3.08	8.94	87.99	0.43	3.20	8.38
Recife	94.08	0.34	1.96	3.61	94.35	0.38	2.24	3.03
Maceió	87.69	0.00	0.00	12.31	88.30	0.00	0.00	11.70
Salvador	95.45	1.54	0.44	2.57	94.60	1.96	0.66	2.79
Belo Horizonte	92.59	1.50	1.16	4.74	92.11	1.30	1.32	5.27
Vale do Aço	98.09	0.00	0.24	1.67	94.45	0.00	1.59	3.96
Grande Vitória	97.45	0.17	0.91	1.48	97.94	0.16	0.61	1.28
Rio de Janeiro	99.18	0.01	0.22	0.59	99.21	0.01	0.23	0.56
São Paulo	97.24	0.61	1.04	1.11	95.83	0.69	1.75	1.72
Baixada Santista	99.57	0.00	0.04	0.39	99.60	0.00	0.04	0.36
Campinas	92.26	2.84	0.59	4.31	93.10	2.72	0.56	3.61
Curitiba	90.32	1.21	0.00	8.47	90.00	1.56	0.17	8.27
Londrina	91.65	0.14	0.00	8.21	93.33	0.14	0.00	6.53
Maringá	91.29	1.45	0.00	7.26	91.45	1.95	0.00	6.60
Florianópolis	89.74	0.63	0.36	9.27	89.40	0.55	0.38	9.68
Vale do Itajaí	84.53	0.00	1.80	13.67	83.21	0.00	2.17	14.62
North/Northeast Catarinense	95.91	0.00	1.23	2.86	92.72	0.00	4.36	2.92
Porto Alegre	93.65	1.08	0.52	4.75	93.60	0.57	1.60	4.24
Goiânia	94.96	1.28	0.00	3.75	95.08	3.19	0.00	1.73
Fed.District and surrounding area	88.79	0.97	0.25	9.98	87.93	2.25	1.23	8.60

Urban includes 'non-urbanized areas', IUA Isolated urban areas, RAUE Rural agglomerate of urban extension, Rural includes 'isolated rural agglomerate'.

Source: FIBGE, Demographic Censuses.

In sum, the limitations of current approaches to measuring urbanization and urban growth in Brazil are being increasingly recognized. In the metropolitan context, unlike in less urbanized areas, official statistics generally underestimate the true urban population. Migration flows from central municipality to other parts of metropolitan areas and beyond are very unlikely to be reinforcing the rural nature of these destination areas and their communities. It is therefore suggested that the eightfold classification of settlement, just described, represents an important step forward in the study of urbanization in Brazil and, at the very least, provides a useful starting point for developing a more sophisticated approach.

Limitations of the Data

Any analysis of a reality as complex as urbanization in times of globalization and restructuring of production, especially in a country with the size and territorial diversity of Brazil, will inevitably run up against methodological difficulties related to the characteristics of the information used, the form in which it was collected, and even the total lack of certain data.

In the Brazilian context, one of the first difficulties results from the conceptual bases within which the data are gathered. In fact, the analyst's problems begin with the fact that the definitions of categories such as 'urban' and 'rural', or 'metropolitan area', are determined by the municipalities and states, respectively. That is, they may not only vary from one region to another, but they also depend on decisions and interests that, as is known, are not always in consonance with rational or functional logic. Egler (2001) emphasizes 'the fragility of the political-administrative concept of urban population employed for statistical purposes in Brazil' (p.11). As noted above, this concept can sometimes result in settlements of under 100 inhabitants being considered cities, regardless of the functions they carry out or the collective services they possess.

In a country like Brazil, it would be very useful to have a more consistent definition of urban and rural areas that could be used to allow comparisons. There is a need to assess the appropriateness of using criteria employed by other countries, such as the size and density of places, even while recognizing that this can be problematic in the case of Brazil because the meaning of these indicators would certainly differ from one regional context to another. Beyond this, several authors (Sawyer, 1986; Egler, 2001; Veiga, 2002) argue that functional criteria can also play an essential role. As Sawyer cautions, 'By adopting criteria in relation to size, one must consider that on the fringe, where distances are greater, some smaller settlements (in population terms) can play important urban functions at a local level' (Sawyer, 1986, p.43). What this study claims is that, even among Brazilian scholars, there is no consensus about such definitions using demographic criteria.

A similar situation is encountered in the task of defining a 'metropolitan area'. As Negreiros (1993, quoted by Patarra, 2000, p.40) warns:

> If, on the one hand, the lack of conceptual definitions of the new spatial categories provided greater autonomy to the states in organizing their regionalization, ...on the other, this process may lead to a distortion of the characterization of the Brazilian urban system, ...since the characterization of the metropolitan regions shows distinct urban dynamics.

For example, how is one to compare situations as diverse as the Metropolitan Area of São Paulo, with over 16 million inhabitants and 38 municipalities, with the Metropolitan Area of Belém, with 1.8 million and only five municipalities? What type of data would be necessary for one to adequately apprehend the heterogeneity and internal dynamics of these areas?

One possible answer to the first of these two questions was given in a report on the characterization and trends of the urban system in Brazil (IPEA/IBGE/NESUR, 1999). There, five criteria were used for distinguishing the 49 urban agglomerations identified: degree of centrality/area of influence; existence of decisionmaking centers and international relations; scale of urbanization; complexity/diversification of the urban economy; and diversification of the tertiary sector. Thus, as well as quantitative criteria such as density, size, pace of demographic growth (built into the scale of urbanization), and integration (centrality/area of influence), other aspects were considered in order to better identify both the function and importance in the settlement hierarchy of each of these places.

Regarding the question about the type of data needed, the single most important development for Brazil would be in the fuller and more methodologically effective use of data at the census tract level.[2] In fact, in the case of both the metropolitan dynamics and the characterization of the rural and urban populations discussed earlier, any type of advance in conceptual terms or in data collected would be inadequate by itself, given the usual practice of releasing the data only at the municipal level. Even for less complex human settlements, some important features could be better seen and understood with data at a more disaggregated spatial level. Studies carried out using data available at the census tract level have shown their great potential for acquiring an understanding of intraurban dynamics (including Torres, 1997; Lago, 2000; Cunha and Bitencourt, 2000; Sposati, 2000, 2001).

Another operational type of problem that exists in regard to Brazilian data is the difficulty of establishing a historical series at the municipal level, due to the continuous dismemberment of municipalities that have taken place, especially during the last two decades. Besides the great complexity of the Brazilian urban system in any census, researchers must also be attentive to changes in the territorial division of the country. Apparent changes in the population size of municipalities, in the spatial distribution of population, or in the composition of the Brazilian urban system, may actually be the result of the simple subdivision or reconstitution of municipalities. Sometimes, areas are reorganized in extremely picturesque ways, such as the case of Holambra, in the State of São Paulo, whose current territory was established from parts of four other neighboring municipalities.

Improvements are also needed in the information collected and the methods of handling it. It would be useful if censuses and other surveys could collect more data on people's work patterns and other day-to-day activities (e.g. social, educational, consumption, etc.), as these would reflect interactions between the urban and rural sectors, or within integrated areas. For one thing, we need to

---

[2] In Brazil a census tract includes, on average, a group of 300 households in urban areas and 150 in rural areas, with territorial units being used as 'minimum sample area' for the demographic censuses. The information usually made available at this level corresponds to that gathered in the questionnaire applied to the entire population, containing only basic data, such as sex, age, kinship with the head of household, illiteracy and educational level, and income of the head of household.

understand better the type and location of activities carried out by individuals and the time dedicated to each, as well as information on the locations of consumption and/or satisfaction of basic needs. Such work can build on the example set by the Census Bureau (FIBGE, 1987), which used data on the flows of people for reasons of work and consumption in order to measure the centrality of municipalities and their hinterlands. Also highly desirable is more spatially disaggregated data on changes of residence, these being fundamental for a more a precise demographic analysis of the process of expansion of the large cities.

As regards data handling, developments in Geographical Information Systems (GIS) and the diversification of spatial data (e.g. satellite images, aerial photography, GPS mapping) make it ever easier to manipulate data at the household level. The latter can then be linked to a digital cartographic base containing physical features such as terrain and land use and aspects of infrastructure such as utilities and transport. This type of information would greatly assist the characterization of spaces, enabling them to be more adequately classified.

Though the number and quality of the specialists in this new technology in Brazil has increased considerably in recent years, there is still very limited application of such techniques in the field of social and urban studies. In the past, this would appear to have been due much more to difficulties regarding access and harmonization of data than to resistance to this technology or lack of expertise. Fortunately, things are starting to change, thanks to a policy of openness in terms of data access – especially regarding the universities. Nevertheless, acquiring access, for example, to the Brazilian census data obtained from census tracts, to digitized cartography of these data or – which is even more complicated – to satellite images, still requires researchers to pay high prices, in view of the considerable commercial value of such information.

Concluding Remarks

As can be seen throughout this book, all around the world there is now much questioning about the concepts of urban and rural. Brazil is no exception. This chapter has called attention to some of the most important methodological challenges to be faced if there is to be a clearer understanding of the complexity of urbanization and metropolitanization processes in Brazil. It is all too easy to be misled by official data published for spatial units that owe more to historical inertia and political factors than to a careful reading of settlement patterns on the ground. Also challenging is the regional diversity within Brazil, which means that a set of criteria adopted for one context may not be nearly so appropriate for another.

Several proposals have been presented for lessening the serious problem that the normative nature of the definitions of urban-rural and metropolitan-nonmetropolitan have imposed on demographic analysis. Brazil must move ahead in its search for new ways of presenting information on the characteristics of people's place of residence. In particular, the current, basically 'administrative', approach must be replaced, even at the cost of a break with a tradition of almost six decades. New categories of settlement are needed that go beyond the dichotomy between the rural and the urban, and thus may disclose other important dimensions of settlements in Brazil. Allied with the opportunities opened up by new data collection and handling technologies, these steps can help to bring us much closer to the reality and complexity of contemporary human settlements.

Annex 1   Urban and Rural Definitions Used by Brazil's Census Bureau

Urban Categories

A.1.   **Urbanized area** - An area legally defined as urban and characterized by buildings, streets, and intense human occupation. This category includes areas that have been affected by transformations resulting from urban development and those reserved for urban expansion. The following are therefore urbanized areas:

> a - areas intensely occupied by buildings, streets, squares, etc.;
> b - areas where there is less intense use of the land than that found in the preceding category, identified by the presence of reservoirs, leisure areas, cemeteries, experimental agricultural projects, warehouses related to industrial or commercial activity, wastefills, etc., forming a continuous space with areas in the preceding category;
> c - areas reserved for urban expansion, that is, idle land with no rural use and not yet occupied by buildings or urban equipment but contiguous to areas included in Categories 1 and/or 2.

A.2.   **Nonurbanized area** - An area legally defined as urban but with occupation of a predominantly rural nature:

> 1 - Areas occupied by farming (agriculture and animal rearing in general), and extractive activities;
> 2 - Unused, idle land not near urbanized areas of Types 1 and 2.

A.3.   **Isolated urban area** - An area defined by law and separated from the main town (or municipal) districts by rural areas or by some other legal boundary.

Rural Categories

B.1.   **Rural agglomerate** - A place located in an area legally defined as rural and characterized by a complex of permanent and adjacent buildings, forming a continuously constructed area with streets that are recognizable or arranged along a connection road.

> B.1.1)   **Rural agglomerate of urban extension** - A place that has the defining characteristics of a rural agglomerate and is located less than 1 km from an actually urbanized area of a city or village or from a rural agglomerate already defined as urban extension and contagious to same. It consists of an extension of the actually urbanized area with subdivisions into already occupied lots, housing complexes, agglomerates of so-called subnormal housing or developed around centers of establishments for industry, commerce or services.
> B.1.2)   **Isolated rural agglomerate** - A place which has the characteristics of a rural agglomerate located 1 km or farther from the actually urbanized area of a town or village or from a rural agglomerate already defined as urban extension.

a) **Isolated rural agglomerate village** - A place having the defining characteristics of an isolated rural agglomerate and having at least one commercial establishment for frequently used consumer goods and two of the following three services or equipment: an establishment for primary education from the first to fourth grades, operating regularly; a health center operating regularly; a religious church or equivalent of any persuasion to serve the occupants of the agglomerate and/or nearby rural areas. It corresponds to an agglomerate without a private or business nature or that is not related to a single owner of the land and whose occupants exercise economic activities of either primary, tertiary or even secondary nature at the location itself or elsewhere.

b) **Isolated rural agglomerate nucleus** - A place that has the defining characteristic of an isolated rural agglomerate related to a single owner of the land (farming or industrial companies, factories, plants, etc.), that is, that have some private or business nature.

c) **Other population agglomerates** - Places of a nonprivate or nonbusiness nature that have the defining characteristic of an isolated rural agglomerate and partially or completely lacking in services or equipment for the dweller's use.

B.2. **Rural area (excluding rural agglomerate)** - An area outside the city limits but excluding rural agglomerate areas.

Chapter 11

Changing Urbanization Processes and *In Situ* Rural-Urban Transformation: Reflections on China's Settlement Definitions

Yu Zhu[1]

China's urbanization has experienced tremendous changes since the 1980s. One of the most important of these has been the emergence and development of *in situ* urbanization. Here the term '*in situ* urbanization'[2] refers to the phenomenon that rural settlements and their populations transform themselves into urban or quasi-urban ones without much geographical relocation of the residents. This new urbanization pattern in China has been caused by the massive development of rural nonagricultural activities, and the permeation of urban and quasi-urban facilities into the Chinese countryside since the 1980s. A similar process has been identified in '*Desakota* Regions' or Extended Metropolitan Regions in other Asian countries (McGee, 1991), but China's *in situ* urbanization seems much more developed, as it is not confined to the surrounding areas of large cities (Zhu, 1999). Another important change has been the emergence and development of China's 'floating population', which bears many similarities to circular migrants identified by Hugo (1982) in Indonesia in terms of their double (urban and rural) residential identities.

However, these important changes in China's urbanization process have not been well understood. This has much to do with the inadequacy of China's settlement definitions, and has been further complicated by administrative and definitional changes since the 1980s (Zhu, 1998). Although these problems have,

[1] The author wishes to acknowledge the Wellcome Trust, which provided the fellowship to fund this research (Reference number 055950), and the Provincial Census Office of Fujian Province in China, which provided him with the preliminary results of the 2000 census.
[2] The term '*in situ* urbanization' is closely related to the more widely used term 'reclassification' in the sense that both concern the transformation of settlements from rural to urban. However, while 'reclassification' mainly concerns the final result of the transformation usually achieved by rural-urban administrative changes and implies that there are clear distinctions between urban and rural, the term '*in situ* urbanization' here refers to the whole transitional process from rural to urban regardless of whether this involves any administrative changes, and does not have the connotation of rural-urban dichotomy.

to a certain extent, been solved by the adoption of the new urban definition for the 2000 census, China's urban definitions and statistics are still confusing and subject to misinterpretation. More importantly, China's current settlement definitions, including the one adopted in the 2000 census, are still based on the traditional builtup-area-based dichotomous approaches, and hence do not fully reflect *in situ* transformation of rural areas. There is a need to further improve China's settlement definitions and ways of monitoring settlement changes, and a close examination of China's new urbanization patterns and related definitional, statistical, and conceptual issues is a necessary step.

This chapter starts with a brief account of urban definitions before the reform era. This is followed by a review of China's urbanization process since the early 1980s, and of the relevant definitional and policy changes. Then the process of *in situ* urbanization is examined in more detail, focusing on various dimensions of the new rural-urban transformation pattern and the blurring of the rural-urban distinction. The last part of the chapter will discuss the implications of these changes for the modification of settlement definitions and for ways of monitoring settlement evolution.

The *Hukou* System and Urban Definitions before the Reform Era

China was a typical dichotomous society in terms of rural-urban relations before the reform era. Its unique '*Hukou* system' (the household registration system)[3] divided people into two distinct groups: the agricultural population and the nonagricultural population. As most of the nonagricultural population lived in the urban areas, the agricultural-nonagricultural dichotomy was identical to the rural-urban dichotomy. While those people with nonagricultural *Hukou* status enjoyed many privileges provided by the State, including subsidized food and housing, free education, medical care, old-age pensions and other services, it was extremely difficult for ordinary rural residents to have their *Hukou* status transferred from agricultural into nonagricultural, and hence to move from rural to urban areas. In a similar way, the *Hukou* system also effectively prevented people moving from a place lower in the urban hierarchy to one higher in the urban hierarchy, especially large cities (Zhu, 1999, pp.102-4).

As the *Hukou* system effectively separated the rural population from the urban population and controlled rural-urban migration and urbanization, China's rural-

---

[3] China's *Hukou* system can be dated back to the Northern and Southern Dynasties some 1,500 years ago (Liu, 2001, p.100). Its recent form was shaped through the promulgation of the 'regulation on household registration' in 1958. This regulation stipulates that all citizens must register themselves to relevant authorities at the places of their permanent residence, with the household as the basic registration unit; all births, deaths and migrations are required to be registered by the same authorities; and the transfer of one's household registration from a rural to an urban place needs to be approved. All people are assigned a registration status as either 'agricultural' or 'nonagricultural' in the registration system.

urban settlement definitions were relatively simple before the reform era. According to the definitions promulgated in 1963 (Central Committee, 1986), a place could qualify as a town (the urban unit lowest in the urban hierarchy), if it had a minimum of 3,000 persons with 70 per cent being nonagricultural, or a population of between 2,500 and 3,000 with more than 85 per cent being nonagricultural. An urban place with a population of more than 100,000 could qualify as a designated city, but its suburban area should be restricted to a certain scope so that the agricultural population of the city would not be more than 20 per cent of the total population of the city.[4]

Before the 1980s China's urban population was primarily composed of the nonagricultural population of cities and towns (NAPCT) as its main part and the rest of the population in the urban administrative areas (including the suburban areas outside the cities and towns proper) as its subsidiary part. The urbanization process was mainly determined by the growth of the NAPCT, which was sponsored, and strictly controlled, by the State. However, the total population of cities and towns (TPCT), which includes the total population within the administrative areas of designated cities and towns (including the agricultural population of suburban areas), was usually regarded as a more reasonable representation of the size of the total urban population, because most of the agricultural suburban population were often considered either involved in nonagricultural activities or/and using the urban infrastructure intensively in the urban administrative areas (Chan, 1994a; Zhu, 1999). Both series of data on urban population (TPCT and NAPCT) were published annually, with the former referred to as the 'first definition' (National Bureau of Statistics of China, 2001, p.103).

China's Urbanization since the 1980s: Changing Definitions and Processes

New Urbanization Processes vis-à-vis New Urban Definitions

The adoption of economic reform and open door policies since the late 1970s have fundamentally changed the simplicity of the urbanization process and the applicability of the pre-existing urban definitions in China in at least three ways.

Firstly, since the 1980s, the invisible wall created by the *Hukou* system between the cities and the countryside has gradually been eroded by the implementation of various reform policies. In the early 1980s, China introduced the household responsibility system and abolished the People's Commune system in rural areas. These measures revealed and intensified the once seemingly invisible problem of rural surplus labor. According to Taylor and Banister (1991), the number of rural surplus laborers was 132.3 million in 1980, accounting for 42.5 per cent of total agricultural workers. Since they could no longer rely on the People's Communes for employment and livelihood, many of them entered cities to seek

---

[4] Places commanding special administrative, strategic, or economic importance could qualify for city or town designation with a smaller population. For more details, see Zhu (1999, pp.211-8).

work. At the same time, the government gradually loosened control over rural-urban migration and the agricultural to nonagricultural transfer of *Hukou* status. Meanwhile, economic development and reform undermined the effectiveness of the *Hukou* system in controlling rural-urban migration. In 1984 the State Council decided to allow farmers and their dependents to move to designated towns (excluding county-level government seats) for permanent settlement, provided they met certain conditions so that they did not create a financial burden to the State (China, State Council, 1984a). In 1985 the Ministry of Public Security promulgated 'Interim regulations regarding the management of temporary residents in cities and towns', symbolizing the beginning of legal residential status of migrants without local *Hukou* registration (Liu, 2001, p.102). Since then, it has become increasingly easier for rural residents to enter urban areas as temporary residents and seek employment and living opportunities, mostly in the non-State sectors, although they are still disadvantaged in the urban areas because of their agricultural, non-local *Hukou* status.[5]

Thus, since the 1980s an increasing number of temporary residents, more commonly referred to as the 'floating population', have entered Chinese cities and towns. According to some widely accepted estimates, the volume of the floating population increased from 30 million in 1982 to between 80 and 100 million in the mid-1990s, with most in urban areas (Zhu, 2001a). A majority of the floating population leave some members of their families behind at home, change their living and working places frequently, and do not have the intention of settling down in the destination, due not only to their unfavorable *Hukou* Status, but also their household strategies to minimize costs and bring back as much saving as possible to their hometowns (Zhu, 2001a). The emergence of, and the fast increase in, the floating population makes *Hukou* no longer a useful criterion for defining the rural-urban populations, and raises the issue of how to categorize the floating population to reflect their double identities in relation to both their hometowns and their temporary places of residence.

Secondly, since the 1980s there have been several rural-urban administrative and definitional changes, and *reclassification* has become a major component in China's urban growth and urbanization. Contrary to the policies restricting urban development before the reform era (Chan, 1994a), the Chinese government has taken an increasingly more positive attitude towards the designation of cities and towns, and the number of officially designated cities and towns has increased substantially since 1978 (Table 11.1). However, this has been accompanied by profound rural-urban administrative and definitional changes which have had major implications for the accurate definition of urban and rural areas and populations (see Figure 11.1 for the current rural-urban administrative structure).

---

[5] These disadvantages include restricted access to certain kinds of employment, extra fees for children's education, no entitlement to some social benefits such as subsidized medical insurance and low price housing, etc. However, these disadvantages have been also increasingly reduced in the process of reform. In fact, many cities in China introduced new regulations to allow temporary residents to have local urban *Hukou* status in 2001, and it seems that more and more cities will follow suit.

Table 11.1   Number of cities and towns in China, selected years

Year	1949	1956	1978	2000
Cities	132		193	667
Towns		3,672	2,173	20,312

Source: Dai (2000, pp. 174-9, 210); *Xinhua Net News*, 30/10/2001 and 4/11/2001.

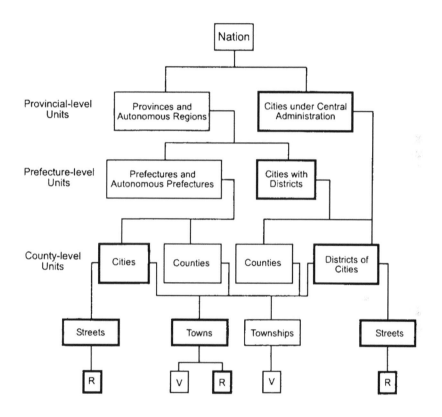

R Residents' Committees
V Villagers' Committees

The heavy boxes are urban administrative units at various levels. Please note that not all areas within the urban administrative units (except streets and residents' committees) are necessarily urban, nor are all areas within non-urban administrative units necessarily rural. The latest boundaries of the *bona fide* urban areas are delineated by the 2000 census urban definition (see the relevant text for more details)

Source: Substantially modified and updated from Chan (1994b, p.250).

Figure 11.1   China's urban and nonurban administrative systems

The most important change of this kind was that in 1984 China's State Council relaxed the criteria for the designation of town status, and adopted policies of 'abolishing townships and establishing towns' (*chexiang jianzhen*) and 'town administering village' (*zhengguancun*). As a result, townships were abolished and their territories and populations were placed under the jurisdiction of towns[6] in order to promote the development of small towns, which were regarded as reservoirs of rural surplus labor and focal points of rural development (China, State Council, 1984b; Ma and Cui, 1987; Lee, 1989). According to the new criteria and administrative structure, townships with a total population of less than 20,000 can be designated as towns if the nonagricultural population in the township government seat is more than 2,000. Also, townships with a total population of more than 20,000 can be designated as towns if the nonagricultural population in the township government seat is more than 10 per cent of the total township population. It is important to note that the conversion of a township into a town involves not only the territory and population of the builtup area but the entire territory and population of the administrative area, making the total population of the town (TPT) a serious overrepresentation of the really urban population there.

Similarly, in 1986 the policy of 'abolishing counties and establishing cities' (*chexian jianshi*) was introduced.[7] This had the effect of exaggerating the urban population because the statistics of the total population of cities (TPC) included many agricultural areas and their populations. The city definitions have also experienced major changes since the 1980s. Apart from changes in the criteria on the size of population and the share of nonagricultural population, the new definitions introduced economic criteria for the designation of cities, and made the criteria dependant on population densities. The latest version of the city definition promulgated in 1993 is summarized in Table 11.2.[8]

Thirdly, since the 1980s there have been some important processes which have made a great contribution to China's urbanization, but are not fully reflected in conventional urban definitions and statistics (Wang and Zhou, 1996; Zhu, 1998). The most important has been the rapid development of rural enterprises, which were named commune and brigade enterprises before 1984 and have been renamed township and village enterprises (TVEs). Contrary to former policies restricting

---

[6] Note that although 'town' and 'township' have very similar meaning in English, they are widely used as the translation of the Chinese terms '*zhen*' and '*xiang*' respectively, whereas '*zhen*' is an urban administrative unit while '*xiang*' is a rural administrative unit in China's administrative system (see also Figure 11.1).

[7] This does not preclude some economically developed towns that have become the regional economic centers being designated as cities and separating from the rest of the counties.

[8] Counties meeting the criteria in this table are designated as county-level cities. They can be further designated as prefecture-level cities if meeting certain higher criteria (Dai, 2000, p.77). Besides, cities are also classified into small cities (with a nonagricultural population smaller than 200,000), medium-sized cities (with a nonagricultural population of 200,000 to 500,000), large cities (with a nonagricultural population of between 500,000 and one million), and very large cities (with a nonagricultural population of more than one million), according to their size of the nonagricultural population.

rural areas from participating in the industrialization and urbanization processes, since the 1980s the Chinese government has actively encouraged the development of TVEs in rural areas. These were expected to play major roles in absorbing rural surplus labor, promoting the development of small towns, and preventing too many farmers from entering cities. This kind of rural-urban transformation poses great challenges to the traditional dichotomous settlement definitions and statistics, because it involves a blurring of any clear-cut transition from a definitely rural to a definitely urban settlement and the acquisition of a degree of urbanity in rural settlements. China's official rural-urban boundaries and the *Hukou*-based rural-urban classifications have become even less meaningful, as more and more rural settlements and their residents perform urban functions and take on urban characteristics without having nonagricultural *Hukou* status or relocating to urban areas.

Table 11.2   Criteria for the designation of cities promulgated in 1993

Population density of the county (persons/km²)		>400	100-400	<100
The town where the county government is situated	Nonagricultural population (000s)	120	100	80
	Nonagricultural population by *Hukou* (000s)	80	70	60
	% of the population using running water	65	60	55
	% of paved road	60	55	50
	Fairly good public facilities, infrastructure and drainage system			'
The county	Nonagricultural population (000s)	150	120	80
	% of nonagricultural population in total population	30	25	20
	Gross output value of industry at the township level and above (million *Yuan*)	1,500	1,200	800
	% of gross output value of industry in the total gross output value of industry and agriculture	80	70	60
	GDP (million *Yuan*)	1,000	800	600
	% of tertiary sector value in GDP	20	20	20
	Local budgetary financial revenue — Total value (000s)	60,000	50,000	40,000
	Local budgetary financial revenue — *Yuan* / per capita	100	80	60
	A certain part of the revenue is turned over to the higher authorities			

Source: Dai (2000, p.77).

As a result of these three sets of changes, the statistical series of NAPCT, although still used for urban and other social economic planning by the government, has become a serious underestimate of the real scale of urban-rural transformation in China.

Modifications to Urban Definitions in the 1990 and 2000 Censuses

To tackle the problems of inconsistency between the administrative boundaries of cities and towns and the boundaries of their actually urbanized areas, two modifications were made to the official urban definitions in the 1990 and 2000 censuses. The first, commonly called the 'second definition', was adopted in the 1990 census (National Bureau of Statistics of China, 2001, p.103). According to this definition, the city population refers to the populations in the districts of designated cities, and the populations in the streets administered by designated cities which are not subdivided into districts; the town population refers to the populations in the residents' committees of designated towns administered by the cities not subdivided into districts and designated county towns. In other words, the modified definition takes TPCT as the urban population for cities subdivided into districts, and the NAPCT as the urban population for cities not subdivided into districts and for towns.[9] The census result on China's urbanization level based on this definition was widely accepted (Chan, 1994a), and since then China's National Bureau of Statistics has produced a third series of data on China's urban population for the period between 1982 (the year of the third national population census) to 1999[10] (the first two being TPCT and NAPCT) based on the second definition.

However, this modification does not solve the fundamental problem with China's urban definitions, as it is actually a mixture of the old administrative-boundary-based definition and the *Hukou*-based definition. In fact, it makes the problem even more complicated, because it takes different criteria for the population of cities subdivided into districts and the population of cities not subdivided into districts and towns, so that the urban population data are not comparable between cities and towns (Zhou and Sun, 1992). The definition is internally inconsistent, exaggerating the city population while underestimating the town population.

Thus in the 2000 census, a more sophisticated modification was made to the urban definition and a new definition referred to as the 'third definition' was made (National Bureau of Statistics of China, 2001, p.103). According to the third definition, cities only refer to the city proper of those designated by the State Council. In the case of cities subdivided into districts, the city proper refers to the whole administrative area of the district if its population density is 1,500 people per square kilometer or higher; or the seat of the district government and other areas of streets under the administration of the district if the population density is less than 1,500 people per square kilometer. In the case of cities not subdivided into districts, the city proper refers to the seat of the city government and other areas of streets under the administration of the city. For the city district with the population density below 1,500 people per square kilometer and the city not subdivided into districts, if the urban construction of the district or city government seat has extended to some part of the neighboring designated town(s) or

---

[9] The nonagricultural population of a city not subdivided into districts or a town by *Hukou* status is almost identical to the population of its streets in the former case or the population of its residents' committees in the latter case, as pointed out by Zhou and Sun (1992, pp.21-7).

[10] In China Statistical Yearbook (2001) the 2000 census definition rather than the 1990 census definition is used for the data of the year 2000 urban population (National Bureau of Statistics of China, 2001, p.91). It is still not clear whether a new adjusted series of data on urbanization based on the 2000 census definition will be created.

township(s), the city proper should include the whole administrative area of the town(s) or township(s).

Similarly, the new definition stipulates that towns only refer to designated towns proper. The town proper in turn refers to the seat of the town government and other areas of residents' committees under the administration of the towns; if the urban construction of the town government seat has extended to the seat(s) of the neighboring villagers' committee(s), the whole area(s) of the villagers' committee(s) should be included in the town proper. All areas outside cities and towns are referred to as the countryside. It is noteworthy that a distinction is made between market towns and the rest of the countryside, with the former referring to the seats of townships and nondesignated towns serving as certain economic, cultural and service centers and acknowledged by the county governments.

Hence there are three official urban definitions in China: the total population of cities and towns (TPCT), with the nonagricultural population of cities and towns (NAPCT) as its supplement, the 1990 census definition, and the 2000 census definition. The first two definitions have little credibility. The new definition adopted at the 2000 census is essentially builtup-area-based. It solved the overbounding problem of TPCT, the underbounding problem of NAPCT, and the problems of inconsistency and incomparability of the 1990 census definition by introducing the population density criterion for cities subdivided into districts, conforming to the conventional international practice in urban definitions. Because the new definition covers the neighboring townships or villages extended to by the urban construction of cities or towns, it also reflects some evident results of *in situ* urbanization in the rural areas. This seems to be a satisfactory definition.

Urbanization Trends Reflected in Statistics Based on the Urban Definitions So Far

Having reviewed China's urban definitions and their changes since the 1980s, we can now look at Figure 11.2 to examine China's urbanization trends since the late 1970s measured by different urban definitions. The urbanization trend before 1982 can be represented by the series of TPCT. As the figure shows, China's urbanization started from a low level, with urban population only 17.9 per cent of the total population, and increased to 21.1 per cent in 1982. The series based on NAPCT is parallel to that of TPCT but lower due to its more restrictive definition, indicating the fact that China's urbanization trend at the beginning of reform was still closely associated with the growth of *Hukou* based nonagricultural population of cities and towns (NAPCT). However, since 1983 the urbanization level as measured by TPCT has increased dramatically, and differed widely from that measured by NAPCT, reflecting the overbounding problem of the TPCT caused by the urban administrative changes. As can be seen from the figure, the series of TPCT became increasingly unreasonable, with its proportion of the total population surpassing 50 per cent in 1989, and even 73 per cent in 1999.

The new series based on the 1990 census definition created for the period between 1982-99 is also presented in the figure. According to this series, China's urbanization level increased from 21.1 per cent in 1982 to 26.4 per cent in 1990, and further to 30.9 per cent in 1999. Although this is a better estimate on China's urbanization level and its changes than the series of TPCT, by comparing the urbanization level of the year 1999 from this series with the urbanization level of the year 2000 from the 2000 census (36.1 per cent), it is obvious that the series based on the 1990 census definition is on the lower side of the real urbanization level.

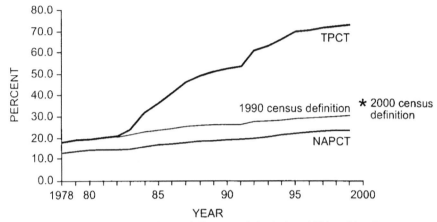

Source: Based on data from National Bureau of Statistics (2001, p.91); Department of Population (2000, pp.447-8).

Figure 11.2   China's urban population as a proportion of the total population by different definitions, 1978-2000

The author has been able to get the relevant data from the 2000 census for Fujian Province processed to obtain the data of urban population by the 1990 and 2000 census definitions respectively to demonstrate quantitatively the problem with the 1990 census urban definition. The result shows that, if the 1990 census definition were used to replace the 2000 census definition, Fujian's urbanization level would be 34.3 per cent rather than 41.6 per cent. Its town population and city population would be 2.44 million and 9.26 million rather than 6.35 million and 7.83 million respectively. This means that Fujian's urbanization level would be underestimated by 7.3 per cent; Fujian town population would be underestimated by 3.91 million; however, Fujian's city population would be overestimated by 1.43 million. This exercise indicates again that the 2000 census urban definition is a significant improvement over the 1990 version. By this new definition, China's urban population was 455.9 million, and its proportion of the total population was 36.1 per cent, on November 1, 2000.

Blurring of Urban-Rural Distinctions and *In Situ* Urbanization

In terms of conventional builtup-area-based dichotomous approaches, the adoption of the 2000 census urban definition seems to have solved most of the problems with China's urban definitions. However, from the perspective of *in situ* urbanization, many problems still remain. In fact, the 2000 census definition only reflects visible changes in the extent of urban builtup areas: it does not adequately reflect the other quasi-urban changes beyond those areas, which have become increasingly important since the 1980s. Therefore China's settlement definitions need to be further improved to take consideration of these changes.

Figure 11.3   The location of Fujian Province and its municipalities and selected cities

As relevant data at the national level is not easily available, the following discussion is based mainly on the case of Fujian Province. Figure 11.3 shows the location of Fujian Province, its 9 municipalities and selected cities. Located in southeast China, Fujian is a coastal province where *in situ* rural-urban transformation is well developed.[11] Fujian is fairly representative of China's coastal areas (especially coastal provinces), whose urban population accounts for 52.2 per cent of China's and whose rural nonagricultural labor force accounts for 49.6 per cent of China's, and for China as a whole it represents a case on the higher side of development.

Development of TVEs and Functional and Physical Changes in the Countryside

The most important dimension of China's *in situ* rural-urban transformation has been the functional changes of rural settlements brought about by the development of township and village enterprises (TVEs). In 1978 the number of TVE employees was only 28.3 million, but this number increased to 130.5 million in 1997 (National Bureau of Statistics of China, 1998, p.420). In 1999 the value-added of TVEs amounted to 2,530 billion Yuan, accounting for 30 per cent of the national GDP. TVEs' export value also accounted for one third of China's total export value this year (*People's Daily*, 21 September 2000). In general, it is well accepted that TVEs account for one third of the national economy in China, and half of the economy or even more in some coastal provinces like Fujian.

In the current context there are two important characteristics of TVEs. Firstly, they include nonagricultural, especially manufacturing enterprises. As can be seen from Table 11.3, only a small fraction of TVEs and their employees are engaged in agricultural activities. The largest sector is manufacturing, followed by wholesale, retail sales and trade contrasting with the traditional image of rural areas being identified with agricultural activities.

Table 11.3   Industry structure of TVEs in Fujian, 1999

Industry	TVEs		TVE employees	
	Number	% of total	Number	% of total
Agriculture	11,874	1.5	188,700	3.6
Manufacturing	243,878	30.9	3,111,347	58.7
Construction	37,496	4.7	529,031	10.0
Transport	141,149	17.9	336,708	6.4
Wholesale, retail sales and trade	237,673	30.1	689,994	13.0
Tourism and catering trades	92,710	11.7	326,815	6.2
Other	25,736	3.3	114,272	2.2
Total	790,516	100.0	5,296,867	100.0

Source: Calculated from 1999 statistics of TVE Management Bureau of Fujian Province.

---

[11] For a detailed study on Fujian's urbanization patterns since 1978, see Zhu (1999).

A second important characteristic of TVEs is that they are small, and mostly township- or village-based. The average number of employees for each TVE in Fujian in 1999 was only 6.7. In fact, many TVEs in Fujian are households or joint-household enterprises. Table 11.4 shows the distribution of 50,178 TVEs covered in the 1996 agricultural survey. These TVEs met the conditions of having at least 8 employees and therefore were larger than average. However, even among these enterprises, nearly 80 per cent of them and 70 per cent of their employees were found to be located in villages. Moreover, TVEs not covered by the survey are mostly located in the villages, as the smaller the enterprises, the more likely they are located in the villages.[12]

Table 11.4 Location of TVEs* in Fujian Province, 1996

Location	Enterprises		Employees		Average number of employees
	Number	% of total	Number	% of total	
Large and medium-sized cities	615	1.2	32,156	2.2	52
Seats of county governments	2,035	4.1	64,987	4.4	32
Industrial and mining areas	168	0.3	5,710	0.4	34
Seats of township or town governments	8,176	16.3	365,198	24.6	45
Villages	39,184	78.1	1,015,172	68.4	26
Total	50,178	100.0	1,483,223	100.0	30

* Only those having at least 8 employees or meeting certain other conditions

Source: 1996 Agricultural Survey of Fujian Province.

One obvious effect of the fast development of TVEs is the transformation of the employment structure of rural areas. As can be seen from Table 11.5, in 1984 agricultural employment still accounted for 83 per cent of the total employment in Fujian's rural areas; but in the 1996 agricultural survey, this proportion dropped to 61.8 per cent. In lowland rural areas, nonagricultural activities were very close to overtaking agricultural activities in the employment structure. In many coastal areas, nonagricultural activities have already become dominant in areas officially defined as rural (Zhu, 2000). In China as a whole, only 7.1 per cent of its rural labor force was engaged in nonagricultural activities in 1978, but this share increased to 31.2 per cent by the end of 2000, largely due to the development of TVEs (National Bureau of Statistics of China, 1998, 2001). In southern Jiansu Province, nonagricultural sectors overtook the agricultural sector as early as the 1980s (Ru *et al.*, 2001).

---

[12] Evidence shows that these two characteristics apply to TVEs in China as a whole as well (Tang and Kong, 2000, pp. 10, 391, 425), although there are some regional variations.

**Table 11.5   Employment structure in the rural areas of Fujian Province,
selected years (per cent)**

	Primary sector	Secondary sector	Tertiary sector
1984[a]	82.83	9.27	7.90
1996, all rural areas[b]	61.78	20.26	17.96
1996, lowland rural areas[b]	52.94	24.57	22.49

Source: a: Statistical Bureau of Fujian Province (1985, pp. 74-5); b: 1996 Agricultural
Survey of Fujian Province.

However, the increase in urban characteristics caused by TVE development
differs from conventional patterns of urbanization. In its early stages, TVE
development was rarely accompanied by spatial concentration of people and
enterprises. Rather, in most cases it brought urban functions, such as industrial and
other nonagricultural activities, down to the lowest levels of the hierarchy of
human settlements. Yet they have not been increasing in a concentrated way, and
therefore cannot be reflected in the builtup-area-based urban definitions. This is
still the case for most small TVEs today. Nevertheless, some TVEs in Fujian, as
well as in many other parts of China, have started more concentrated development
by moving to industrial zones, some of which are part of the builtup areas of
designated towns. The inflow of foreign investment in the rural areas since the late
1980s, which is often connected with TVEs and mostly concentrated in major
development and industrial zones, has further enhanced this. TVE development
and foreign investment have also promoted the development of public facilities,
infrastructure, and service sectors, and the revenues from TVEs and other
enterprises have been the major source of funds financing these developments
(Fan, 1998; Zhu, 2000; Ru *et al.*, 2001).

Hence there has been a significant growth in townships becoming towns and
towns becoming cities in Fujian Province (Zhu, 2001b). However, many TVEs are
still located in central villages (which are at a level in the settlement system
between the town or township and the village), or even in ordinary villages.[13] In
the former case, these zones will be included in the urban areas according to the
2000 census urban definition and reflected in the census result, but in the latter case
many of them will be left out. Even if the seats of their villagers' committees are
extended by the urban construction of the towns proper and therefore included in
the urban areas according to the 2000 census definition, they are obviously a
transitional area between urban and rural, not urban builtup areas in the
conventional sense. This kind of urban or quasi-urban physical changes, together
with the more widespread functional changes in the rural settlement, cannot be
adequately reflected in the current settlement definitions.

---

[13] In an extreme case, there are 40 industrial zones at the municipal or town level, but 535 at
the central village or village level, in Jinjiang Municipality of Fujian Province with a total
area of 649 square kilometers where I conducted fieldwork.

Increasing Universality of Urban Facilities in Rural Areas

A second major dimension of China's *in situ* urbanization is that many urban or urban-like facilities, which are not necessarily related to TVE development, have also permeated into rural areas. As can be seen from Table 11.6, most villages in Fujian Province have access to public electricity, postal services, public roads suitable for motor vehicles, TV communications, primary school, primary health care clinics, and telephones, although the coverage of telephones in the villages is still relatively low. At the town and township levels, more urban facilities can be found. In Fujian Province, 93 per cent of towns and townships have cultural centers, 97 per cent have secondary schools, and 98 per cent have hospitals, according to the 1996 Agricultural Survey. The universality of the above facilities at the bottom of the settlement hierarchy is further enhanced by modern transport and communication facilities, such as highways, IDD telephone and fax services, and even the Internet, which connect many rural settlements closely with major urban centers, not only those in the province, but those in other parts of China too.

Table 11.6   Urban facilities in Fujian's villages, 1996

	Total number of villages	Number of villages with named facilities	
		Number	% of total
Access to public electricity	15,087	15,066	99.86
Access to postal services	15,087	14,572	96.59
Access to public roads suitable for motor vehicles	15,087	14,424	95.61
Access to telephone	15,087	10,845	71.88
Access to TV communications	15,087	14,417	95.56
Primary school	15,087	13,895[a]	92.10
Primary health care clinic	15,087	16,897[b]	112.00

[a] number of primary schools in the rural areas.
[b] number of clinics in the rural areas.

Source: 1996 Agricultural Survey of Fujian Province.

High Population Densities and Improved Transport and Communications: Two Enabling Factors for In Situ Urbanization

Chinese government policies have played important roles in the emergence and development of these dispersed rural-urban transformation patterns. However, other factors have also been influential, such as rapid population growth and the widespread improvement of transport and communication facilities that makes spatial concentration less necessary. The former can lead to very high rural population densities and thus facilitate the creation of urban like settlements on a widespread regional basis. For example, Fujian Province has a total area of 121,400 square kilometers and, by the 2000 population census, its population density had reached 286 persons per square kilometer. The 27 coastal counties and municipalities, where 56 per cent of Fujian's population and 66 per cent of its TVE employees live, have a total area of 29,266 square kilometers and an average population density of as high as 663 persons per square kilometer. This area size is similar to that of Belgium (29,456 square kilometers) or Holland (32,538 square kilometers) at the end of the nineteenth century (Weber, A.F., 1968, p.182), but as Table 11.7 shows, the population density of Fujian's coastal area is several times higher than those of Belgium and Holland at that time.

Table 11.7   Population density of Fujian in the 2000 census and some developed countries or regions at the end of the nineteenth century (persons per square kilometer)

Fujian 2000	
Fujian Province	286
Coastal area of Fujian Province	663
Countries or regions at the end of the 19th century	
Saxony	234
Belgium	206
England and Wales	192
Netherlands	139
Italy	107
Japan	107
Germany	92
France	73
United States	8

Source: Population densities in Fujian are calculated from data provided by Population Census Office of Fujian Province, May 2001; otherwise, Weber, A.F. (1968, p.147).

As a consequence of high population density in the rural areas, rural settlements with the population size and density of an urban or semiurban place are not unusual in Fujian's coastal area. In the 27 coastal counties and municipalities, nearly 50 per cent of towns and townships have population densities higher than 800 persons per square kilometer (calculated from data provided by Civil Affairs Department of Fujian Province, April 2000). This compares to the average population density of 1,000 people per square kilometer of the urbanized areas in USA in 1990 (US Bureau of the Census, 1990, cited in Zhou and Shi, 1995), and 400 persons per square kilometer as the density criterion for identifying urban territory proposed in the US (Lang, 1986). In Jinjiang, a place with a population of 1.5 million and well known for TVE development, the population density was as high as 1,506 persons per square kilometer in 1990, before it was designated as a city. In terms of settlement size, villages with a population of 2,000 persons are not unusual in this area (Zhu, 1999). Many villages have even expanded and connected to each other, forming bigger incorporated villages (Chen and Huang, 1991). Therefore, although many settlements in these areas are still regarded as rural, in a way they have already achieved urbanization in terms of population size and density, and a certain degree of specialization and scale economy is also achievable for nonagricultural industries (Zhu, 2000). What they need to complete the urbanization process is to transform their functions from rural to urban, and the fast development of TVEs served just this purpose.

The role of improved transport and communication facilities in *in situ* rural-urban transformation can also be illustrated by comparing Fujian Province with developed countries in the past. In the period when developed countries were urbanizing, and even in the middle of the twentieth century when many cities developed in developing countries, widespread motor transport was still not available, and dense settlement near the center was required so that people could walk to work and goods could flow easily between manufacturers, wholesalers and retailers, causing huge concentrations of persons in small areas (Hackenberg, 1980; Speare *et al.*, 1988). However, *in situ* rural-urban transformation in Fujian has been occurring under very different transport conditions. Here relatively cheap transport such as motorcycles, buses and trucks has increased rapidly and become commonplace since the late 1978, and the road networks serving these vehicles have also improved tremendously. In Fujian Province as a whole, the number of motor vehicles increased by more than 11 times in the period 1978-2000, from 26,148 to 321,278, and the length of highways increased 75.5 per cent in the same period, from 29,109 to 51,073 kilometers, including 345 kilometers expressways connecting major cities along the coast (Statistical Bureau of Fujian Province, 2001).

Such improved transport conditions have two major effects. On the one hand, they greatly reduce the separating effect of distance between major cities and the rural areas, making geographical proximity to large cities less important in development.[14] The widening use of modern communication services in business activities, especially IDD telephone and fax services, further enhances this effect. On the other hand, easily available and affordable means of transport connect almost all rural settlements together, making internal agglomeration of people and enterprises less necessary. This is confirmed by a survey of 100 enterprises in Jinjiang Municipality and Huian County in Quanzhou, a prefecture-level municipality well known for TVE development This showed that, although none of the enterprises were located in cities, only a small proportion of their managers felt inconvenience in terms of transport (2.0 per cent), acquisition of raw materials or parts (5.1 per cent), and sales of their products (9.1 per cent). There is a newly emergent terminology of the 'half-hour urban agglomeration', which refers to the fact that most major cities and towns in the Municipality can be reached by half an hour's bus drive from Quanzhou's city center (Policy Research Office and

---

[14] It took at least four hours to travel from Fuzhou (the provincial capital) to Jinjiang (one of the field sites) before the expressway was put into use, but now it takes only two hours to complete the same journey.

Department of Construction, 2001).[15] These conditions greatly facilitate the circulation of commodities and people, and increase the accessibility of rural areas to external resources and markets, making it more feasible for nonagricultural activities to be located in rural areas.

China's Settlement Definitions in the Twenty-First Century: A Continuum and Evolutionary Approach

Remaining Problems with the Current Settlement Definitions

Three problems are paramount. Firstly, China's current settlement definitions and categories do not adequately reflect *in situ* urbanization and the blurring of rural-urban distinctions. The widespread functional and landscape changes of rural settlements driven by the development of TVEs, and the permeation into rural areas of urban or quasi-urban facilities, are two major dimensions of *in situ* urbanization. However, it is often not appropriate to categorize settlements undergoing these changes as either rural or urban because, while these settlements possess some urban characteristics, they may not reach full urban standards and/or may not be within, or contiguous to, the boundaries of the builtup areas of designated cities and towns. Therefore a new way of defining, categorizing, and monitoring *in situ* rural-urban transformation needs to be found. This kind of approach is also urgently needed for planning purposes. China's urban planners have increasingly recognized the reality of blurred rural-urban distinctions and the need to incorporate those rural areas undergoing functional and landscape changes within their integrated plans (Zhang, 2000). To meet this end, a more sophisticated, nondichotomous settlement definition and classification covering the whole continuum of rural-urban changes is indispensable for high quality planning practice.

Secondly, a related problem with the current urban definitions concerns the double identities of the floating population in relation to their places of origin and destination. Most urban populations in China now involve a significant number of people working and residing in the city, but keeping close ties with their places of origin in rural areas through the flow of remittances, frequent visits (especially at the time of spring festival), and eventual return migration. In fact, most members of the floating population have their own home and land, and many of them leave some members of their families at the place of origin. This further complicates the rural-urban distinction in China. Obviously, members of the floating population are

---

[15] The improved transport conditions could also give rural laborers the option of commuting to urban jobs without moving permanently to existing urban areas, as pointed out by Hugo's study on rural-urban commuting and circular migration in West Java of Indonesia as early as in the early 1970s (Hugo, 1980). However in Fujian Province, rural-urban commuting and circular migration of local farmers are not significant compared to *in situ* rural-urban transformation, due to the lack of economic strength of the major urban centers and well developed rural nonagricultural activities.

part of their destination cities, but they are less committed to the destination cities and have needs for housing and other urban services that are different from those of permanent urban settlers. At the same time, they are still part of their home communities in many ways. How to properly reflect such double identities of the floating population in China's settlement definitions and statistics remains an unresolved issue.

Thirdly, although high population densities and improved transport and communication facilities are two enabling factors facilitating *in situ* rural-urban transformation, they are not properly taken into account in China's settlement definitions. In fact, under the current urban definitions, transport and communication facilities are not considered as urban criteria.

Apart from these three problems, two other issues also need to be addressed. One is that the current criteria for the designation of cities and towns still carry some legacies of the *Hukou* system, taking the size of the nonagricultural population by *Hukou* status as an important criterion for the designation of cities and towns. This is increasingly irrelevant because of the gradual collapse of the *Hukou* system. In fact, in Guangdong, Fujian and Hunan Provinces, it has been announced that the categories of agricultural and nonagricultural *Hukou* will be completely abolished, and only the place of residence will be registered in the *Hukou* system (*Xinhua Net*, 5/12/2001; *People's Daily*, 24/12/2001, *China Youth Daily*, 1/1/2002).

The second issue is that the criteria for the designation of cities include indicators of the absolute volume of economic activities, such as the GDP, the gross output value of industry, and the amount of financial revenue. It is even proposed that similar indicators should be introduced to the criteria for the designation of towns (Dai, 2000). These economic indicators do not reflect the essence of urban settlements, making the criteria for the designation of cities and towns unnecessarily more complicated. In fact, the function of cities and towns as the places for nonagricultural activities can be adequately reflected by the criterion of employment structure, as used by many other countries in defining urban places. Some Chinese geographers also use structural indicators, i.e. the proportion of nonagricultural employment in the labor force and the proportion of nonagricultural added value in the GDP as the major criteria for determining the outer zones in their study on China's major metropolitan areas (Hu *et al.*, 2000). The official criteria for the designation of cities can be also modified following this line.

Some Guidelines for the Modification of China's Settlement Definitions

The foregoing discussion suggests at least three ways in which China's settlement definitions could be modified and thereby permit the more accurate monitoring and analysis of urbanization and urban change. In the first place, in order to adequately reflect the reality of *in situ* urbanization and blurred rural-urban distinction in China, a more complex classification of settlement types, which goes beyond the simple rural-urban dichotomy, should be introduced. Secondly, the criteria for the designation of cities and towns need to be revised. Thirdly, it is suggested that the

Hukou system be used to reflect the double residential identities of the floating population in settlement statistics. Each of these is now examined in more detail.

Moving towards a more complex classification of settlement types This would reflect the fact that there are transitional settlement types which are between urban and rural. At the same time, a multidimensional scoring system can be developed to measure the degree of urbanity of the settlement units, determine their settlement types, and monitor their transformation. The criteria used in such a scoring system should be able to reflect the following major dimensions of the settlement system: population density; 'urban' facilities, which must include transport conditions to reflect the accessibility of the settlement unit; and employment structure of the population.

A settlement unit can be assigned a score for each of the above dimensions and a total score taking consideration of all the dimensions, and the total score can be compared to a threshold value to decide whether the unit is urban or rural. Those units identified as urban and contiguous to the designated cities and towns can be included in the cities and towns proper (this is actually a more accurate and objective way of delineating the boundaries of cities and towns proper than that adopted in the 2000 census). Those units with scores below the threshold value can be categorized as either rural or a transitional settlement type, according to their scores reflecting their degrees of urbanity. In this way, all settlement units at the village level with different degree of urbanity (or rurality) will be covered by the settlement category system.

This classification and scoring system could be implemented and maintained jointly by the census office and the construction department of the government through China's well-organized administrative network. Details of the system, including the way of assigning score, the selection of 'urban' facilities, the new settlement categories, and the cutting point for urban and rural, still need to be further explored.[16]

Revising the criteria for the designation of cities and towns While China's basic rural-urban administrative structure at and above the town/township level should remain largely unchanged to ensure continuity of the settlement system, the criteria for the designation of cities and towns need to be revised in three main areas:

---

[16] In Indonesia the following facilities are used as criteria for the scoring system: primary school, junior and senior high school, cinema, hospital, maternity hospital, clinic, road negotiable by three- or four-wheeled motorized vehicle, post office or telephone, market, shopping center, bank, factory, restaurant, public electricity, party-equipment renting service (Firman, 1992, p.107). Similar criteria are also used in Taiwan for identifying 'urban localities' (Tsay, 1982, p.212). These can be taken as references for the selection of 'urban' facilities in the scoring system of China. On the basis of this, some more facilities relevant to the accessibility of localities, such as accessibility and distance to expressways and designated cities and towns, should be included.

- Population density should be used as a criterion for the town where the county government seat is situated and for the resident committee where the township government seat is situated when designating a new city or town respectively. Possible criteria could be 1,500 persons per square kilometer for the city and 800 persons/square kilometer for the town. This is not only necessary to make the criterion of population density consistent with its role in the settlement evolution analyzed above, but also a way to avoid the situation where some counties with dispersed population distribution are designated as cities.
- The absolute values of GDP, the gross output value of industry, and the financial revenue should be removed from the criteria for the designation of cities. This will make the criteria simpler and more relevant to urban functions, and conform to international practice.
- Thirdly, changes are needed in relation to the categories of cities and towns. Towns can be categorized into two groups: central towns serving as the center of a county and ordinary towns. The category of city size mentioned earlier should be based on the population of the builtup areas rather than the nonagricultural population by *Hukou* status, and the size criteria for the categories of small, medium, and large cities also need to be adjusted. For example, the upper limit of large cities may be increased to one million people. In some developed regions with high population densities, metropolitan functional regions can be identified and relevant settlement categories can be introduced.[17]

Recognizing double residential identities in settlement statistics   China's *Hukou* system, although having lost its functions in distinguishing urban from rural settlements, can be used to reflect the double residential identities of the floating population in settlement statistics. Some information of this kind is already available from the 2000 census results, and it can be provided on a regular basis by making use of the temporary and permanent registration of the *Hukou* system. Under this system, migrants without local permanent *Hukou* status at the places of their destination need to be registered as temporary residents while keeping their permanent *Hukou* registration at their places of origin. This kind of register can be used to identify members of the floating population as well as their special status in the destination areas, while still reflecting the fact they are still members of their home communities. At the same time, some efforts are needed to tackle the problem of underreporting of the number of such migrants. The availability of such information would make it easier for relevant government departments of migrant destination cities and towns to have better planning for this group.

---

[17] Some studies on this have been conducted in China for the Pearl River Delta Region, the Yangtze River Delta Region, the Beijing-Tianjin-Tangshan Region, and the Central and Southern Liaoning Province (Hu *et al.*, 2000).

Conclusions

Efforts have been made to modify settlement definitions in China since the 1980s in order to capture the fast changing reality of the settlement system. However, this is not a simple task. Even the urban definitions adopted in the 2000 census, which at last present a reasonable estimate of China's urban situation according to the conventional dichotomous builtup-area-based approach, still carry some legacies of the old settlement definitions based on China's planned economic system of the past, and are not compatible with the fast development of market-oriented reforms. More importantly, the important role of *in situ* urbanization in China suggests that more fundamental modifications to current settlement definitions and ways of monitoring settlement changes are needed to reflect the blurred rural-urban distinction and to cover the temporal evolution and spatial continuum of the settlement system. This is not unique to China: many other developing countries, especially those densely populated regions with good transport conditions, also need to consider these sorts of issues.

PART IV
CONCEPTUALIZING SETTLEMENT SYSTEMS

Chapter 12

An Evolutionary Approach to Settlement Systems

Denise Pumain

The concept of a settlement system is important to understanding the urbanization process (Berry, 1964). According to this approach, urban and rural population are disaggregated into elementary and spatially autonomous settlements located within a given territorial framework and which evolve interdependently. The derived concept of relative situation, including size, geographical location and functional specialization of any settlement within the system, is essential for qualifying its status and predicting its potential development.

The dichotomy between 'rural' and 'urban' was meaningful at some stage of the development of settlement systems, when they could differentiate settlement types whose characteristics were opposite on several dimensions. However, more detailed typologies of settlements are needed for analyzing new urbanization trends. Therefore, it is necessary to consider the *evolution* of the dynamics of settlement systems. As settlements have become more complex, the simple rural-urban dichotomy has to be replaced by a number of classes that are defined in terms of their settlement sizes and functional types and represent their evolutionary trajectories within the system.

This approach not only aims at establishing settlement types as geographical contexts having a possible impact on population characteristics and behavior, but also refers to a more circular causality chain where settlement types are considered as being produced by social and spatial interactions among people and may in turn influence them. This systemic approach is not only micro-macro but also multilevel: it considers that each settlement is a subsystem that cannot be described merely in terms of the current characteristics of its own residents. A settlement acquires specific collective properties during its development, and the latter is embedded in a competitive process with other subsystems. Settlement systems may then be interpreted as a spatial mode of the structuration of societies, as well as the product of successive social choices in the patterning of spatial interactions.

Why the Urban-Rural Dichotomy has Never been Totally Relevant

Over the last 30 or 40 years urban sprawl has blurred the spatial limits between urban agglomerations and the countryside. Moreover, the 'urbanization of the society', including the diffusion of the urban economy and urban way of life in the most remote parts of every developed country, has meant that there is nowadays less substance in the traditional distinction, which is still made by national and international statistical institutes, between rural and urban populations.

However, such a dichotomy, although practical, has never been totally relevant. Several formal distinctions were made in the social sciences to qualify the difference in nature between rural and urban places, according to a variety of theories and criteria: economically, there is a break between agriculture and other activities; sociologically, towns and cities are places where new ways of life and status and professions were invented in a more or less continuous process of social division of labor; for political economy, urban places are nodes of power and platforms for exchanges; geographically, rural settlements depend for their survival on resources that are located near their site, whereas towns and cities use more distant resources through the various networks they activate. According to these dichotomy principles, a 'finite' story has been suggested as an analogy with the demographic transition process.

Transition in the History of Settlement Systems

The concept of 'urban transition', as first mentioned by Gibbs (1963) and related to the mobility transition by Zelinsky (1971), is useful to portray the evolution that transformed mostly rural settlement systems – made up of numerous and scattered hamlets and villages relatively homogeneous in their size and functions – into almost entirely urban systems made of elements that are much more differentiated in their sizes and functions and hierarchised in a number of levels, with a much higher degree of spatial concentration of the population. To various degrees and at different timings according to countries in the world, the urban transition is linked with the demographic transition, industrial revolution and increasing speed of transport. Starting as soon as the beginning of the nineteenth century in developed countries, an acceleration in the urbanization rate took place only after 1950 in most developing countries (Bairoch, 1996). Everywhere it has been accompanied by large migrations from rural settlements to urban ones, although in developing countries a larger share of urban growth has been taken by natural increase. When summarized by logistic curves showing the historical progress of urbanization rates, it appears typically as a worldwide spatial diffusion process.

Such a major event could be called a *bifurcation*, analogous to a phase transition in physics. When interpreted in this framework of self-organizing systems theory, the evolution of urban systems may appear as not finished once the whole population has become urban. It does not consider that a specific historical period marked by a diffusion process has ended, nor does it expect reversal trends to take place as in 'counterurbanization' theories (Berry, 1976), nor does it support explicitly cyclical perspectives about urban development (van den Berg *et al.*, 1982). It is necessary to produce a more integrative and continuous theory of our ways of inhabiting the planet, which would include the major episode of the transition from rural to urban, as well as what happened before and also what will happen afterwards.

Difficulties in Measuring Rural and Urban Populations

However important historically, the categories of 'rural' and 'urban' were always difficult to separate with precise limits, especially quantitatively (Goldstein and Sly, 1975b). The existence of a legal definition, separating the status of population on either side of town walls, was by no means universal, and disappeared well before censuses became general. The size of settlements and their densities are attributes which were often used for establishing thresholds between settlement categories but which have no basis in theory. One may, for instance, question the significance over time of a threshold which defines urban population in France by a value fixed to 2000 inhabitants established in 1856. Major uncertainties in the application of definitions come from the emergence of settlements which have contradictory attributes. For instance, are mining towns, exploiting only local resources, truly 'urban'? In which category should 'towns' with population above the urban threshold but with agricultural activities dominant (as for instance in Southern Italy) be classified? Can communes with low population densities and a green environment still be considered rural if they are located on the edges of urban areas and have a majority of their residents involved in urban activities?

Another general property of settlement systems prevents any easy partition between rural and urban populations. Instead of more or less visible thresholds in size between rural and urban settlements, there is always a continuum in statistical distributions of the population size of settlements. The Pareto law describes the distribution of urban settlements, while the lognormal distribution is appropriate to the whole settlement systems (Baker, 1969; Pumain, 1982). All settlement systems share this property, even those in prehistoric times, as shown by Fletcher (1986) for a variety of settlements worldwide. Continuity in size distribution can only be explained in a dynamic perspective. There is no absolute distinction between rural and urban settlements because the latter proceeds from the former. There is no definition of settlement size in theory because the size of a settlement does not result from any optimization process.

An Evolutionary Theory of Settlement Systems

The rural-urban continuum in settlement sizes and the generality of the hierarchical models of size distribution, whatever the historical or regional political circumstances, have suggested many tentative explanations. Among them, several were static, referring to some optimizing principle (as for instance in Christaller's central place theory based on spatial economic equilibrium, or in Zipf's suggestion of an equilibrium between two opposite principles of spatial concentration and dispersion). There is now a general acceptance that it can be better understood when related to the spatial distribution of population growth in settlement systems (Robson, 1973; Pumain, 1982; Dendrinos and Mullaly, 1985). Knowledge of the growth process may provide an explanation of how the system's structure is generated. It may help in making predictions about its future. It can also be related to the dynamics of settlement systems by orienting explanation towards the interactions between the settlements and the way they change.

The statistician Gibrat (1931) developed an explanation and suggested a dynamic process for generating the lognormal distribution. According to this, the size of settlements in any system becomes lognormal if the statistical distribution of growth rates maintains over time the following properties: 1) the absolute growth in population of a settlement is proportional to its initial size (this is equivalent to say that in average the growth rates have the same value, they have a random distribution around a mean value); 2) the value of the growth rates are independent of the initial size of the settlements; 3) growth rates are distributed independently between successive time intervals. Such rules have been tested on empirical processes of urban growth and these hypotheses appear relevant in most of the cases studied (for reviews, see Pumain, 1982; Moriconi-Ébrard, 1993).

A Model of Spatially Distributed Growth

The statistical process suggested by Gibrat can be interpreted as a dynamic model of settlement systems, including their ecological and spatial dimensions. The competition for resources between subsystems is the main driving force shaping these dynamics. Among villages, there is an *ecological constraint* which limits the growth of population of each settlement, according to the available state of agricultural technologies and/or social organization, but there is also a limit to their development which comes from the pressure exerted on local resources by neighboring villages. This *spatial constraint* becomes more obvious once a few among those villages have succeeded in developing commercial activities or crafts and enter in competition with other centers for trade. With time and network building, the competition space expands and more and more distant places are directly connected. Competition means that individuals and groups located in a place are trying to capture the benefits from innovation – by inventing new ways of attracting wealth, by developing their economy, by imitating the innovations which appear elsewhere, or through war. This competition process is the main explanation for the consistency observed in the development of a settlement system. Even well before direct connections are made between distant places, they become integrated in a co-evolving system of places through local interactions (Bura *et al.*, 1996). Another explanation lies in the establishment of a circulation space that is progressively made homogeneous for ensuring a variety of transactions, through imposition of common rules and provision of facilities by political powers.

During the early history of settlement systems, many inequalities are observed between the growth rates of settlements. The sudden emergence of new sites, or the total disappearance of settlements which were prosperous for decades or even centuries, are observed (Archaeomedes, 1998). Over time, the creation of new rural or urban places becomes rarer and the probability of disappearance of settlements becomes more inversely correlated to their size. The regime of 'distributed growth', as described above, becomes the 'normal' way of evolution for the settlement system.

Selection Processes

The stochastic distribution of growth among settlements belonging to the same system is expressive of the interdependences between these subsystems. It is related to the process of spatial diffusion of innovation within the system (Pred, 1977). When changes in urban social life and economy are examined, there are continuous local adjustments to the ongoing transformation of society and technologies, which lead to fluctuations in the growth process in time and space, but create rather little differentiation among the settlements in the long run. This property, which is sometimes interpreted as demonstrating the 'inertia' of settlement structures, on the contrary is proof of their ability to incorporate new developments by ensuring their diffusion throughout all parts of the settlement system. The system is self-organizing through the amplification of some of these fluctuations or asymmetrical interactions.

Diffusion processes are not always instantaneous or ubiquitous. Selection processes contribute to differentiate settlements, usually in connection with innovation cycles. For instance, a selection among towns and cities has occurred each time an important innovation was linked to some ecological resource that was unevenly distributed (as for instance coal mines for industrial revolution of the nineteenth century). Linkages of some uneven urban developments with economic innovation lead in most cases to a cyclical episode of growth for settlements which are specialized in this particular activity. The result is the emergence of what historians and geographers called 'generations' of towns and cities. These settlements keep the mark of their period and type of specialization long after the time when they benefited from this innovation, since the gap which was once created between them and the other settlements is maintained through the 'normal' incremental mode of evolution of the settlement system. That is why typologies of settlements reveal former selection processes: the major differences which are identified by multivariate analysis or classification of settlements can very often be interpreted in terms of traces of the unequal diffusion of former cycles of innovation within the settlement system. For instance, in Europe old industrial centers which emerged at the time of the industrial revolution still appear as very distinct from older administrative and political capitals, while new urban categories like technopolitan nodes have emerged more recently. Such contemporary trends in restructuring the settlement system can be revealed by significant correlations between growth rates at successive time intervals, revealing the persistency of trends in growth and decline in the same places. For instance, during the second half of the twentieth century such a trend has intensified in France, and taken more systematic spatial patterns, opposing regions gaining or losing population in almost all settlement categories (Bretagnolle *et al.*, 2002).

Another example of a more durable selection process refers to the way in which the dichotomy between rural and urban settlements has been created and maintained. When adjusting the total distribution of settlement sizes in France to a statistical model, one gets a better fit if, instead of considering a single lognormal distribution, two different lognormal curves with different means and dispersions are used: the one for the actual 'rural' settlements, typically under 10,000 or 20,000 inhabitants, has a clearly different slope (meaning dispersion) from the 'urban' ones whose distribution is represented by another line on a gausso-logarithmic graph (Figure 12.1). One can explain and simulate the formation of this inflexion between the two distributions by considering that the rural settlements have systematically lost population for more than 150 years, while the average growth rate of urban settlements remained positive. Rural and urban settlements would then have ceased to evolve under the same common growth process, once those which would become towns and cities had been selected.

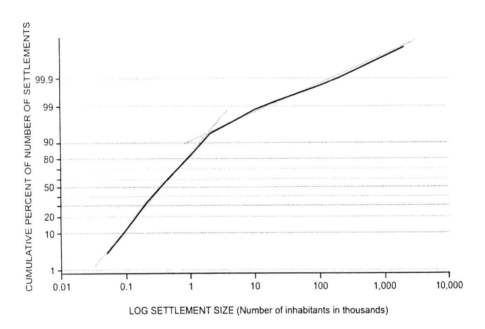

Figure 12.1   Adjustment of human settlement sizes by lognormal distributions

Evolutionary Trends

The human settlement growth process is dominated by clear historical trends that give an irreversible direction to its evolution. This suggests that one can conceive a more specific model for human settlements (Pumain, 2000) that integrates specific features of social and spatial interaction and their transformation over time. For instance, the Gibrat model alone can account for a small part of the reinforcement of settlement-size hierarchies, but it does not explain entirely what is observed. There is a recurrent trend for larger urban settlements to grow faster than the smaller: if not expressed by a high correlation between growth rates and city size the trend appears frequently in statistics when size classes are considered. This is linked with the initial advantage given to the largest places by the hierarchical diffusion of innovation (Pred, 1977).

More generally, the features of settlement system dynamics are governed by major historical trends. This is a combination of two main processes. Firstly, a process of creation of innovation sustains demographic and economic expansion and ensures the emergence of new types of activities and of more complexity in the upper levels of the settlement system's hierarchy. This innovation process is partly incited by the emulation between settlements, and computer simulations have shown that its continuity is necessary for maintaining the structural properties of the settlement system (Bura *et al.*, 1996). Secondly, a process of acceleration in the speed of communication causes an apparent shrinking of geographical space (Janelle, 1969), and contributes to the further hierarchisation of the system in two ways. The hierarchical diffusion of innovation gives an initial advantage to the largest settlements, which are connected first to rapid communication networks and whose more complex society adopts innovation quicker. Meanwhile, smaller settlements are short-circuited as their customers are captured by larger ones, which have become closer in time-space (Bretagnolle *et al.*, 2002).

These evolutionary trends are defined at a macro scale for the entire settlement system. It should be added that they are only probability laws which are compatible with possible reversals in trends at a local scale, for some individual settlements. There is no absolute determinism in the evolution of a specific place, which can always be reoriented by inventing new activities or capturing innovation. The probability of this, however, is much lower for smaller settlements, especially if their geographical location has a low accessibility to urban centers. This is because of what is called 'path dependency' in complex systems theory. Despite postmodernist collages, individual settlements keep their specificity as defined by their original trajectory within the system: they cannot become what they have never been.

Conclusion

The hypothesis is that the same processes which have already occurred during the historical development of settlement systems will continue, even after the end of the 'urban transition'. A few conclusions can be taken from the evolutionary concept of settlement systems for preparing settlement typologies. Firstly, most urban settlements proceed from formerly rural ones: they represent the result of a selection process which is durable but not totally irreversible. No classification of places can then be assessed in definitive terms: always it has to be defined in relative terms. Secondly, there is no permanence in size thresholds or in portfolio of services which would identify a rural or an urban character, because of the continuous changes in the income, needs and traveling capacities of the population. Hence settlement typologies have to be evolving in time, as settlements are. Below we shall see from European examples that they also have to be adapted to the regional context of population concentration and spatial patterns (which themselves reflect the past trajectories of settlement subsystems).

Classical Approaches to Rural-Urban Classification in Europe

The European continent has a long and continuous history in human settlement. When compared to other parts of the world, it is characterized by a high population density (around 100 inhabitants per square km on average), which is not too heterogeneously distributed, at least at regional scale. It also has a high density of towns and cities (average spacing between two neighboring towns with over 10,000 inhabitants is less than 15 kilometers as against 50 in North America). It has a spatial distribution of urban population which is less concentrated than in other parts of the world: the first 30 largest metropolises have less inhabitants than those in North America, and the share of urban population living in agglomerations over one million inhabitants is less than on other continents (Moriconi-Ébrard, 1993). At a local scale, another feature can commonly be observed in the spatial organization of urban settlements (Benevolo, 1993): a concentric model of urban development. This includes: an old core, still highly valued as architectural and urbanistic heritage; densely and continuously builtup 'banlieues' surrounding that core; and more recently developed and less dense suburbs which have spread out in the countryside during the last three decades.

Despite this shared history and many physical and cultural similarities, there is no common statistical definition of what an 'urban' settlement is among European countries, perhaps because there are not uniform representations of a 'town' or of a 'city' throughout Europe. The noncomparability of European statistics, mentioned as early as the nineteenth century (Meuriot, 1897), is a recurrent problem (Bunle, 1934), despite repeated efforts at promoting comparison made by international agencies (International Institute for Statistics, 1962; OCDE, 1988; Eurostat, 1999) and research communities (Hall and Hay, 1980; Pumain and Saint-Julien, 1991; NUREC, 1994). Heterogeneity in statistical sources and noncomparability of urban information remain the rule in Europe (Pumain and Saint-Julien, 1991, 1996; Le Gléau et al., 1996).

Diversity in European Concepts of Urban Settlements

In some countries, the urban concept is based on *administrative or political status*. The definition of a Greater London area – including its recent fluctuations – is a well-known example of this approach, but it is also the case for Copenhagen and Frederiksberg in Denmark, the municipalities of Madrid and Barcelona in Spain and in Ireland. More generally, in Austria as well as in England and Wales and in Germany, urban territories are recognized according to a legal decision. Such administrative definitions may register the existing balance of power between the State and the cities at a given moment, but they do not help much in measuring the progressive expansion of the urban realm. Sometimes, as in Germany, the limits of urban municipalities have been extended to follow the progress of urbanization, but these adjustments of political territories cannot be frequent enough to provide an adequate measurement of the urban settlements. This is made easier in countries where there is a purely statistical concept of urban units.

In most countries, urban character is defined in terms of a *threshold size of the resident population in local units*. In Italy and Spain, the definition classifies as urban all municipalities with a population of over 10,000 inhabitants, whereas all those under this threshold are considered rural. In order to free the definition from administrative boundaries, nine EC countries as well as Switzerland have defined a concept of urban agglomeration (named 'urban areas' in UK) which groups in a single urban entity the local units covered by a continuous builtup zone. The exact definition of the continuously builtup area varies between countries, notably in terms of a separating distance between buildings as well as a threshold size for total population. There have been criticisms of the concept of urban agglomeration as being too restrictive in evaluating the functional extension of urban settlements, especially since the diffusion of automobile and commuting habits and the general trend of urban sprawl in the last three decades. Nevertheless, for historical comparisons at least, the morphological definition has one decisive advantage over all other possible definitions. It offers the certainty of continuity over time. Urban builtup areas vary only in one direction: they expand outwards, by successive accretion of adjacent localities. This will not necessarily be the case with entities defined on functional criteria based on the location of jobs or economic activity, whose center of gravity may indeed shift and whose fringes are more subject to fluctuation over time.

From Urban Agglomeration to Urban Field

However, even when a common definition of urban agglomeration is applied (as for instance in the historical data base on European cities developed by Bairoch *et al.* (1988), or in the data base Geopolis prepared by Moriconi-Ébrard (1994) for the 1950-2000 period consisting of more than 5000 urban units of 10,000 inhabitants over Western and Central Europe), a perfect comparability of the resulting urban entities is not ensured because of the large diversity of the surface (and population) of the local administrative units which form the spatial basis for collecting population figures (Table 12.1). For instance, there is a factor of 80 between the mean surface of French and Swedish communes, and still a factor of 20 between their average populations. Within the same country, there are also variations in the size of local units (as shown by the wide amplitude between mean and median values in Table 12.1).

This heterogeneity in size and shape of elementary administrative or political subdivisions also hampers the comparability of urban settlements whose definition is usually based upon journey-to-work statistics. The first attempts came from researchers. The 'Metropolitan Economic Labour Areas' designed by Hall and Hay (1980) in the 1970s were later converted into 'functional urban regions' by Cheshire *et al.* (1988). They group around an employment core (more than 60,000 jobs) all municipalities and districts sending more commuters to this center than to any other. Such definitions, which completely partition geographical space, however useful for regional studies or spatial planning at a broad scale are too large for permitting a detailed analysis of settlement systems. Official statistical definitions of daily urban systems are still rare in Europe. The recent design of *aires urbaines* in France is worth mentioning since it may help in correcting the underestimation of French urbanization, which arises from both the too restrictive concept of urban agglomeration and the very small surface of the communes. The new definition gathers around a center of at least 5000 jobs all communes (adjacent to the core or to an already formed cluster) sending more than 40 per cent of their labor force to work in communes belonging to the core or cluster. The resulting units number about 350. The name which was chosen for them may lead to misunderstandings when translated into English (the concept of 'urban area' being closer to the definition of an agglomeration).

Table 12.1 Size of local units in Europe (NUTS 5)

Country and local units	Population (thousands)		Surface (square km)	
	Mean	Median	Mean	Median
Sweden (kommuner)	30.3	16.0	1437	676
The Netherlands (gemeenten)	22.6	11.9	61	35
Denmark (kommuner)	18.7	9.8	156	143
Belgium (communes)	17.1	10.7	52	40
Finland (kunnat)	11.1	5.0	743	364
Italy (comuni)	7.0	2.3	37	22
United Kingdom (wards)	5.2	3.9	22	6
Germany (Gemeinden)	5.5	0.9	24	13
Spain (municipios)	4.8	0.6	62	35
Austria (Gemeinden)	3.4	1.6	36	25
Luxemburg (communes)	3.3	1.4	22	19
Portugal (freguesias)	2.3	1.0	22	11
Greece (demo)	1.7	0.4	22	15
France (communes)	1.6	0.4	17	11
Ireland (DEDs/wards)	1.0	0.5	20	n.d.

NUTS   Nomenclature des unités territoriales statistiques (nomenclature of statistical territorial units).

Source: Eurostat, after Le Gléau *et al.* (1996).

How can the agglomeration and the daily urban system concepts be evaluated in an evolutionary perspective? The latter is obviously narrowly related to time-space, as it has been shown that, if commuting distances have considerably increased, the time which is devoted each day to commuting has remained remarkably invariant during the recent decades. Thus the conception of urban settlements as urban fields where spatial interaction can occur in a given time favors a definition based upon frequency and length of daily movements around an urban center. On the other hand, the continuously builtup area which lies under the definition of urban agglomeration and seems to refer solely to topographical space cannot be considered as totally obsolete. Agglomerations have a fractal structure in the spatial arrangement of their buildings and open spaces, as well as in transportation or technical networks or even daily circulation flows: this means together heterogeneity, fragmentation, but also self-similarity (repetition of the same morphology at different scales). It has been shown that the external limit of the agglomerations, which is variable over time, corresponds more or less at each date to the extension of the urban space which has a fractal structure, compared to the outward rural or urbanized space that remains more homogeneous (Frankhauser and Pumain, 2001). So according to the question under consideration, either oriented towards behavioral or morphological space, one definition may be preferred to the other, but both have an evolutionary character.

Intermediary Settlement Types between Urban and Rural

An interesting aspect of the new definitions used by INSEE since 1996 is that they identify several types of settlements, for which population and other statistics are regularly computed and published: central communes are the cores; the *banlieues* comprise all other municipalities belonging to the urban agglomeration; the suburban communes (*couronnes péri-urbaines*) are those belonging to the *aire urbaine*, but not to the agglomeration; together with the 'urban centers' (made of cores and *banlieues*), they form the 'space with a dominant urban character'; other communes where commuting is important but not directed toward a single urban center are considered as 'multipolarized urban'; the remaining communes are considered as 'space with a dominant rural character' and are divided in four categories: 'communes under a weak urban influence' (sending 20 per cent of their labor force to work in an *aire urbaine*), 'rural nodes' (little towns with 2000 jobs or more and a job ratio greater than one), their peripheries (communes sending 20 per cent of their labor force to work in the rural nodes) and 'remote' or 'isolated' rural (10,000 communes). Plate I shows a subset of this classification around the city of Toulouse in southern France, illustrating cases of more or less large urban cores (Toulouse, Albi, Castres, Carcassonne) as well as rural nodes (Lavaux, Revel, Limoux) with their respective polarized areas and examples of isolated rural zones (on the eastern and southern sides of the map).

Several European countries have defined classifications of settlements including intermediate categories between urban and rural. They provide useful information by completing the description of a settlement system but they are hardly comparable between countries.

Population Densities or Corine Land Cover?

One may question the relevance of settlement classifications which are based upon the concept of density (Eurostat, 1999). First from a theoretical point of view, density is a concept borrowed from physics and related to an idea of homogeneity, whereas the spatial distributions of population densities, especially urban, can never be referred to this uniform distribution and on the contrary are of a fractal type (Batty, 1995; Batty and Longley, 1994; Frankhauser, 1993). Secondly, from a more practical view (since at the moment data on densities are easily computed and available while fractal measures are still in discussion among specialists), there is a systematic bias when assigning a density threshold for a geographical classification purpose, because, within a particular settlement style, urban densities are correlated to city size (Bussière and Stovall, 1978).

A solution towards harmonization of definitions in Europe could be found by using satellite image information. The European Agency for Environment already provides spectacular maps of the 'artificialized' lands, which mainly correspond to human settlements (Plate II). According to experts, the criteria which have been used in different countries for identifying the categories of urban land use are not yet fully comparable. More detailed tests have been made for delimiting large metropolises as from Landsat and Spot images (Eurostat, 1995) and research is continuing on evaluating the capabilities of this new source of information, especially when it will provide images of higher resolution with Spot 5 (SCOT, 1997). In any case, this apparently objective and strictly comparable instrument reflects the morphology of settlements, but it has to be complemented with population statistics in order to meet more general objectives of the classification of human settlements. This integration with data relating to administrative and other spatial units can be achieved using Geographical Information Systems.

If progress toward harmonization of definition is desirable for comparative studies, it could be argued, especially from the point of view of an evolutionary theory of settlement systems, that a variety of definitions is not to be necessarily rejected, because it may partly reflect the diversity of configurations of national settlement systems or specific national representations of settlement types.

A Regional Typology of Human Settlements Styles in Europe

As we have seen, the cacophony among European definitions of settlement types is partly arbitrary, partly explained by specific features of the historical structuring of political and administrative territories, whose consequences in terms of shaping contemporary social and spatial interactions could still remain relevant, but also could be partly depending upon more general aspects of the geographical context of development of the settlements. It will now be shown how a limited number of neatly distinct settlement styles can emerge from a simple crossnational comparison of the most apparent characteristics of settlement systems. Since it is not possible to rely on national definitions of rural and urban settlements in Europe, original dedicated data bases have to be built for providing such a comparative view.

A basic level of resolution has to be chosen. It should not be too large because wide spatial subdivisions would erase the internal differentiation of settlement systems (which are very heterogeneous in size and spacing), but also it cannot be too fragmented because too small spatial units would not take into account the geographical consistency in urban systems. Different types of indicators are needed for describing characteristics of settlement systems which are more continuous in their spatial distribution such as rural population densities – whose regional variations may be related to constraints of natural milieu (altitude, climate) – and those that are discontinuous. The latter are linked to the various forms of local urban concentrations, which may be either grouped in a single major center or distributed into several nodes, and can be surrounded by high densities or isolated inside much less densely populated areas.

Such considerations lead us to a list of indicators combining two main data bases: the Eurostat regional data base for basic information about rural and total population densities and the Geopolis data base (Moriconi-Ébrard, 1994) for a more detailed description of the upper part of the settlement system's hierarchy (including about 5000 urban agglomerations of 10,000 inhabitants and more, following the UNO definition, for Western and Eastern European countries, excluding countries of the former USSR). The NUTS 3 level has been chosen as a grid representing a good compromise. It has been a little modified when subdivisions are not homogeneous in size compared to other countries (former Yugoslavia excepted). All data have been computed in this spatial framework, excepted for the indicators which characterize the spatial influence of a large metropolis, which were smoothed over broader surfaces. The indicators describe the size, spacing and intensity of human settlement at three spatial scales. The class size of the largest city, the urbanization rate and the average rural density are measured at the local level (NUTS3 region); the urbanization rate and rural density in neighboring regions also are considered; the regional urban context is defined for a circle of increasing distance with largest city size and is characterized by the class size of this metropolis, the inequality index of town size distribution and a primacy index computed inside the circle.

The classification identifies six main types of regional human settlement (Figure 12.2). A few regions are dominated by a very large metropolis. Paris, Madrid and Athens, belong to this type. Their development has been associated with a highly centralized political and administrative organization of the state, which has been maintained in the long run. It has concentrated the population in one primate city and generated around this another type of region, which consists of rural areas deprived of major towns because this metropolitan attractiveness hampered any other urban development.

A different style of settlement hierarchy consists in a less contrasted distribution of sizes, including neighboring metropolises in a polycentric pattern, scattered in high population density areas, both rural and urban, as observed in the Rhine region, or mainly urbanized as in central England. Here there are two different explanations for similar forms. In England the precocity and strength of the industrial revolution explain the emergence of very large industrial cities and conurbations which considerably increased and spread out the density of urbanization in a formerly rather centralized country. Political fragmentation combined with strong economic development explain the persistence of cities, not so large as Paris or London but numerous and close to each other on the corridor linking the Mediterranean and North Sea regions through the Alps and Rhine valley in continental Europe.

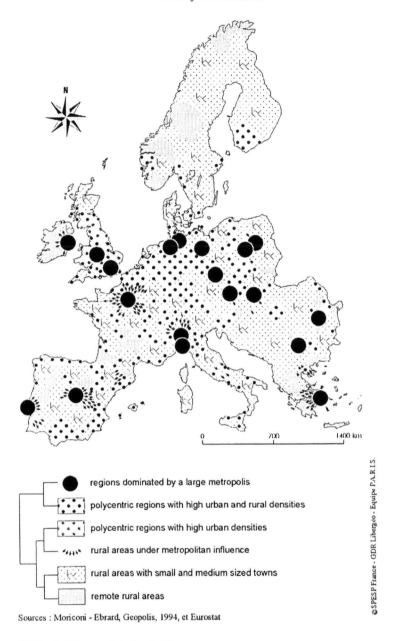

regions dominated by a large metropolis

polycentric regions with high urban and rural densities

polycentric regions with high urban densities

rural areas under metropolitan influence

rural areas with small and medium sized towns

remote rural areas

Sources : Moriconi - Ebrard, Geopolis, 1994, et Eurostat

©SPESP France - GDR Libergéo - Equipe P.A.R.I.S.

Figure 12.2   Settlement styles in Europe

A third style of settlement is made up of less urbanized areas, that either include regular networks of smaller towns, as in Eastern European regions, or are essentially rural, as in some peripheral regions. The former have had a regular pattern of settlement from a systematic colonization which occurred rather late (thirteenth century) when compared with other parts of Europe, while the latter have always been sparsely populated because of natural or accessibility constraints.

The importance of the exercise is not so much in the resulting typology, which could be subdivided in more different styles and perhaps give rise to slightly different regional limits if different indicators had been chosen (the relative stability of the map on Figure 12.2 was however tested with respect to these considerations). What is more significant in these results is the weight that the historical circumstances of the development of the settlement systems take in the definition of the settlement styles of today. Not only the main three families of settlements brought forward by the classification but also smaller groups can be explained in such a way.

Suggestions from Evolutionary Theory for Settlement Systems

There is a need to depart from a too simple dichotomy and to move towards more detailed and flexible typologies of settlements, according to the purpose these classifications are aiming at. This attitude would rejoin an old tradition of political economy which used to distinguish for instance not only between hamlets, villages, towns and cities but which introduced nuances between borough, market towns, and towns, or between mining towns and administrative centers (de Vries, 1984).

At the local level, at least four main concepts could be defined in order to typify the nature of the spatial interactions which are involved in the constitution of a settlement entity:

- The municipal level is an elementary territory for political representation and decision making.
- The urban agglomeration corresponds to a physical unity covering several municipalities, which is defined by the spatial continuity of the builtup area and necessitates a minimal coordination of municipal policies for the management of basic urban services functioning on physical networks like water provision, sewage, transportation.
- The metropolitan or functional area, or daily urban system, also includes surrounding municipalities which have a mainly residential character and send more than a specified proportion of their labor force (and /or customers) toward the agglomeration on a daily basis.
- The polycentric metropolises have two main origins: either they are conurbations which joined together several former towns and cities or agglomerations into a single continuous urban entity, or they are former monocentric agglomerations which integrated other urban centers in the course of their spatial development or where new centers emerged on the edges of the already urbanized area for various reasons (deconcentration of certain economic activities, emergence of new technological poles, especially close to airports or to research campuses, swarming of real estate investment for a part of urban bourgeoisie and offices, and so on).

At the upper scale of national or continental territories, other typologies may be interesting for a range of purposes from making purely academic distinctions (Geyer, 2002) to the application of European policies (see for instance Cattan *et al.*, 1999; Conti and Spriano, 1990; ESDP, 1999). One should keep in mind that if size thresholds are the main indicator for the design of such typologies, they have to be defined in relative ways: firstly, in respect to the national or regional territory they belong to (which represents the main subset of local interactions through which their development took place) and, secondly, according to the spatial context (including location, human densities and physical constraints) in which they are embedded. For instance, if a category of settlement as 'remote rural' can be defined everywhere, the population size and spacing threshold which will be used are different if the settlements are located in the plain of Berry in the middle of France or the central Apennine mountains in Italy or the northern part of Sweden.

For analytical reasons, including the evolutionary perspective on settlement systems (Pumain, 2000), as well as for planning purposes, consideration of the *relative size* of a settlement is essential, since it represents together a summary of the historical trajectory of the settlement in a competitive context, a proxy for a level of complexity of its society in terms of the number and diversity of the activities that it can supply, and a rather good approximation of the limits of its development perspectives in a given period of time.

At this scale, it is more difficult to suggest *a priori* a number of settlement types which should be taken in consideration. In the context of most European countries, rural settlements could be (and often are) identified as settlements which do not provide all services which would be sufficient for family life on a daily basis. For instance, they do not supply schools for secondary education. According to the countries and the geographical context, the typical sizes for such settlements may vary between 200 and 2000 inhabitants. Distinctions can be made between them for instance on a functional basis between those offering no services at all and those supplying a few basic services, and others having specialized in activities like forestry, fishing, mining, tourism, which enlarge the range of their relationships compared to the other types of rural settlements but also always include at various degrees the risks and constraints on further development which are associated to a mono activity.

Among towns and cities, the range of variation in size and functions is much more important and there is a much wider list of relevant criteria which may be used for classification. If the same logic is applied as was used previously for rural settlements, i.e. the level of autonomy in their functioning and development, the portfolio of their activities may be relevant if taken in a dynamic perspective. Towns and cities depending upon a dominant sector of activity – whether manufacturing or tourism or public funding (redistribution) – may be more fragile in the long run than more diversified urban settlements. It is obvious that, for planning considerations, mushrooming urban settlements (for reasons as varied as residential boom or discovery of a new resource or building of a new communication infrastructure) could be profitably established as one category.

The subset of the largest urban settlements is often referred to. The identification of a metropolitan level is easy if using the classical models of statistical distribution of city size such as the model of a Paretian form known as Zipf's rank size rule, since it amplifies the 'anomalies' which appear at the top of urban hierarchies. How can it be made acceptable for politicians? First it has to be remembered that a metropolis should be defined not by absolute terms in size or panel of activities but in relative terms, inside a given territorial context and relatively to the other elements of the settlement system it belongs to. Second, one has to admit that such a level is already clearly apparent in Europe, for instance when policies for polycentric development are put forward (ESDP, 1999). More or less implicitly, the relevant subset of cities which could benefit from such a policy includes all state capitals (which range in size from 10 millions to 100,000 inhabitants) and the major regional capitals of the largest countries. It may also include a few towns or cities without any political prerogatives or major economic role but which have an international visibility through their specialized activities like tourism or finance.

Typically, the typologies which are suggested at the European scale and for spatial planning purposes include five or six levels which are not too different from the results of the classification that we have presented above.

Conclusion

It has been shown in this chapter that, considering the long history of human settlement, the two categories of rural and urban could be expanded into broader typologies. The relevant types and criteria may be provided by referring to an evolutionary theory of settlement systems. It relates different classes of human settlements to selection processes which introduce discontinuities in their development and explain the major size and functional differentiation between settlements.

So the main suggestion is to extend the list of the criteria suggested in Chapter 1 of this book, i.e. settlement size, density and accessibility to services, through the addition of information about the type of trajectory of the elementary settlement, relative to its context in the settlement system. Although examples in this chapter have been restricted to Europe, application of the theoretical approach can be readily made for other continents, taking into account different stages and phases of the urbanization process.

Chapter 13

The Conceptualization and Analysis of Urban Systems: A North American Perspective

Larry S. Bourne and Jim Simmons

Urbanization is generally considered to be the dominant process of social, economic and territorial transformation in the twentieth century in most of the developed world. The same will hold true for the developing world in this century (UN, 1996, 2003). Yet, surprisingly, there is little or no agreement on how to approach the study of this complex process or its outcomes. What are the most appropriate conceptual and theoretical frameworks? Specifically, how do we approach the measurement and analysis of the outcomes of recent trends in the urbanization process at a broad macrogeographic scale when those trends and outcomes are so varied?

Our view is that our conventional ways of approaching the urban question are now, in many regards, inadequate and outdated. These approaches have been overwhelmed by the increasing complexity and diversity of the global urbanization process, by the fact that the rural-to-urban transition in the developed countries is now largely complete, and most recently by another phase of global integration, as well as by the complex transformations of space and place attributable to economic restructuring, social and demographic change, and the introduction of new technologies. In highly urbanized societies, such as those of the world's most advanced economies, the traditional urban-rural partition is now largely irrelevant. Between 80 and 90 per cent of national populations now reside in designated urban areas, and within the residual territories – rural or nonurban – most people live urban life styles.

This chapter argues the case for an urban-system perspective – indeed a continental or even global perspective – and approach to contemporary urbanization. It comprises five parts. The first outlines the concept of the urban system, while the second reviews its evolution, and its strengths and weaknesses. The third part describes, with special reference to North America, those recent trends that are redefining the urban process writ large and reordering the challenges posed for researchers and government agencies. The fourth reviews questions of the definition, measurement and classification of urban settlements and urban systems, the importance of links to urban nodes outside the system, and the information gaps confronting researchers. Finally, looking beyond definitional questions, we provide suggestions on emerging issues and future directions of research and outline the information improvements that are required to address them.

The Urban System: Concept and Dynamics

The concept of an 'urban system', or system of cities, is based on notions of interaction and interdependence. It offers an approach to, and a way of thinking about, the urban process at a broad geographical scale, rather than a specific technique or methodology. The approach argues that individual cities do not exist in isolation. These cities grow and change primarily because of their roles in the larger urban systems, the rate of growth (or decline) in their hinterlands, and specifically through their interconnections – both complementary and competitive – to other urban places. Cities therefore are simultaneously part of several larger sets of urban places – within a local region, the nation and internationally.

The premise, simply stated, is that in an urban society most human activity is located in, organized by and controlled through urban areas, particularly the larger metropolitan areas, and the agents, institutions and corporations that dominate urban economies and societies. Urban areas serve as organizing nodes within a nation's economy, culture and demography, and as the primary gateways to the rest of the world. The characteristics of urban places thus reflect their relative position in larger urban systems at different geographical scales. For present purposes we can assume, under this conceptual umbrella, a nested and increasingly fluid or 'elastic' urban hierarchy consisting, first, of a continental or global settlement system, and within that a national urban system, and within that a set of regional and metropolitan systems.

The logic of the urban system approach is relatively straightforward, and is supported by an extensive literature dating from the 1960s onwards (Bourne, 1975; Berry, 1976; Pred, 1977; Simmons, 1986; van den Berg, 1987; Borchert, 1987). It argues that the first place to look in understanding growth and change in settlement patterns is to start with the urban system of which that settlement is a part. For any particular place we are then required to examine the relative position of that place within the larger system – in effect, to examine its external relations. The economic and demographic fortunes of individual urban areas are then determined by the sum of their interconnections – the flows and linkages – with other places. These linkages in turn define a hierarchy of urban places, based on systematic differences in population size and the range of social services and economic functions performed, with each level in the hierarchy subservient to all levels above.

The existence and importance of interurban connectedness, of course, is not a new idea or a recent phenomenon. For example, consider the extensive colonial urban systems maintained by and centered on ancient Rome, or Athens, or the trading networks created by the Hanseatic League of cities, or the system of military and trading posts that guided the European colonization of the Americas.

Urban systems, in other words, have been around a long time. They have been shaped and reshaped by the relative advantages of geographical concentration, the costs and difficulties of transport and communications, and the varied challenges posed by shifts in economic competition and political control. The parameters of these relationships have continued to evolve, however, as cities continue to increase in size and to expand their influence over wider geographical areas. As the constraints imposed by distance and accessibility have weakened over time, and the larger urban nodes continue to grow, the influence of nearby nodes gives way to the influence of larger nodes further away. Cities now connect with, and compete with, places that are farther and farther away; and the networks of linkages and spheres of competition involve many more cities than before.

The importance of urban areas, and of metropolitan areas most notably, in the process of national development is both well known and well documented. The literature on the attributes of metropolitan areas, individually or collectively, is enormous. What is generally lacking in the available statistical data and the empirical analyses, however, is an explicit image of what an urban or 'metropolitan system' actually is; that is, a model (or models) of how these urban concentrations fit together, and empirical evidence sufficient to test these ideas and models. What kinds of social, demographic and economic linkages bind these urban areas together? How do migrants, and capital, flow through the urban system? How do these cities compete with each other for investment, for people, and for economic and political rewards?

The reasons for this relative lack of theoretical engagement and empirical research on systems of urban and metropolitan areas are twofold in the North American context. First, the federal political structure of government in both Canada and the US has led to an emphasis on the provision of data at the provincial or state level. This is especially the case for research on economic activity, demographic change, and public sector activities. With a few exceptions, these spatial units cannot be used to infer urban or metropolitan conditions or linkages. An obvious example is foreign trade and investment. Data on exports and imports and capital movements are almost totally absent for any level of the settlement system, even for large metropolitan areas. Most urban researchers, and almost all geographers, find that the crude subnational spatial template provided by ten provinces and 50 states in North America is a totally inadequate framework for the analysis of change in urban societies.

Second, metropolitan regions have never been permitted to evolve into meaningful political entities, and thus they lack the political clout necessary to influence the spatial architecture of information collection and distribution. Certain kinds of subnational data, such as public revenues and expenditures, are generally available only for political entities such as municipalities or counties. These entities, however, have become increasingly inappropriate as units of analysis since the urban development process has spread far beyond their typically static and outdated boundaries. In short, metropolitan regions are seldom actors in any real sense; they are typically simple aggregates of many small and often diverse entities. In effect, no one speaks for the entire New York region or for the Greater Toronto area.

Urban Systems: The Evolution of an Idea

Although the concept of the urban system itself is not new, its analysis and interpretation have changed over time. A generation ago the national urban system was viewed as a set of distinct nodes that together formed a more-or-less distinct hierarchy of places. Each urban area organized the region that it served, attracting population and commodities from that regional hinterland and shipping them to other places in the urban system. Each node also imported goods and services from other places and redistributed them to the local hinterland, serving in a kind of middle management role. The urban system itself was often considered to be closed to imports or immigration. Defining the individual nodes that constituted the urban system was less important than defining levels in the urban hierarchy and measuring the cross-sectional attributes of the nodes.

Growth and change were also viewed in a rather simplistic fashion. Urban population growth was driven primarily by local or endogenous demographic components – that is, by natural population increase and domestic inmigration. Economic linkages tended to be hierarchical and, depending on the type of good or service provided, to be contained largely within the region and/or the nation state. Urban growth then tended to reflect the expansion (or contraction) of the immediate trade area and localized competition for markets. In effect, the economy of the trade area supported the urban economy. In this context, research focused on the attributes of the cities in the urban system, and classifications of cities based on these attributes were commonplace. At the time, these studies were novel and informative. Over time, however, they have become more restrictive and less useful.

In the last few decades, urban systems have evolved in rather dramatic ways, and our perceptions of the structure and dynamics of these systems have changed accordingly. We have increasingly come to see cities and metropolitan regions not as isolated nodes but as points of organization, management and control, set within highly competitive spatial economic and demographic settings (Simmons, 1986; Borchert, 1987; Bourne and Simmons, 2002). The nation state is now seen not simply as a detached regulator that is outside, or external to, the urban system, but as an active player internal to the social and economic processes of change. The nation state, even if weaker than previously, has not been hollowed-out, as some have suggested. It is, for example, still preeminent in setting taxes and macroeconomic policy, collecting and redistributing public revenues, making location decisions for infrastructure investments, devising equalization strategies and development programs (e.g. regional policy), defining boundary conditions (e.g. immigration policy), and negotiating external exchange arrangements (e.g. trade policy). Moreover, as growth processes at the continental and global scales become more important, the nation state becomes the crucial gatekeeper in managing the influence on urban and regional growth of external forces of change (e.g. through trade, immigration and international treaties).

Recent Urban Trends

Urban systems have continued to evolve in complex and often unpredictable ways, and this in turn poses challenges for both urban researchers and for those agencies with the responsibility to monitor and regulate settlement patterns. Specifically, the roles of individual cities and metropolitan areas, and the spatial organization of the entire settlement systems, are being redefined by a series of simultaneous events, policies and processes. These include widespread economic and corporate restructuring, the postwar demographic transition, massive transnational migrations, new information technologies, trade agreements, and by the globalization of culture, commodity flows, labor and capital investment. We illustrate the imprint of these changes with examples from recent urban experience in North America. This experience can be characterized, allowing for wide variations between the two countries and among their regions, as one of continued growth and rapid spatial reorganization, combined with considerable volatility. The result has been a highly uneven geography of urbanization and urban development (Sternlieb and Hughes, 1988; Knox, 1993; Bourne, 1999; Bunting and Filion, 2000).

Metropolitan Concentration, Rural Decline

In both the US and Canada, population and productive capacity have continued to concentrate in the larger metropolitan regions, despite an earlier tendency to counterurbanization (Berry, 1976). Over 78 per cent of the US and Canadian populations now live in large metropolitan areas, and in the US over 50 per cent reside in metropolises of over one million population. Although many of the largest metropolises are not growing as rapidly as some of the medium-size metropolitan areas, this should not be interpreted as a symbol of extensive deconcentration or of the decline of large cities. Neither the US or Canada are abandoning large cities. Instead, they are building new ones, typically in newer regions, and on a very large geographical canvas. The next generation of megametropolises will emerge from the current second tier metropolitan areas – such as Atlanta, Dallas-Ft. Worth, Houston, Tampa-St. Petersburg-Orlando, San Diego, Phoenix and Toronto-Oshawa-Hamilton.

The physical scale of these metropolises also defies simple classification and boundary measurement. Some metropolitan statistical areas, such as New York (19 million) and Chicago (9 million), sprawl into three different states. Others represent the consolidation of several metropolitan areas; for example, Los Angeles-Long Beach (15 million) and San Francisco-Oakland-San Jose (8 million). Still others straddle national boundaries (e.g. Buffalo-Toronto; Seattle-Vancouver; San Diego-Tijuana). Others may not transcend political boundaries but represent the coalescence of several previously independent metropolitan areas – e.g. Miami-Ft. Lauderdale-West Palm Beach.

One obvious outcome of these trends is that urban areas now dominate their local trade areas, rather than the reverse. Rural markets, in much of North America, are relatively small and rapidly declining. Urban places at the lowest level in the traditional urban hierarchy, especially in those low-density agricultural and resource-based regions, will likely continue to decline, or even disappear, unless they are transformed to become servants to specific niche markets (e.g. tourism, recreation) or particular population groups (e.g. retirement). The only small places that will likely prosper will be those located near, and thus accessible to, growing metropolitan regions and/or those with special features, such as distinctive environmental attractions, heritage and cultural amenities, or research and educational facilities. The growth of the service sectors, especially financial and producer services, relative to manufacturing and the primary sectors, has also further increased the dominance of metropolitan regions.

The New Demography

The demographic basis of urban growth has also shifted. Canada and the US both witnessed a series of dramatic demographic transitions over the postwar years. The 'baby-boom' of the initial postwar period (1947-1963), involving exceptionally high fertility and marriage rates, was followed by an equally dramatic decline in fertility (the 'baby-bust' period from 1963). The 'echo-boom' anticipated as the baby-boomers reached reproductive age never materialized. The result of this sequence is a demographic profile in which there are huge differences in the size of individual age cohorts. These differences, in turn, are magnified in and by the unevenness of urban growth and change, and underlie the variable pressures shaping housing and labor markets and the demands for public goods and services in different urban regions (Foot and Stoffman, 2000).

In North America, as in most western countries, rates of natural population increase are at or below replacement level. National populations, as a result, are ageing rapidly, and in the near future the populations in many regions could begin to decline in absolute terms. Moreover, domestic (internal) migration rates – the principal means of population redistribution in the past - are stable or declining. As a consequence, urban growth in most locations is now driven, not by natural increase or domestic inmigration, but by immigration. In the last decade immigration has accounted for 50 per cent of total national population growth in Canada (Figure 13.1), and over 35 per cent in the US. Moreover, in Canada, over 70 per cent of recent growth in the labor force is attributable to new immigration.

At the same time, the destinations of these immigration flows are highly concentrated geographically. Most of the major immigrant flows in North America are focused on a number of 'gateway cities' – notably, New York, Los Angeles, Chicago, Miami, Vancouver and Toronto. In some instances, the proportions of the metropolitan population that are foreign-born now exceed 40 per cent. Ironically, many of these same gateway metropolises have recorded net losses in the exchange of domestic migrants over the last decade or two. These losses, however, have been more than compensated for, to date, by overseas migration. What is not clear here is whether high levels of foreign inmigration are related to the net outmigration of domestic residents from the gateway cities, and if so how and why? Jencks (2001), for example, suggests that immigrants may drive down wages and thus reduce the attractiveness of certain metropolitan areas for domestic migrants. It is also possible that outmigrants are simply 'equity-migrants' taking advantage of the demand for housing provided by inmigrants from overseas by moving out.

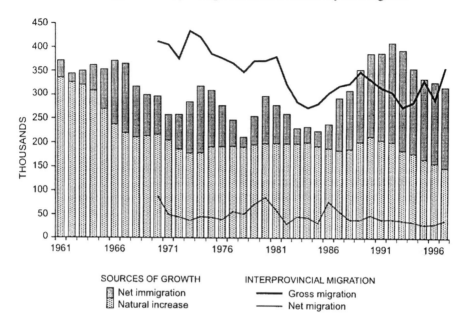

Figure 13.1   Components of population growth, Canada, 1960-1997

For many of those cities and regions in North America that have not received substantial numbers of immigrants, or an inflow of domestic migrants, the future appears bleak. Their populations will remain relatively homogeneous, their age structures will be truncated, and the ageing process will accelerate as members of the younger age cohort leave and are not replaced. Areas of population decline will become even more widespread than in the past. The social distances and demographic contrasts among urban areas, especially between those receiving immigrants and those that are not immigrant destinations, will also increase. Given the dramatic transformation in the source countries of these immigrants, and thus in their ethnocultural, religious and racial attributes, the gateway metropolitan areas are beginning to look and feel much more diverse and less and less like other cities and nonurban communities in their respective nations.

Politically, the tensions between areas of growth and decline are exacerbated by the increasingly sharp cultural differences between the places of immigrants and places populated by earlier immigrants and the native-born. The politics of the United States balances the slow-growth Midwest and western interior against the complex growth dynamics of the high immigration cities along the two coasts and along the Mexican border. In Canada, cities such as Toronto and Vancouver appear to be moving further apart from their respective regions as the imprint of globalization through immigration transforms their social character, economies and political institutions.

An additional dimension of the demographic transition that is changing settlement dynamics is the transformation in social organization, notably in terms of family structures, household types and choices of living arrangements (Bourne and Rose, 2001). Key aspects include the increasing proportion of nontraditional families and the shrinkage in average household sizes. Rates of household formation increased dramatically over the postwar period in all developed countries. As a rule of thumb, there are now some 50 per cent more households than there would have been on the basis of family and household sizes prevailing in the 1950s and 1960s. These changes not only influence income distributions, housing demand, domestic consumption and life style patterns, but the locations within the urban system at which those demands are expressed. A geographic area with a fixed housing stock would have lost, on average, between 25 and 30 per cent of its population over this period due to this process of demographic thinning. These trends in turn call into question the use of traditional measures of density based strictly on population.

The New Economy

While the basic outline of the new demography is relatively visible, except for its future geography, the degree of change in the economy is more subtle, and more difficult to monitor. In terms of employment, primary sector activities have almost disappeared. Even the jobs involved in manufacturing, long the central component in urban growth models, are melting away. Instead, almost all of the new employment is found in services, some of them serving local markets but more and more of them being oriented to national or international markets. The location attractions for sophisticated (and high-wage) services and employees have shifted to amenities based on city size, culture and life style, and away from natural resources and transportation crossings. The mix and success of local firms becomes an important element in the process of competing with other urban regions. Indeed, the competition increasingly involves cities in other countries. Corporations play off cities in the US against places in Canada, Mexico or China. Moreover, a small number of corporations in each sector generate most of the movements of goods.

The problem for urban researchers is how to differentiate among urban places when two-thirds or more of all jobs are in services, and many of the interactions are within corporations. What kinds of services are most important; and how do these services connect to other sectors of the economy? How important is the role of the key corporation in urban growth and in defining urban characteristics? Seattle and San Francisco, for example, are leaders in the digital economy, but for different reasons. New York continues to lead in financial services; and Los Angeles provides a place name for the media. But how do these specializations shape the rest of the economy? The challenge is much more complicated than just counting jobs; and in each case the international market has become at least as important as the market contained within the national urban system.

In parallel, the longstanding hierarchies linking urban places are giving way to much more complex and varied hierarchical arrangements. Well-established national urban systems seem to be fragmenting. International flows are increasing almost everywhere in relation to domestic flows, measured in terms of trade, migration, capital and the exchange of ideas and culture. Capital, for example, now moves quickly and effortlessly around the globe. Transborder migration has increased despite tighter border controls on legal movements. Corporations market their goods and services in markets in several countries at the same time, and can rapidly relocate their production and management functions to another jurisdiction as needed.

Linkages among places – the glue of the urban system concept – have become less hierarchical and more focused on institutional, government and corporate links. They have also become more closely defined by the particular economic specialization of each place rather than by distance alone. Many services are now delivered electronically. As the overall effects of distance and proximity have declined in the face of the telecommunications revolution, this decline has had uneven impacts on different urban regions and types of activities. For some regions and activities, including those in the new digital economy, intense local clustering effects are still evident. In other words, distance may still be crucial, but not as intensely or as universally as previously.

The immediate impacts of these shifts, on the settlement system in general and the metropolitan system in particular, are still under negotiation. Nevertheless, several generalizations are relevant here. One is that as national systems fragment, the ties among the nation's urban areas are weakening, and new transborder or global urban systems are being constructed, or in some cases reconstructed. Individual metropolitan areas are now dependent on, and competing with, not only their national counterparts, but also cities and urban systems in other countries. For most of the larger metropolitan places the field of competition is now international; in fact, in many regards these international metropolises are now much closer together than before – in terms of time-distance, sociodemographic character and economic attributes. As one example, Figure 13.2 illustrates how the air passenger flows from Toronto, Canada's largest airport, have been reoriented over time from domestic connections to those with US cities and especially further afield.

Whether these linkages and similarities suggest the emergence of a continental or world urban system is unclear. Surprisingly, much of the literature on global cities is strangely detached from the urban system perspective that it would seem to require. There is, for example, little empirical evidence provided on the relative strength of linkages between these global cities and their national and regional hinterlands. Until such evidence is provided the hypothesis that some cities are more global than national or regional remains untested. What is true is that certain cities have emerged as relatively strong world leaders and command points for the global economy (Knox and Taylor, 1995; Beaverstock, Smith and Taylor, 2000; Sassen, 2000).

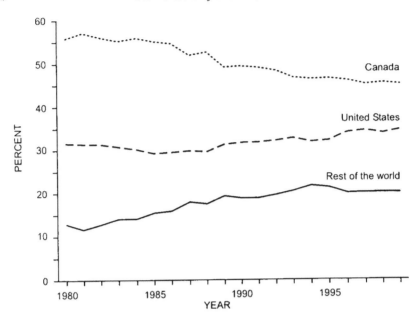

Figure 13.2   Destinations of air passengers from Toronto, 1980-1999

For smaller urban places, in contrast, the competitive field has remained national, regional or local, depending on their size and functions, and indeed may have become even more localized. Among several other side-effects, increasing global integration, whether of capital sources, supply chains, final markets or labor supply (including immigration), means greater volatility in rates of population growth and economic activity. Although increased cross-border economic flows and social integration may augment national growth rates and improve standards of living, they also import risk and uncertainty.

For students of urbanization, and specifically for researchers of urban systems, the consequences of these trends require a rethinking of conventional approaches and measurement indices, and a shift in the focus of research. As examples, research on urban and regional change has shifted from an emphasis on criteria that measure activities to those that measure flows, from studies of patterns to studies of processes, from narrowly deterministic analytical models to more holistic models, from an emphasis on the role of agglomeration economies in location decisions to more flexible paradigms, from the role of initial advantage to that of competitive advantage, from a focus on physical capital to one on human and social capital and innovation capacity, from the study of the individual firm to corporate networks or clusters of corporations, and from examinations of localized government stimuli to wider studies of the governance of corporate activity, immigration and international trade. All of these shifts render the usual questions of settlement conceptualization, classification and measurement all that more difficult.

Data Problems and the Definitional Conundrum

With these trends and conditions as background, we can return to the questions of the conceptualization, definition and measurement of urban systems and the role and importance of new information sources. Here we examine a series of issues relating to the concepts, statistical building blocks and basic information required to analyze urban systems under diverse national and global conditions, systems that are themselves undergoing rapid transformation.

Traditional Urban Definitions

The core of any definition of urban places typically involves three tasks. The first is the establishment of a minimum threshold total population and density for a settlement to be classified as urban. In North America this generally means a minimum population of 1,000 in Canada and 2,500 in the US, living at a density of at least 400 persons per square kilometer. All other areas are considered rural, or at least nonurban.

The second task is identifying a geographical boundary or limit for those settlements. This issue is obviously more acute in defining the extent of the larger metropolitan regions than for smaller urban places. Almost everywhere those regions are now defined as extensive 'functional urban regions', combining city and adjacent suburbs into regional labor and housing markets, typically using threshold population densities and the journey to work as the principal criteria.

The third task is the identification of a hierarchy of urban settlements, usually based on population size groupings. These groupings are often treated differently by national statistical agencies. The most obvious example is the partition of an urban system into metropolitan areas (over 100,000 population) and nonmetropolitan areas (under 100,000), under the assumption that size is an important indicator of urban attributes. The larger size grouping is typically favored with relatively rich information sources, the smaller size category with very little information. In Canada there is an additional category of functional regions, the census agglomeration, for places between 10,000 and 100,000 populations (Statistics Canada, 1997).

All three of these criteria are open to serious debate and criticism. For example, the widespread use of GIS and the increased availability and flexibility of census data and special surveys have reoriented much of this debate. The researcher is now able to redefine urban boundaries and urban groupings to satisfy particular research problems, using combinations of measures such as distance, density, social character or political jurisdiction. The primary issue then shifts to questions of spatial scale and the content of the data. What measures are required to understand growth and change in urban systems at different spatial scales? At the same time, the data content is affected by conventional metropolitan area definitions since the latter are the most common units for the release of disaggregated data, at least initially.

The Information Gaps

It follows from the above discussion that different kinds of information are required for the analysis of contemporary urban systems. There are three areas of particular concern. Two involve domestic data sources, while the third requires international cooperation.

The first issue derives from the extension of urban systems analysis to incorporate information on economic, demographic and political processes. National statistical agencies generally do an excellent job in measuring and compiling information on individual households and economic activities that have been the main source for the description of urban systems to date. These same agencies now provide a variety of other measures that are relevant to our research but are seldom available in spatially disaggregated form. For example, it should be possible to generate more information about the institutions that now play such a major role in the growth of urban systems, and to obtain data on the activities of various levels of government (e.g. incomes, expenditures) at the metropolitan level. And, ideally, it should be possible to describe corporate concentrations within an urban region – e.g. the relative proportions of local, state, national or international control of the economy.

Second, aside from data on commuting and migration, largely derived from censuses, we have little information about the patterns of interaction among urban areas. Why not release more information about the movement of goods and information among cities – the phone calls, mail, highway usage, and so forth – so that researchers can monitor the evolution of interurban relationships? It might only be possible, at least initially, to release such data for large and well-defined metropolitan regions, but over time the information base could be expanded in order to enrich our understanding of the changing relationships between large and small urban places.

Finally, we need to support efforts leading to the standardization of urban and other classifications whenever possible so that we can begin to trace the development of international urban systems. If urban places are defined in reasonably comparable ways – which will require the translation of national urban definitions into a set of generalized international standards – researchers will be in a better position to compare rates and locations of urban growth. If, for example, economic activities are classified in the same way it should then be possible to analyze changes in levels of economic specialization in various national urban systems and relate those to shifts in international trade. The recent agreement to standardize the industrial classifications used in Canada, Mexico, and the US provides a good example.

Definitional Questions

The above discussion raises a number of questions related to the premises underlying traditional definitions of urban places and the nature of the relationships between urban and nonurban places. In this section we examine five sets of questions: the issue of 'urbanity'; the importance of the contrast between low and high density settings; the question of whether our focus is on people or territory; whether urban classifications should involve top-down or bottom-up approaches; and whether such urban classifications should be inclusive of all national territory or exclusive.

Urbanity and Rurality

The first question relates to the concept of 'urban' itself. Why should 'urbanity' be defined by a certain size clustering of people, or 'rurality' by the absence of such clusters, especially when goods and information move more freely than in the past and many services are now widely available by electronic means? Why should the extent of metropolitan regions be limited to daily routines and the daily cycle of commuting to work when people participate in many increasingly varied behavioral fields ranging in scale from the neighborhood to the region? And why should urban regions over 100,000 be treated differently in the provision of information from those of smaller size? Has anyone actually tested the idea that this size threshold is meaningful in terms of social, demographic and economic attributes? Are these differences reflected in the processes shaping metropolitan form and structure and the processes shaping entire systems of cities?

High- versus Low-Density Settings

The definitional challenge is also very different in countries – urban systems – that have markedly different histories and geographies, notably those with high or low population density. The former tend to have a long history of settlement, a dense population outside of urban areas, and a settlement system that as a result is also relatively dense. Towns and cities are often located in close proximity, reflecting inherited population geographies and the historical difficulties of travel and service delivery. In contemporary times, the greater ease and lower costs of travel have resulting in many of these towns and cities being absorbed into the same overlapping labor markets and service hinterlands as part of a multinucleated urban landscape. Drawing single lines on the map to separate these market areas into discrete zones is now almost impossible, or at best highly arbitrary.

In lower density countries, often those with more geography than history, settlements tend to be smaller and further apart. Population densities outside the urbanized core in much of North America, as well as in Australia and Latin America, also tend to be relatively low. Indeed, vast regions of those continents are either unpopulated or can be characterized as sparsely populated islands of settlement in a sea of wilderness. In such cases there are fewer overlapping labor markets and more distinct service hinterlands. Drawing boundaries in these situations may appear to be a simpler and less ambiguous task. However, even this simple exercise poses its own methodological problems, not the least of which is that in very low-density settings with extremely high transportation costs, daily commuting and service delivery are replaced by weekly patterns and, increasingly, by internet-based service provision (e.g. for medical assistance).

It should also be acknowledged that there are wide differences in the size hierarchies and spatial distribution of cities within national urban systems. In smaller countries variations in the locations of, and distances separating, urban centers may not be that significant, but in larger continental-scale countries those variations are highly significant. How, for example, does the spatial distribution of cities in an urban system influence growth rates within that system? How do such differences in the macrogeography of urbanization influence our approach to urban definitions and classifications? And, how do we overcome these differences in establishing an international classification system?

Focus: People, Territory or Life Style?

Urban analyses are also complicated by the fact that there is frequent confusion in the literature between the identification of urban and rural (or nonurban) territories and urban and rural (or nonurban) populations, and between the definition of the urban and metropolitan populations. The two pairs of concepts are not the same. Nonurban can refer to where people live or what they do, or both. A territory described as rural may contain urban populations and settlements; a metropolitan area that is broadly defined on functional criteria will invariably contain zones and populations classified as rural. To illustrate the extent of possible confusion, in both the US and Canada, 'metropolitan areas' – a functional definition applied only to larger urban centers – hold about 78 per cent of their respective national populations. This is almost exactly the same proportion cited for the total urban population which includes settlements too small to classify for metropolitan status (see Table 13.1 for Canada). How can both figures be correct?

Table 13.1 Urban, rural and metropolitan populations of Canada, 1996

Functional area	Populations (000s)			%
	Urban	Rural	Totals	
Metropolitan (CMAs) >100,000	16,707	1,157	17,865	61.9
Census Agglomerations (CAs) 10,000-100,000	3,704	881	4,585	15.9
Large urban subtotal (CMAs+CAs)	20,410	2,038	22,449	77.8
Small town (<10,000) and rural	2,050	4,347	6,397	22.2
Totals	22,461	6,386	28,847	100.0
%	77.9	22.1	100.0	

Source: Statistics Canada.

One explanation lies in the fact that metropolitan areas are functional constructs that use entire urban municipalities or counties as their basic building blocks and thus cover extensive geographical areas.[1] As a consequence, they also contain substantial populations classified as rural – not because they are necessarily engaged in typical rural pursuits, most are not, but because they live in areas designated as rural fringe. The latter designation means they do not meet the minimum population density threshold, and are not adjacent to the urbanized core. But appearances can be misleading. In contrast, non-CMA/CA areas in Canada contain significant urban settlements and populations – those living in small towns and villages that meet the size and minimum density criteria for urban areas but tend to be isolated and have fewer than 10,000 people.

These examples suggest the need to revisit the measurement, indeed the relevance, of density measures. It also confirms the need to clarify whether urban and nonurban labels refer to territory or to people or to type of activities, or to some combination of these attributes. How useful are the partitions of urban-rural and metropolitan-nonmetropolitan under these conditions?

Top-down or Bottom-up Approaches

A related methodological question is whether one approaches the construction of an urban system, or more broadly a settlement system, from the local or microlevel and then aggregate upward, or from the macrolevel and work downward to the local. In the latter case, the researcher might begin with a definition of the national urban system and then disaggregate that system into its various component parts. The level of disaggregation could be based on the purposes and data at hand, but with the requirement that one disaggregation does not exclude any other. The actual construction of components could be based on criteria such as proximity (adjacency), demography, socioeconomic characteristics, life styles and functional linkages.

In the bottom-up case, researchers would start with the smallest possible building block, being the smallest units of the settlement geography. They would then systematically combine these units – again using measures such as proximity, demographic structures, social homogeneity, economic attributes and especially, flows and spatial linkages – into larger regional units, including metropolitan areas and hinterland regions beyond. For small and isolated settlements, there would be few, or no, aggregations involved. For larger metropolitan regions, there would

---

[1] Table 13.1 illustrates the two different but parallel interpretations of the terms 'urban' and 'rural' that are used in Canada. The first is the typical definition of what is urban – a minimum concentration of 1,000 or more population – with everywhere else as rural or non-urban. The second is a functional-based definition of urban areas that comprises all those living in census metropolitan areas (CMAs) and census agglomerations (CAs). This includes all places of 10,000 or more: everywhere else (i.e. rural and small towns) is non-CMA/CA. Of course, CMAs and CAs also contain people classified as rural, and there are urban populations (e.g. isolated small towns with populations between 1,000 and 10,000) that are not included in the CMA/CA definitions. Ironically, the proportion that is CMA/CA and the proportion that is urban are roughly equivalent, but the actual populations are not the same.

likely be many possible aggregations of the building blocks. These aggregations might vary from the narrowly defined urbanized core (the builtup landscape) to the commuter-sheds of the MSAs to larger consolidated metropolitan regions (CMSAs) to even more extensive regions based on other types and cycles (e.g. recreation) of human behavior and interaction (Dahmann and Fitzsimmons, 1995).

Inclusive or Exclusive Approaches

In both the top-down and bottom-up cases a decision has to be made as to whether the approach is inclusive of all of the national territory or whether it excludes some or all nonurban areas. The first approach assigns all territory, settled or not, to one or more of the urban nodes in the system; the second leaves nonurban areas as residuals, thus maintaining the traditional urban-rural dichotomy. The advantages of the former are twofold. First, there is no residualized territory or population. No one is excluded. Second, rural (or nonurban) areas are then fully incorporated into a broader settlement system that is focused on urban nodes as the points of territorial organization. In the Canadian example this approach requires a considerable stretch of the imagination, and presents a challenge to existing data sources, since many of the resource-based and native (First Nation) communities in the far North are extremely remote, accessible often only by airplane.

The disadvantages of the inclusive approach are that the special needs of low-density, nonurban populations and peripheral territories may be overlooked simply because they are incorporated into, and swamped by, the larger system. The obvious disadvantages of maintaining the urban-rural partition are precisely that it tends to isolate nonurbanized regions and populations, and that it perpetuates a simple dichotomy that appears to be increasingly irrelevant. Recent research on the Canadian urban system, for example, confirms the existence of multiple kinds and levels of peripheries rather than a single core-periphery dimension (Bourne, 2000). Many parts of the country's traditional core region have taken on the usual attributes of the periphery (e.g. low income, high unemployment, net outmigration), often because of a decline in basic manufacturing activity, and new core regions have emerged in the far west (Vancouver-Victoria, Calgary-Edmonton). In contrast, many of the country's higher income and well-educated communities, albeit mostly small places, are located in isolated regions in the northern resource periphery. Parallel trends are evident within the US urban system as growth shifts from cities in the northeast manufacturing belt to a diversity of urban nodes in the south and west.

Challenges and Opportunities

At the same time, the explosion of new GIS technologies and data sources allows, indeed challenges us, to undertake research outside the usual constraints imposed by the rigidity of traditional data sources, such as the census. We no longer have to think exclusively in terms of partitioning the settlement landscape into distinct units with clear and uniform boundaries designed to serve all purposes and all regions. Instead, it is possible to allow for individual localities to be simultaneously

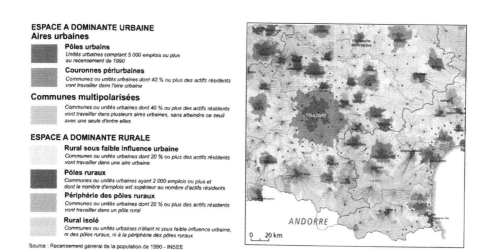

I   INSEE classification of settlement types in the Toulouse area of Southern France

VARIETES DES INSCRIPTIONS LOCALES URBAINES
Catégorie "surfaces artificialisées" dans une classification de l'utilisation du sol d'après les images satellitales de Corine Land Cover

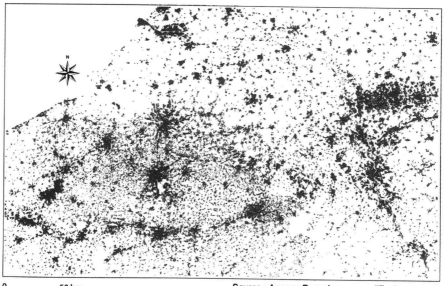

Source : Agence Européenne pour l'Environnement

II   Diversity of European settlements after CORINE

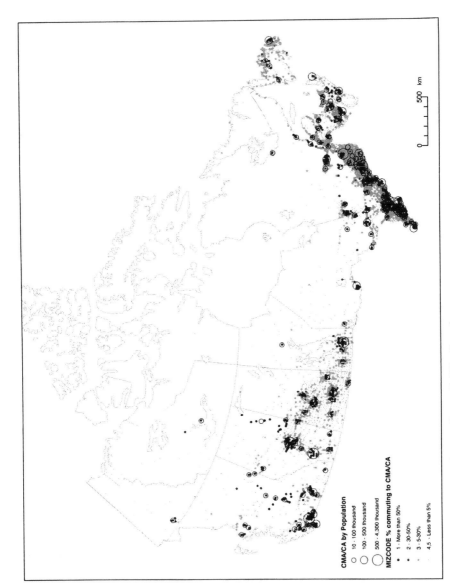

CMA/CA by Population
○ 10 - 100 thousand
○ 100 - 500 thousand
◯ 500 - 4,300 thousand

MIZCODE % commuting to CMA/CA
· 1 - More than 50%
· 2 - 30-50%
· 3 - 5-30%
· 4,5 - Less than 5%

0   500 km

III   Metropolitan Influence Zones (MIZ), Canada

part of several larger settlement regions, with the attributes of these localities taking on the character of those urban regions with which they are most closely connected. In this sense, 'geography' becomes a variable of analysis, not simply a static framework within which we examine differences in urban attributes (Mendelson, 2001).

The challenge in each of these cases is the relative absence of appropriate data at any geographic scale below the nation state, and certainly for the settlement system in general and the metropolitan system in particular. What this requires are not only flexible spatial units, but units that can be constructed and reconstructed as needed. We also require data on economic, demographic and public sector activities for the same spatial units. In particular, we need better data on the wide variety of linkages and interactions among cities, including economic linkages, migration flows, telecommunications and public expenditures and revenues. An additional component of this requirement is improved information on institutions and on corporate structures and behavior. Finally, it is also important that we repackage data and measures used nationally into acceptable international standards to ensure consistency and comparability.

Metropolitan Influence Zones (MIZ): A Canadian Example

One example of recent efforts at measuring the areas of influence of metropolitan areas and smaller urban agglomerations on nonurban regions, based on interaction criteria, is the delineation of Metropolitan Influence Zones (MIZ) by Statistics Canada (McNiven *et al.*, 2000). The approach is interesting precisely because it illustrates the importance and challenge of incorporating geographical settings in settlement classifications, and the unique attributes of the Canadian urban system. In a country as vast as Canada, the entire set of 25 census metropolitan areas (CMAs), with over 100,000 population, and 112 smaller census urban agglomerations (CAs) with between 10,000 and 100,000 population, accounts for nearly 80 per cent of the national population but only five per cent of the nation's territory (Statistics Canada, 1997; Bourne and Simmons, 2002). There has, to date, been no effective method of differentiating and classifying communities located outside of this CMA/CA system – that is, in most of the country.

The purpose of the MIZ methodology is to classify all nonmetropolitan areas outside the urbanized territory in terms of the degree of influence exerted by urban areas. The degree of influence is measured by physical accessibility (e.g. road connections) and by the level of commuting flows into a metropolitan or large urban area. The approach parallels recent studies undertaken in the US using county-level data. It also replicates and then extends an earlier OECD classification of rural areas, but instead it employs interaction criteria (commuting) rather than population density alone, and uses a much finer spatial grid (ICRRC, 1995).

The methodology begins with a set of basic building blocks – in this case some 3,400 Census subdivisions (CSDs, e.g. municipalities, townships, Indian reserves) outside the CMA/CA system – and a series of georeferenced data sets. The analysis identifies the percentages of the employed labor force in each CSD working in the urban core of any CMA or CA. Thus MIZs represent not the zone of influence of any single urban area but rather the role of multiple centers of attraction. Several measures of accessibility were explored, but in the present analysis the emphasis is on commuting data. Based on an empirical examination of significant breakpoints in the frequency distributions of outbound commuters, CSDs were then classified into four discrete categories: those showing strong, moderate, weak and no metropolitan influence (over 30, 5-30, 1-4 and 0 per cent outcommuters, respectively). A separate category for the far North was subsequently added.

In each case the time and distance of commuting were calculated, and the socioeconomic characteristics of the communities in each category were identified and compared. Not surprisingly, the characteristics of places in the four categories differed systematically. For example, average household and family sizes, income levels and house prices were highest in the group characterized by strong metropolitan influence and lowest in those displaying weak or no influence. The differences, however, were not as large as might have been expected from past experience – an indicator of the trend to a convergence of many social attributes of urban and rural (nonurban) areas generally.

The result is a classification of the rural or non-CMA/CA portion of the nation's settlement system based on a universal measure of interaction (commuting to work) and on the degree of connectiveness of each CSD to the core of an urban region. One example of the national pattern is reproduced in Plate III. Essentially the approach identifies the country's ecumene, the settled portion of the country with daily access to urban-based services, housing and employment opportunities. The rest of the territory, based on lower levels of accessibility to the urban system, is considered to be nonecumene.

The disadvantages of this approach are probably obvious. One is that it tends to replicate earlier maps of the local hinterlands of metropolitan areas; second, it still relies on physical access and travel (e.g. commuting) on a daily basis; and third, it does not address the issue of how to deal with areas of overlapping metropolitan influence in the more densely populated parts of the country. The advantages of the approach are also threefold: that it relies primarily on interaction data, it builds upward from the smallest common measurement unit available in the Census of Canada (Census subdivision), and it provides a useful characterization of the nonurban portion (non-CMA/CA) of the country's settled landscape.

Conclusions

There is no single best method of either conceptualizing or defining a settlement system at regional, national, continental or global scales, or identifying the urban components of those systems. The method that is most appropriate in any particular case depends on the kind and quality of information available and the purpose and potential uses of the exercise. Nor are uniform criteria, such as population density, necessarily the most appropriate. A low-density region in one country would qualify as medium or high-density in another. The settlement systems of individual countries are so widely varied, based on history, politics and geography, not to mention level of economic development, that any approach must be both conditional and flexible, as well as readily understood and easily applied. In other words, the criteria of classification and analysis must be contingent on conditions in the setting under study, and be sensitive to differences in spatial scale.

We have argued that the concept of the urban system provides a suitable and practical approach. It attempts to shift our focus from static descriptions of the attributes of individual urban places to the dynamics of those places set within a larger and more flexible systems framework The core of the concept rests on the importance of linkages and interaction – the flows of people, goods, capital, profits, information and ideas among urban nodes. These flows, either tangible or intangible, both reflect and define the relative position and stature of any particular urban area within an urban system. For one area to grow it must expand its linkages to other places and increase the size of its immediate hinterland region.

The urban system concept also offers the potential advantages of flexibility and inclusiveness. The concept can be applied to local or regional urban systems, in which there are several cities in close proximity, as well as to provincial (state), national and eventually global scales. Individual urban nodes can simultaneously be treated as part of several urban systems. New GIS technologies allow us to use more flexible definitions than has been feasible in the past. The boundaries of those systems are porous, and subject to constant change, as is the case in the real world. For individual countries, the urban system can be limited to the urbanized core or continuously settled ecumene, leaving out low-density populations and rural territories. Or it can be inclusive of all populations and territories. In the latter case, all areas would be assigned to an urban node, however distant that node might be. In this framework rural areas become not second-order residuals but integral parts of an urbanized society and of a spatial demography and economy.

The difficulty facing widespread application of the concept is that it requires much better, and more flexible, information sources than are currently available at a subnational level, especially with respect to flows and interactions of all sorts. We also need to be able to integrate information from sources that are traditionally incompatible, such as economic and demographic data, as well as data from public and private sources. One challenge for national and international agencies involved in the comparative assessment of settlement change is to address the issues of what particular kinds of flow information is currently, or should be, collected, in what form, at what spatial scale, and by whom. The challenge here is that the interests of (and constraints on) information-gathering agencies and firms are not necessarily the same, indeed, they may be in conflict.

In sum, traditional approaches to the study of urbanization, and specifically to the conceptualization, delimitation and analysis of change in human settlement systems, are now increasingly inadequate and outdated. The scale and rapidity of sociodemographic and economic change have overwhelmed those methods and the concepts that underlie them. We need to rethink our approach, seeking concepts, techniques, measurement criteria and information sources that are both more imaginative and more appropriate to the new realities of the urban process in an era of increasing global integration and uncertainty.

Chapter 14

The Nature of Rurality in Postindustrial Society

David L. Brown and John B. Cromartie[1]

Urbanization is a dynamic social and economic process that transforms societies from primarily rural to primarily urban ways of life. Few would dispute this definition, but how useful is it for examining the spatial reorganization of population and economic activities in postindustrial societies where, by any definition, a large majority of people, jobs, organizations, and power are already urban? The answer to this question hinges on our ability to distinguish that segment of the population still at risk of becoming urban.

At the most basic level, urbanization cannot be understood without also examining the nature and extent of rurality. Notions of what constitutes urban and rural communities that were formulated in the era of industrialization may no longer offer a reliable lens through which to view contemporary settlement structures. In particular, guidance is needed for examining places that fall into the rapidly-expanding intermediate zone between what is clearly urban and clearly rural.

Hence, our purpose in this chapter is two-fold. First, we argue that the rural-urban dichotomy, whether defined traditionally or using the metropolitan-nonmetropolitan division, needs to be further subdivided, and that newly modified, official US classification schemes represent an important step in recognizing rural diversity. Second, we present a multidimensional approach for conceptualizing rurality that reflects economic, institutional, and cultural realities, alongside standard ecological criteria based on population size, density, and accessibility. We feel this expanded approach is needed as rural-urban distinctions in postindustrial society become increasingly blurred. Our case is the United States but we believe that the situation we describe in the US is similar to that in most other postindustrial societies.

[1] The views and opinions expressed in this chapter do not necessarily reflect the views of the Economic Research Service or the US Department of Agriculture. We acknowledge the helpful comments of Calvin Beale, William Kandel, Kai Schafft, Laszlo Kulcsar, Tony Champion and Graeme Hugo. Brenda Creeley prepared the manuscript.

Background

Early social scientists saw urbanization and industrialization as being reciprocally related. One process could not proceed without the other. While most scholars understood that urban and rural were not entirely discrete categories, relatively clear lines could be drawn to distinguish urban from rural communities and distinct ways of life associated with each. In addition, early social scientists were convinced that the transformation from rural to urban-industrial society would be accompanied by a wide range of negative social outcomes. In fact, this concern is generally credited with motivating the rise of the new discipline of Sociology (Marx, 1976; Durkheim, 1951; Weber, M., 1968; Wirth, 1938).

The social and economic organization of community life has been thoroughly transformed by the automobile and other technological and institutional changes since the beginning of the twentieth century. Agricultural communities supported by 50 square-mile trading areas have been replaced by 'Wal-Mart' towns serving 5,000 square miles. Many areas once entirely rural have been absorbed into metropolitan regions, and economic activity within those regions has deconcentrated into suburbs and coalesced into 'edge cities.' In the context of these on-going transformations, traditional classification schemes must be modified in order to maintain a reliable delineation of what is rural. Otherwise, we may not be able to determine whether the level of urbanization is advancing, declining, or remaining constant and so our analyses of population redistribution may bear little connection to the reality of spatial reorganization. The large literature on counterurbanization, to which we are both contributors, may be missing the mark because it depends on data systems and geocoding schemes that reflect a prior era of socio-spatial organization. Hence, we need to carefully scrutinize and expand definitions of rural.

How do we Define what is Rural in Postindustrial Society?

We agree with Halfacree (1993, p.34) that '...the quest for an all-embracing definition of the rural is neither desirable nor feasible,' but we believe that social science can and should develop conceptual frameworks and geo-coding schemes to situate localities according to their degree of rurality. Since rurality is a multidimensional concept, the degree of rurality should be judged against a composite definition that includes key social, economic and demographic attributes. The operationalization of rurality should be flexible enough to differentiate urban from rural, while recognizing and appreciating the diversity contained within each category.

Our approach to defining rurality involves the material aspects of localities, but we acknowledge the validity of other approaches. As many highly-respected scholars have observed, rurality can be defined as a social representation (Moscovici, 1981; Giddens, 1984). As Halfacree puts it, 'the rural as space, and the rural as representing space' should be distinguished (1993, p.34). We do not propose to debate the relative merits of the material and representational approaches here (see Chapter 15 for a fuller discussion). Both have respected traditions in social science. We feel that they are complementary rather than competitive. As Martin Lewis has observed, 'In the end, only by combining the insights of the *new* geography with those of the traditional approaches may human relatedness be adequately reconceptualized' (1991, p.608). However, we emphasize the socioeconomic approach because of its utility for informing statistical practice essential to the quantitative empirical study of urbanization.

Our multidimensional approach elaborates and extends earlier work by Cloke. Cloke's (1977) objective was to develop a quantitative statement of rurality that could be used as a basis for comparative studies among rural areas, and between them and urban areas. He used principal components analysis to identify nine 1971 Census variables associated with rural-urban location. Principal components loading scores were then used as weighting criteria to form an index of rurality. The resulting scores were arrayed in quartiles ranging from *extreme rural* to *extreme nonrural*, and each of the administrative districts of England and Wales was assigned to one of these four categories. In a second study that updated the index using 1981 Census data (Cloke and Edwards, 1986), it was found that the variables differentiating rural from urban areas then were somewhat different from those in the initial analysis. In particular, population decline and net out migration were important rural attributes in 1971, during a period of population concentration, but not in the 1981 analysis after the relative rates of rural-urban population change and net migration had reversed in favor of the periphery. This suggests that the nature of rurality had altered to some extent over the intervening decade.

Why do we Need to Know what is Rural in Postindustrial Society?

It is important to point out at this juncture that neither Cloke nor we are geographic determinists, i.e., we do not contend that the type of environment people live in has an independent causal effect on their attitudes and behavior. On the other hand, we believe that spatial locality is more than simply a setting in which social and economic relationships occur: a person's place of residence in a nation's settlement system can shape social and economic outcomes and have a profound impact on life chances (Brown and Lee, 1999). While a growing number of social scientists agree that space should be incorporated into social theory and research, there is little agreement on the manner in which space enters into social behavior. The debate hinges on the question of whether spatial arrangements are an elemental cause of social behavior, or whether space acts in a more contingent manner. Our position is consistent with the latter view: that space has an important but contingent causative role in social relations. Hence, we see value in distinguishing rural from urban areas because variations in socioeconomic status, for example, can only be understood by taking into account how characteristics of rural and urban places modify the access to opportunities. As geographers have long been arguing, local social structure contextualizes social and economic behavior (Massey and Allen, 1984; Johnston, 1986). For example, research shows that education is positively related to income in all locations, but the strength of this relationship varies across local labor markets depending on their industrial and occupational structures. Education matters everywhere, but returns to education are higher in some spatial contexts than in others depending on the availability of well paying jobs and on the nature of the stratification system (Duncan, 1999). Hence, while the relationship between education and economic rewards is fundamental, it can be modified by spatial variability in social and economic context.

In addition, what we believe about rural people and communities sets the agenda for public policy. The American public, for example, holds a strong pro-rural and/or anti-urban bias that provides continuing support for agricultural and rural programs (Kellogg Foundation, 2002; RuPRI, 1995; Willits *et al.*, 1990), and quite possibly promotes population deconcentration (Brown *et al.*, 1997). However, research has demonstrated that this pro-rural bias is based on nostalgic positive images of rural places, and a misunderstanding of the social and economic realities of rural life (Willits *et al.*, 1990). What people value in rural communities is often formed 'at a distance,' through literature, art and music, not through actual experience. As John Logan (1996, p.26) has observed, 'A large share of what we value is the mythology and symbolism of rural places, rather than their reality.' Accordingly, more reliable research-based information about the social and economic organization of rural areas, their role in national society, polity and economy, and their relative share of a nation's population and economic activity will provide a stronger basis for public policy. Bringing beliefs about rural areas into closer connection with empirical reality will improve the fit between rural problems and opportunities, public priorities, and the targeting of public investments.

Changing Rural Definitions in the United States

In defining what is rural in a rapidly urbanizing world, difficulties arise primarily in the intermediate zone along the constantly-shifting urban-rural boundary. There are two thresholds to select in drawing a line through this intermediate zone: the city-size threshold determines what size a city or town must be to be considered urban, and the edge threshold determines where the city or town ends and the rural countryside begins. In the context of rapid urbanization, it is necessary to adjust the city-size threshold upward to mark changing city-size functions. A metropolitan center of 100,000 people now provides central-place functions that once were available in the county seat town of 10,000 or less. Similarly, as suburbanization proceeds, it is necessary to move the edge threshold outward to reflect urban incorporation of previously rural territory. Changes to official rural-urban definitions, and changes in the use of them by researchers and policymakers, reflect an attempt to keep up with the upward and outward thrust of urbanization.

In the United States and some other postindustrial countries, two residential categorizations have evolved side-by-side: rural and nonmetropolitan. Even though their respective shares of the nation's total population have tracked quite closely during recent decades, nonmetro (as we will refer to it from now on) is a newer and quite different concept. In light of rapid urbanization over the past several decades, it was inevitable that nonmetro counties came to replace rural territory in defining what is 'rural' in most research and in many policy contexts as well. The choice of using nonmetro to study rural issues was originally a practical consideration because nonmetro areas are comprised of counties, and a broader array of data is available at the county level than for lower levels of aggregation. However, it eventually led to a changed perspective on rural that, in effect, caused both an upward and outward shift of the urban-rural boundary. The rural category is still an important and viable classification, not for defining 'rural' necessarily but for defining the 'built-up' area in studying land use issues such as urban sprawl (Heimlich and Anderson, 2001; see also Table 13.1 and associated comments by Bourne and Simmons in Chapter 13).

Limitations of Rural and Nonmetropolitan Definitions

Before 2000, rural areas in the US were defined as places (incorporated and unincorporated) with fewer than 2,500 residents plus open territory, except where such areas are included in Urbanized Areas. Urbanized Areas consisted of one or more places that, together with the adjacent, densely-settled territory, had a minimum population of 50,000 persons. The density criterion for the surrounding census blocks or block groups was 1,000 people per square mile. Urbanized Areas appropriately classified as urban the 'spillover' population from large, rapidly suburbanizing cities where the built-up territory reached beyond municipal boundaries. However, they left as rural any population in similarly built-up areas situated further away from these cities.

Before 2000, nonmetro areas were defined as counties that fell outside (1) core counties with one or more central cities of 50,000 or more residents, or an urbanized area (and a total metropolitan population of 100,000 or more), and (2) fringe counties that were economically tied to the core counties (as measured by daily commuting). The shift to a nonmetro perspective on what is 'rural' pushed the size threshold up from 2,500 to 50,000, and pushed the edge threshold out from the built-up area of a city to the county edge of a larger, economically integrated region. The difference in the two concepts is large enough that, in 1990, nearly half of the rural population lived in metropolitan areas. The nonmetro population was only two-thirds rural in 1990.

The problem with nonmetro as 'rural' before 2000 was not the often-heard criticism that it was a residual category, left over after metropolitan areas were defined. Rather, it was a problem of having a large and diverse territory left undifferentiated. Nonmetro counties cover 80 per cent of US land area and counties range in size from fewer than 100 to over 150,000 people. As early as 1975, the US Department of Agriculture was recommending that the nonmetro category be disaggregated according to the degree of urbanization, in effect creating a multidimensional view of 'rural' by combining the metro-nonmetro and urban-rural classifications. In a major publication released in that year, Hines, Brown and Zimmer showed that more populous nonmetro counties, especially those adjacent to metropolitan areas, were more similar to metropolitan areas than to their nonmetro counterparts. Since then, the Department's Economic Research Service has created a series of demographic, economic, and social typologies aimed at capturing the diversity of nonmetro counties (Cook and Mizer, 1994; Butler and Beale, 1994; Ghelfi and Parker, 1997; McGranahan, 1999). These classifications are widely used for conducting research, developing public policy, and implementing government programs.

Recent Changes to Rural and Nonmetropolitan Definitions[2]

The US Census Bureau now classifies as rural all territory outside Urbanized Areas (UAs) and Urban Clusters (UCs). Both UAs and UCs encompass densely settled territory, consisting of core census block groups or blocks that have a population density of at least 1,000 people per square mile and surrounding census blocks that have an overall density of at least 500 people per square mile. In addition, under certain conditions, less densely settled territory may be part of each UA or UC. The difference between UAs and UCs is in their total population size. UAs remain as entities of 50,000 or more people, while UCs contain at least 2,500 people but fewer than 50,000 people. Whereas, before 2000, place boundaries were used to determine the rural classification of territory outside of UAs, rural is now a completely statistical concept based on population size and density. Not only are densely-settled, spillover suburbs outside smaller cities now removed from rural, but sparsely settled territory within places of 2,500 or more is now classified as rural.

Under the new rules, the total urban population was 222 million in 2000, 192 million in 452 UAs and 30 million in 3,158 UCs. The rural population was 59 million, two million less than in 1990. Data are not yet available to identify the effect of the criteria change on how people are classified. Rural population most likely would have grown by 3-5 million in the 1990s under the old criteria, so a 5-7 million net shift of population from rural to urban due to changing criteria is a best guess. The city size threshold of 2,500 people or more was not moved upward. However, by removing the connection to place boundaries and lowering the density criteria to 500 people per square mile, new rules once again caused an appropriate outward shift in the edge threshold between rural and urban.

The Office of Management and Budget (OMB) also modified its metropolitan area classification in 2000, to recognize that both metropolitan and nonmetro territory can be integrated with a population center. OMB has instituted a Core-Based Statistical Area (CBSA) system that establishes a *micropolitan* category as a means of distinguishing between nonmetro areas that are integrated with centers of 10,000 to 49,999 people, and nonmetro territory that is not (OMB, 2000b).[3] CBSAs based on 2000 census data are not yet released, but it is possible to develop a 1990 version of CBSAs by applying the already-published criteria to 1990 data. Those results show that 1990 CBSA-based metropolitan counties contained 79 per

---

[2] For a fuller account of the new US metropolitan standards, see Chapter 18.

[3] Social scientists have objected to the use of counties as building blocks for the nation's metropolitan geography (Morrill *et al.*, 1999), but the new OMB standards have retained counties in the new classification system.

cent of the US population in 2000 and 21 per cent of its land area, while the percentages are exactly reversed for nonmetropolitan territory (Table 14.1). The nonmetro population is almost evenly split between micropolitan and non-NCBAs, although the former category contains 582 counties while the latter has 1,668.

Comparison of Core-based Statistical Area Categories

The data in Tables 14.1-14.3 show substantial diversity between micropolitan and non-NCBAs, and demonstrate the importance of distinguishing between these two types of counties. To begin with, the average micropolitan county had 50,923 persons in 2000 compared with only 18,581 persons in the average non-NCBA. The data in Table 14.1 also show that micropolitan counties had 47 persons per square mile, while only 14 lived on each square mile of non-core-based territory.

Table 14.1   Population, land area and density by CBSA[a] category, 2000

CBSA category	No. of counties	Population		Land area (sq miles)		Population per sq mile
		000s	% US	000s	% US	
US	3,141	281,422	100	3,536	100	80
Metro	891	220,792	79	737	21	299
Large[b]	606	193,228	69	488	14	396
Small[c]	285	27,565	10	249	7	111
Nonmetro	2,250	60,630	21	2,799	79	22
Micro	582	29,637	11	625	18	47
Non-CBSA	1,668	30,993	11	2,174	61	14

[a] 1990 version of CBSA categories.
[b] 250,000 persons or more.
[c] 50,000-250,000 persons.

Source: Economic Research Service, US Department of Agriculture, using data from US Census Bureau.

Table 14.2 compares social and economic characteristics of persons living in various types of US counties. In each instance these data show *regular patterns of decline* as one moves from the largest metropolitan counties to non-core-based counties. For example, 55 per cent of all metropolitan persons attended college compared with 41 per cent of nonmetro residents, but only 39 per cent of non-CBSA adults had been to college compared with 44 per cent of persons living in micropolitan counties. Metropolitan workers are more dependent on jobs in service industries while their nonmetro counterparts depend more heavily on farming and manufacturing, although these differences are not strikingly large. Within the nonmetro category, however, dependence on farming is over twice as high in non-CBSAs compared with micropolitan areas, and consistently smaller percentages of non-core-based employees work in manufacturing, retail and services jobs. Similarly, professional, technical managerial and administrative occupations comprise a much larger share of metropolitan than nonmetro jobs, and a larger share in micropolitan than in non-core-based counties. Data on earnings per job (displayed in the bottom panel of Table 14.2) show that non-CBSA workers earn less than their micropolitan counterparts, especially in services where recent job growth has been concentrated.

Table 14.2 Comparative profile of metropolitan, micropolitan and non-core-based counties,[a] US, 2000

Characteristic	Metropolitan			Nonmetropolitan		
	Total	Large[b]	Small[c]	Total	Micro	Non-CBSA
Educational Attainment						
% Less than High School	18	18	18	24	22	25
% High School	27	26	31	35	34	36
% College	55	56	51	41	44	39
Industry of Employment (selected)						
% Farm	1	1	2	6	4	9
% Manufacture	11	11	15	15	16	15
% Retail	16	16	18	17	18	16
% Other services	33	34	29	24	25	23
Occupation of Employment (selected)						
% Management, Professional	35	36	31	27	28	26
% Sales, Office	27	28	26	24	24	23
% Construction, Maintenance	9	9	10	12	11	12
Earnings Per Non-farm Job						
All Jobs ($000)	37	38	29	25	27	24
Manufacturing ($000)	51	53	37	34	37	32
Retail ($000)	20	20	16	15	16	15
Other services ($000)	33	34	25	20	22	14

[a] 1990 version of CBSA categories.

[b] 250,000 persons or more.

[c] 50,000–250,000 persons.

Source: Economic Research Service, US Department of Agriculture, using data from the US Census Bureau and the Bureau of Economic Analysis.

We have also examined whether micropolitan areas are more 'metropolitan' than non-CBSAs with respect to the presence of various services and facilities typically associated with metropolitan status (Beale, 1984). Cornell University conducted a mail survey of the heads of county government in a 10 per cent random sample of non-NCBAs, and in 20 per cent of the central counties of micropolitan and small metropolitan areas. Our overall response rate was 75 per cent, and response was essentially equal across the three county type samples. These survey data reveal that central counties of small metropolitan areas are clearly differentiated from both nonmetro categories. In all twelve instances the presence of these 'metropolitan functions' is most prevalent in small metropolitan counties, and least available in non-NCBAs. Micropolitan areas, however, appear to be more similar to small metropolitan areas than to non-NCBA counties.

Table 14.3   Presence of services and facilities by county type, 2000[a]

Service or facility	Per cent provided in county		
	Small metro[b]	Micro	Non-CBSA
Scheduled Passenger Air Service	52	23	7
Scheduled Inter County Bus Service	97	69	38
Local Bus Service	97	59	30
Museum[c]	73	52	25
Daily Newspaper	97	88	22
National or Regional Hotel Franchise	100	94	41
Four Year College	81	40	10
Library with Multiple Branches	78	57	34
Commercial Television Station[d]	73	37	11
General Hospital[e]	100	100	72
(N)	(33)	(92)	(129)

[a]   Ten per cent sample of non-core-based counties; 20% samples of small metropolitan and micro counties. Current response rate = small metro: 73%; micro: 74%; non-core: 77%. 1990 version of CBSA categories.
[b]   50,000-250,000 persons.
[c]   Art, science or natural history with focus beyond local county.
[d]   With local news and advertising.
[e]   With at least two of the following services: emergency room, physical therapy, cardiac care/MRI.

Changes in official US classifications seem to be steps in the right direction. The Census Bureau's new urban-rural definition does a better job distinguishing the settlement system's built-up area. OMB's Core-Based Statistical Areas continue the metropolitan and nonmetro distinction, but have added the micropolitan and non-NCBA distinction within the nonmetro category. While we applaud these changes toward recognizing rural diversity, we now go on to recommend that social science researchers examine further the multidimensional nature of rurality in order to enhance understanding of the extent of urban and rural settlement and urbanization in postindustrial societies, and to guide future modifications of official statistical geography.

A Multidimensional Approach to Conceptualizing Rurality in Postindustrial Societies

Data from censuses and other nationwide surveys provide significant opportunities for inquiry by university-based and government researchers into the extent and nature of rurality in postindustrial societies. In effect, analysts can design their own residential categorization schemes to examine various aspects of settlement structure and change, with innovative research eventually contributing to changes in official statistical practice. The previous section demonstrated that it is not necessary to be restricted to an unchanging, undifferentiated perspective on the rural. What follows is a proposal to augment the definitional dimension based on ecological factors with economic, social, and cultural variables so as to better distinguish urban from rural and to better understand the variability of social and economic organization, especially in the broad zone between the extremes of urban and rural.

As mentioned earlier, Cloke's approach to defining rurality in Great Britain was largely inductive. His choice of variables was not shaped by a clearly defined theoretical framework for distinguishing rural from urban, although they were suggested by the literature as being important aspects of the socio-spatial environment. Neither do we claim that our approach emanates from a well-crafted theory of rurality, but we do start with a clear premise about four distinct dimensions that comprise rural environments in postindustrial societies. The concept of rurality we are proposing involves ecological, economic, institutional, and sociocultural dimensions. In this section of the paper we discuss each of these four dimensions in turn, and propose a set of indicators that could be used to empirically develop a composite measure of rurality. We follow Willits and Bealer (1967) in observing that a composite definition of rurality involves both the attributes of rural areas themselves, and the attributes of persons residing in such areas. For each of the four dimensions, Figure 14.1 shows the relevant indicators and the contrasting rural vs. urban situations for these. Our approach indicates the attributes that define rurality, and it does so in a comparative framework vis-à-vis urbanity. Moreover, our contention that rurality should not be treated as an *undifferentiated* residual complements the social-representational approach in which rurality is defined by how people imagine rural life in everyday discourse. Both approaches focus attention on the complexity of contemporary rural life and its continuing distinctiveness in comparison with urban areas.

The Ecological Dimension

Population size, population density, spatial situation within a settlement system and natural resource endowments are included in this dimension. As indicated earlier, conventional statistical practice typically emphasizes this approach. Urban-rural delineations are defined by size and density thresholds, while metropolitan-nonmetro delineations use size and density criteria to identify central cities and employ measures of geographic access such as commuting to signify the interdependence of peripheral areas. Hope Tisdale's (1942) influential article provides one of the clearest theoretical statements for the size/density delineation, while central place theory is the primary theoretical basis for considering geographic location vis-à-vis other places in a settlement system (Berry, 1967).

Dimensions of Rurality	Rural areas or populations are more likely to be:	Urban areas or populations are more likely to be:
Ecological Dimension Population Size Population Density Situation in Settlement System Natural Environment	Small Low/Scattered Peripheral Rich in Natural Resources	Large High/Concentrated Central Lacking Natural Resources
Economic Dimension Dependence on Industrial Activities Size of Local Economy Diversity of Economic Activity Autonomy of Local Economy	Extractive Nondurable Manufacturing Consumer Services Small Workforce Small Establishments Undiversified Low/Dependent	Producer Services Professional Services Durable Manufacturing Large Workforce Large Establishments Diversified High
Institutional Dimension Local Choice Public Sector Capacity	Narrow/Constrained Limited/Modest	Wide High
Sociocultural Dimension Beliefs/Values Population Diversity	Conservative Homogeneous	Progressive Heterogeneous

Figure 14.1 A multidimensional framework of rurality in postindustrial society

The ecological dimension also includes a consideration of the natural environment. As shown in Table 14.1, 79 per cent of land in the United States lies outside the officially recognized metropolitan areas, and 61 per cent is located in the non-CBSAs. While this tells volumes about density, it also indicates that most of America's natural resources are located in its rural territory. Energy, minerals, land for agricultural production, water, and habitat for wild life are all found disproportionately in the rural sector, and together constitute an important aspect of the nation's rurality during the postindustrial era.

The Economic Dimension

This dimension concerns the organization of economic activity in local economies. It focuses on what people do for a living, the size and composition of local economies, and the linkages between local economic activities and national and global capital. Until the mid twentieth century, the rural and agriculture – while not synonymous – were very closely related, and definitions of rural were heavily influenced by measures of dependence on agriculture and other extractive industries. Rural economies were small and undifferentiated both in terms of establishments and workers, and localities had a relatively high degree of economic autonomy.

Many people continue to view rural areas through this archaic lens, even though local economies have been fundamentally restructured during the past 50 years. Direct dependence on agriculture, forestry, mining and fisheries has declined to less than one in ten nonmetro workers although extractive industries continue to dominate economic activity in particular regions of the US (Cook and Mizer, 1994). There is no denying that economic activities in rural and urban America have become much more similar since World War II. Not only has dependence on extractive industries declined throughout the country, but so has dependence on manufacturing, with most economic growth now being accounted for by services. However, the jobs available in rural labor markets continue to be significantly different from urban jobs. Rural manufacturing is more likely to be nondurable than urban manufacturing, and well paying producer services jobs are seldom available in rural economies. Moreover, research shows that full time rural workers earn less than urban workers regardless of their industry of employment, and that rural employment is significantly more likely to be part time and/or seasonal (Gale and McGranahan, 2001). While these rural-urban differences in employment do not adhere to the traditional farm-nonfarm contours, they show that opportunities available in rural labor markets are clearly inferior to those available in urban America, and that rural and urban areas can be differentiated with respect to how people make a living.

Rural economies have traditionally been smaller than urban economies in terms of number of workers, the number and size of establishments, and the gross value of products or services sold. Of the four indicators of rural economic activity, this one has changed the least over time even though the decentralization of urban based branch plants has brought some large employers to particular rural areas. Moreover, rural economies have been much more dependent on one or a few types of economic activity than urban economies, and this lack of diversity remains an important rural-urban difference.

The 'protection of distance' enjoyed (or suffered) by rural economies has clearly diminished in recent decades, leading to a reduction in their autonomy. Technological changes including all weather roads, the interstate highway system, virtually universal telephone service (now including cell phones), and the internet have greatly reduced rural isolation. This is not to deny that some important inequalities in transportation and communication infrastructure persist between rural and urban areas, but for the most part the effect of physical distance has been substantially leveled by technological advances.

Institutional changes, especially the increased mobility of capital, have further diminished rural economic independence. The deregulation of banking means that capital now flows easily to and from metropolitan bank centers and the rural periphery. This has both positive and negative implications for particular rural communities, but the clear result is that rural economies are increasingly integrated within national and global structures (see also Chapter 13). With this change comes a resulting decline of local autonomy and increased dependence on extra-local firms and organizations. This makes rural areas at the same time more attractive sites for certain types of external investment, and more likely to lose traditional employers because of financial decisions made elsewhere. There is little room for sentiment in the globalized economy, including sentiment for rural communities as valued 'home places'. When the bottom line demands it, capital flows across national borders to production sites with low costs and few regulations, locating and relocating according to the demands of the market.

The Institutional Dimension

Communities are institutionalized solutions to the problems of everyday life. Accordingly, some social scientists view communities as configurations of institutional spheres including education, religion, governance and the economy. (Rubin, 1969). While we do not necessarily subscribe to this functionalist view of community organization, there is no denying that institutions are a critical aspect of local social structure, and that human beings would have little use for communities if they did not serve recurring needs. Both urban and rural areas have formal institutional sectors. Most places have some form of politics and local governance, organized religion, education, and voluntary and service organizations. Moreover, as discussed in the preceding section, sustenance and economic activity are important aspects of locality.

Rural and urban areas are not so much differentiated by the presence or absence of particular types of institutions as by their diversity and capacity. For example, schools, churches and newspapers are widespread, but most rural communities offer a narrower range of choices as to where one's children may be educated, where to worship, and the media from which one obtains local news. School consolidation in rural America has resulted in fewer and larger schools. Students are often bussed long distances to school. Similarly, while churches are present in most rural communities, the range of denominations and congregations is narrow. Clubs, service organizations, and voluntary associations are also an important part of rural community life, but the choice of organizations to join is more constrained than in urban environments.

Rural institutions also tend to have more limited capacity than their urban counterparts. Rural governments, for example, are often constrained by part-time leadership, insufficient fiscal resources, ineffective organizational structures, limited access to technical information and expertise, and limited ability to assess changing community needs (Kraybill and Lobao, 2001; Cigler, 1993).

The Sociocultural Dimension

Moral traditionalism is one of the most consistent themes subsumed under the term 'rural culture' (Willits and Bealer, 1967). Rural persons are often considered to be more conservative than their urban counterparts, and data from national surveys certainly indicate this for the United States. Calvin Beale (1995) has shown that 49 per cent of rural respondents to a 1993 National Opinion Research Center (NORC) national survey regard themselves as religious fundamentalists compared with 33 per cent of urban respondents. Similarly, a much lower percentage of rural respondents believe that abortion should be available for any reason (26 per cent vs. 44 per cent), and a much higher percentage of rural persons believe that homosexuality is immoral (84 per cent vs. 62 per cent). Beale also observed that rural voters have been more likely to support conservative candidates in recent elections even though rural persons are slightly more likely than urban persons to describe themselves as belonging to the Democratic party.

A related idea is that rural conservatism is often associated with the homogeneity of the rural population. According to Wirth (1938), increased population diversity was one of the dominant effects of urbanization, and one of the reasons why informal social control was likely to break down in cities. Ironically, Fischer (1975), among several critics of Wirth, argued that ethnic diversity, rather than contributing to a weakening of the social order, was a main reason why the strength of social relations did not diminish in cities, and why community was not 'eclipsed' in urban environments. While the association between ethnic and other aspects of population diversity and social and political attitudes is still an open question, research clearly indicates that rural populations in the US, while increasingly diverse, remain significantly more homogeneous than urban populations (Fuguitt et al., 1989). In addition, the rural population's racial and ethnic diversity is not spread evenly across the landscape, but tends to concentrate in particular regions and locales (Cromartie, 1999). Hence, even though about one out of ten rural Americans is African American, few rural communities are 10 per cent Black. Rather, Blacks tend either to comprise the majority or large minority of a rural population or an insignificant percentage. The same tends to be true with respect to other racial and/or ethnic populations.

Much has been written to suggest that primary social interaction is more prevalent and more intense in rural areas, and that rural areas have a higher level of informal social control than is true in urban areas. However, these contentions, if ever true, are not supported by contemporary empirical evidence. Copious research has shown that urban persons are involved in regular and intense interaction with family, friends and neighbors, and that community has not been eclipsed in urban America (Hummon, 1990; Fischer, 1975). Moreover, Sampson et al. (1999), among others, have shown that social networks are quite effective in regulating social behavior in urban locales. Accordingly, primary social interaction and effective social control do not differentiate rural and urban areas in contemporary American society, and are not components of the sociocultural dimension of rurality.

Conclusion

How urbanized are postindustrial societies? How rapidly is the remaining rural population being incorporated within the urban category? How do rural people and rural areas contribute to and/or detract from the social and economic wellbeing of highly developed nations? We contend that answering these questions accurately is contingent on the availability of theoretically-informed definitions of rural and urban areas.

Changes in a nation's urban-rural balance have significance that extends beyond purely academic curiosity. Misinformation about the social, economic and institutional organization of rural and/or urban areas, and about the size and composition of a nation's population living and working in rural and urban places, will result in misinformed policies. For example, if policy makers believe that most rural persons are farmers, agricultural policies will be seen as a reasonable response to rural poverty and income insecurity. Yet agricultural policies will not have much of an effect on rural poverty in postindustrial societies where most rural persons do not depend on farming for their livelihoods (Gibbs, 2001). Similarly, if research indicates that the size of a nation's rural population has held constant over time, as is the case in the United States where about 55-60 million persons has been classified as rural since 1950, then significant public investments for rural development will be legitimized (at least from an equity perspective). But, if the measurement of rurality is too permissive, and the population that is genuinely rural has actually declined, then public resources may be targeted to the wrong populations.

Rural-urban classification in most national statistical systems typically involves only two mutually exclusive categories. It is not that government statisticians do not understand that rurality is a variable not a discrete dichotomy, that the rural-urban distinction is somewhat arbitrary regardless of the population size or density threshold chosen, or that neither the rural nor the urban category is homogeneous. However, given their responsibilities for monitoring basic aspects of social organization and social change, and for providing data tabulations to the public, to businesses, and to other government agencies, the elemental need is to develop a geographic schema that makes intuitive sense, and where between-category variability exceeds internal differentiation. It has not been realistic to expect statistical agencies to adopt a complex multidimensional delineation of rurality given the realities and politics of statistical practice in which budget constraints, and competition between stake holder groups determine which items are included on censuses and other large scale public surveys, and which variables are routinely included in tabulations and data products. However, the development of GIS techniques, together with new advances in small area data collection and availability, suggest that more flexibility and variability in geocoding is possible. Hence, while we do not necessarily expect statistical agencies to adopt our multidimensional approach, we believe that it raises important questions about conventional methodologies for assessing the level and pace of urbanization in highly developed nations.

Chapter 15

Rethinking 'Rurality'

Keith Halfacree[1]

For a subject repeatedly dismissed as a figment of our analytical imagination, the *rural* world has an unruly and intractable popular significance and remains a tenaciously active research domain. ...conceptual disinvestment has never sat comfortably with everyday reminders of the forcefulness of the idea and experience of rurality. (Whatmore, 1993, p.605)

The aim of this chapter is, firstly, to describe and explore the content and the consequences of recent academic discussions of how the rural is defined and understood. As Whatmore notes, in spite of receiving a battering from numerous commentators who feel that the 'rural' is an outdated concept, residualized and perhaps totally transcended by the spatial hegemony of (urbanized) capitalism, it simply does not go away. Indeed, its social and cultural significance today may be as great as it has ever been. The need to consider these definitional debates in this book brings us to the chapter's second aim, namely to think about the placement of a 'rural' category within any conceptualization of human settlement systems of the new millennium. In general, it is argued that we must move away from considering the rural as necessarily 'residual' and see it instead as an integral part of such settlement systems.

In the first section of the chapter the debates surrounding the issue of 'defining the rural' are introduced. A contrast is drawn between attempts to maintain a largely 'material' understanding of the term, rooted in the presence or absence of a relatively distinct rural 'locality', and efforts to dematerialize the concept, through placing it within the realm of the imagination. The second section shows the significance of recognizing the importance of imaginative understandings of the rural with respect to the United Kingdom and other European contexts. The third section builds on the idea that imaginative definitions of rurality are not just of significance at the level of ideas but also have clear material affiliations. This is illustrated by synthesizing the two main approaches to defining the rural into *networks of rurality*. The playing-out of the networks presents alternative models for the future geography of the countryside. Moreover, the fourth section then suggests that imaginative dimensions of rurality incorporated within these networks embody deeper currents/emotions/intentions/anxieties concerning daily life. The concluding section of the chapter draws out the importance of this

---

[1] This chapter is dedicated to the memory of Pierre Bourdieu (1930-2002), whose work demonstrated the everyday importance of making a 'distinction'.

rethinking of rurality for any new conceptualizations of human settlement systems. It presents a range of issues that must sensitize any specific conceptualization that is established.

Defining the Rural

In a critical intervention in the journal *Rural Sociology* in 1982, Gilbert argued that the search for an adequate definition of the rural had been going on for at least 70 years. This search had been hampered by the lack of theoretical sophistication apparent within rural studies relative to its urban counterpart (see also Cloke, 1980, 1989). Instead of dealing with the definitional question directly, attention had focused either on the best ways and techniques for measuring rurality or on mimicking urban work by considering the rural to have its own 'way of life' analogous to Wirth's famous 'urbanism as a way of life'. Both perspectives assumed the rural to be a relatively separate spatial realm. Thus, from the first perspective, we had Cloke's efforts to produce an 'index of rurality' (Cloke, 1977; Cloke and Edwards, 1986; Harrington and O'Donoghue, 1998; see also the previous chapter in this book) and, from the second perspective, anthropological work in the community studies tradition (Symes, 1981; Wright, 1992).

Gilbert's critique remained on the books and, three years later, Cloke (1985) again raised the definitional issue in his introductory editorial to the *Journal of Rural Studies* as a suitable line of academic enquiry. Subsequently, several studies sought to examine the topic in depth, aided this time by a much stronger theoretical armory. In particular, two traditions have emerged (Halfacree, 1993). On the one hand, there are attempts influenced strongly by Marxian political economy that have sought to identify the presence of rural localities. On the other hand, there are efforts more influenced by symbolic interactionist and poststructuralist ideas, which have located the rural more firmly in the imaginative realm. These will be considered in turn.

The Rural Locality

The idea of a 'locality' as a precise conceptual category comes out of structuralist concerns not to fetishise space but to see it as being constantly produced, reproduced and (potentially) transformed. More specifically, localities are spaces inscribed by social processes or, less passively, spaces both inscribed and used by social processes; product and means of production. For rural localities[2] to be identified, at least two conditions have to be met (Hoggart, 1990, p.248):

- There are *significant* processes in operation that have a local scale, and

---

[2] The understanding of locality used here is more theoretically reflexive and correspondingly more restrictive than that promoted by the 'localities research' literature of the 1980s. This is because the latter ran the danger of, *inter alia*, a return to excessive empiricism (Smith, 1987).

- The resulting spatial inscriptions enable us to distinguish a 'rural' from one or more 'nonrural' environments.

As Hoggart (*ibid.*) put it, we must 'identify locations with distinctive causal forces' that we can label 'rural'.

In an era with a *zeitgeist* of 'globalization', it is unsurprising that many authors doubt whether socially significant rural localities can be identified, certainly in the so-called developed world. Indeed, this doubt within rural studies has been longstanding, as expressed in Copp's self-excoriating critique from thirty years ago:

> There is no rural and there is no rural economy. It is merely our analytic distinction, our *rhetorical device*. Unfortunately we tend to be victims of our own terminological duplicity. We tend to ignore the import of what happens in the total economy and society as it affects the rural sector. We tend to think of the rural sector as a separate entity... (Copp, 1972, p.159, my emphasis)

Hoggart reiterated the essence of Copp's argument 18 years later:

> Undifferentiated use of 'rural' *in a research context* is detrimental to the advancement of social theory... The broad category 'rural' is obfuscatory... since intrarural differences can be enormous and rural-urban similarities can be sharp. (Hoggart, 1990, p.245, my emphasis)

It thus seems that we can 'do away with rural' (*ibid.*) as it is a theoretically unsound figure of speech.

The Rural Dematerialized

When Copp described the rural as being only a 'rhetorical device' and Hoggart bemoaned the incautious use of the term in a research context, they were both suggesting another way in which it can be understood, namely as a concept utilized in everyday life or what Sayer (1989) terms a 'lay narrative'. One way in which the rural can be defined as a lay narrative draws on the work of Moscovici and others influenced by the symbolic interaction school. Moscovici (1984) argued that, in order to deal with the perpetual complexity of the world around us, we are forced to simplify it into a series of 'social representations'. In detail, these are understood as:

> organizational mental constructs which guide us towards what is 'visible' and must be responded to, relate appearance and reality, and even define reality itself. The world is organized, understood and mediated through these basic cognitive units. Social representations consist of both concrete images and abstract concepts, organized around 'figurative nuclei'. (Halfacree, 1993, p.29)

Applying the theory of social representations to the issue of defining the rural we can thus say that the rural may fruitfully be described as a 'social representation of space' (Halfacree, 1993, 1995; Jones, 1995). The rural is thus shifted from the material sphere of the locality to the more dematerialized realm of mental space; it becomes a virtual structure. As Mormont (1990, p.40, 22) puts it, even if one doubts the continued existence of any distinct rural society or locality:

> The rural is a category of thought. ... The category [is] not only empirical or descriptive; but it also [carries] a representation or set of meanings, in that it [connotes] a more or less explicit discourse ascribing a certain number of characteristics or attributes to those to whom it [applies]. (*cf.* Copp, above)

The idea that the rural is a social representation of space can itself be dematerialized further. This comes from a critique of both elements within Moscovici's term 'social representation'. First, we can take issue with its *social* character, which assumes a degree of group-specific consensus in composition. Whilst this clearly facilitates both communication and understanding (Potter and Wetherell, 1987; Halfacree, 2001), it is hard to know where to draw the line with respect to the boundary between social and more 'individual' representations (Potter and Wetherell, 1987). Second, we can question the *representational* or cognitive character of Moscovici's concept. For example, Shotter (1993) suggests that our everyday 'conversational realities' attain their fullness through playing themselves out within specific discursive situations, without there being a need for them to be grounded through 'any reference to any inner mental representations' (p.142).

Following such critiques, Potter and Wetherell (1987) suggest that the role of language has been underplayed and propose an alternative concept to social representation, namely the 'interpretative repertoire'. This fundamentally poststructuralist concept is emergent within discourse[3] alone and comprises 'a lexicon or register of terms and metaphors drawn upon to characterize and evaluate actions and events' (Potter and Wetherell, 1987, p.138; Shotter, 1993). Within debates about defining the rural, Pratt (1996) has taken up the interpretative repertoire direction most directly with his call for a more wide-ranging 'critique of unitary notions of the rural' (p.75). For Pratt it is paramount to stress 'the variability, contradiction and variety of representation and articulation of rural discourses' (p.76). Indeed, in an earlier paper with Murdoch he went as far as to call for replacement of the term rural with that of the 'postrural'. This would highlight the rural's constant 'reflexive deployment' and indicate how *'the point is there is not one* [rurality] *but there are many'* (Murdoch and Pratt, 1993, p.425).

Overall, whilst accepting the fundamental contextual fluidity of a dematerialized rurality and the diversity of meanings taken by the term in a whole host of contexts (Halfacree, 1993, 1995), I argue for a middle road between ideas of social representation and the more radically dematerialized alternative. The idea of interpretative repertoires tends towards an overstated idealism, overemphasizing the productive potential of language through doing away with the referent altogether. This seems to miss the tension between the material and the ideational

---

[3] *Not* to be understood 'merely' as language – see later in the chapter.

that is drawn out in the section after next. First, however, some idea of the content of rural social representations needs illustration, plus the association between this and human actions.

Rural Social Representations and their Links to Human Agency

Social Representations of the Rural

Imaginative geographies of space feature strongly within British, especially English, culture and this is a very good place to start when considering the content of social representations of the rural. A strong historical element is apparent here. For example, in Fisher's 1933 essay on *The Beauty of England*, ideas of Englishness, landscape and rurality were intrinsically interwoven (see also Wright, 1985; Short, 1991; Matless, 1998):

> The unique and incommunicable beauty of the English landscape constitutes for most Englishmen [*sic.*] the strongest of all the ties which bind them to their country. However far they travel, they carry the English landscape in their hearts. As the scroll of memory unwinds itself, scene after scene returns with its complex association of sight and hearing, the emerald green of an English May, the carpet of primroses in the clearing, the pellucid trout-stream, the fat kine browsing in the park, the cricket matches on the village green, the church spire pointing upwards to the pale-blue sky, the fragrant smell of wood fires, the butterflies on chalk hills, the lark rising from the plough into the March wind, or the morning salutation of blackbird or thrush from garden laurels. (Fisher, 1933, p.15)

Immediately, we see the contours of an idealized or even an idyllic rural landscape (see Halfacree, forthcoming). This imagination has been updated to the present day into a well-known family of representations: the 'rural idyll'. The contours of this idyll can be defined as:

> physically consisting of small villages joined by narrow lanes and nestling amongst a patchwork of small fields where contented... cows lazily graze away the day. Socially, this is a tranquil landscape of timeless stability and community, where people know not just their next door neighbours but everyone else in the village. (Halfacree and Boyle, 1998, pp.9-10)

Or, similarly:

> The countryside as contemporary myth is pictured as a less-hurried lifestyle where people follow the seasons rather than the stock market, where they have more time for one another and exist in a more organic community where people have a place and an authentic role. The countryside has become the refuge from modernity. (Short, 1991, p.34)

Of course, it is vital to stress that, in spite of its clear historical legacy, the 'rural idyll' today is in no sense some fixed or all-encompassing representation

(Cloke, 1994; Matless, 1998; Pratt, 1996). Geographically, it varies between the parts of Britain. For example, Cloke *et al.* (1998) talk of interlinked and overlapping national-level, regional-level and local-level constructs of rurality. It also varies socially, as reflected in the particular association between the rural idyll and fractions of the middle class, notably the service class, by some commentators (see later). Which elements of the idyll appear of major importance will also be context dependent (for example, in terms of what activity is being undertaken), inconsistent and even idiosyncratic. These last aspects of variation point to the contradictions and incoherence within our discourses so beloved of poststructuralist deconstruction. In no way, therefore, should the 'rural idyll' be regarded as a totalizing representation (see also Little, 1999), as is clear when any two examples are considered in any detail. Nonetheless, a shadowy, elusive, indeterminate core can still be glimpsed.

Elsewhere in Europe, social representations of the rural can both overlap and vary considerably from those dominant within Great Britain. Images of the peasantry – both positive and negative – tend to feature much more strongly than in Britain. For example, Mathieu and Gajewski (2002) argue that France still represents itself very much as a peasant society. For many people in all social classes 'la campagne' (countryside) is a positive representation (Hervieu and Viard, 1996). As with the rural idyll this representation is linked with physical characteristics (pure air, natural environments, fresh smells, scenic landscapes) and with moral qualities (liberty, conviviality, sociability, local democracy) (Mathieu and Gajewski, 2002). Historically, these positive associations were boosted after 1968 by a 'neoruralism' (Léger and Hervieu, 1979). More recently, rurality in France has become associated with the worldwide radical struggle against capitalist globalization. The Confédération Paysanne struggles against declining food quality and rural employment and for the creation of 'sustainable' or 'peasant' agriculture (Confédération Paysanne, 1994). This struggle has been encapsulated in their leader José Bové's ongoing war on 'American' *malbouffe* (crap food) (Bové and Dufour, 2001).

In Germany, by contrast, the very concept of the rural is more problematic. Laschewski *et al.* (2002) consider it to be a 'secondary concept', usually subordinated to ideas such as 'region', 'peasant' or 'periphery'. For example, within regional policy, rurality has tended to be associated with both backwardness and with the national periphery. There is also, as in France, an association with peasant life styles, although again this has been shifted culturally and politically in recent decades from only being valorized through *blut und boden* (blood and soil) echoes of National Socialism to being associated with the post-1968 'alternative' movement.

Taking a slightly different tack, a positive affirmation of the rural as a place of reality, calm and hope emerges from one of the former Czechoslovakia's foremost writers, although exile in the West might have refined this perspective:

> Tereza looked into the farm worker's weather-beaten face. For the first time in ages she had found someone kind! An image of life in the country arose before her eyes: a village with a belfry, fields, woods, a rabbit scampering along a furrow, a hunter with a

green cap. She had never lived in the country. Her image of it came entirely from what she had heard. Or read. Or received unconsciously from distant ancestors. And yet it lived within her, as plain and clear as the daguerrotype of her great-grandmother in the family album... She wanted to tell Tomas that they should leave Prague. Leave the children who bury crows alive in the ground, leave the police spies, leave the young women armed with umbrellas. She wanted to tell him that they should move to the country. That was their only path to salvation...

'What about the country?' she said.

'The country?' he asked, surprised.

'We'd be alone there. You wouldn't meet that editor or your old colleagues. The people there are different. And we'd be getting back to nature. *Nature is the same as it always was*'. (Kundera, 1985, p.168, 170, 233, my emphasis)

Here we can see an Eastern European variant of the rural; a simple, natural and caring country being opposed to the complex, alien and cruel experience of the city. The idea of nature via rurality as a timeless refuge is a theme to which I return in the penultimate section.

Rural Social Representations and Human Agency

In reflective mode, Cloke (1994, p.164) has observed that:

people do make decisions (where to live, where to set up business, where to go for recreation or leisure, etc.) which presume a category rural, whether this is imagined, or experienced in a material sense.

Crucially then, social representations of the rural are not just important passively, as a badge of English culture (*sic.*), for example. They are not just concerned with conventionalization but are also *used* as behavioral resources by human agents. As Moscovici (1984, p.65) put it: 'it is our representations which... determine our reactions, and their significance is... that of an actual cause'. This typically has the effect of bringing social representations 'down to earth'. Such a Foucauldian discursive combination of the mental and the material is expressed well in Shields's (1990, p.31) allied concept of social spatializations. These:

designate the ongoing social construction of the spatial at the level of *the social imaginary* (collective mythologies, presuppositions) *as well as interventions in the landscape* (for example, the built environment). (my emphasis)

For example, besides having a very strong communicative presence within marketing (Thrift, 1989) and the popular media (Phillips *et al.*, 2001), the English rural idyll is directly associated with human behavior. Seeing a role for personal preference (agency), I have argued elsewhere that it is of considerable importance for explaining counterurbanization. This was demonstrated in a case study of migration in the late 1980s to villages in rural England (Halfacree, 1992). Both the 'physical' and 'social' features of the (rural) destination emphasized by the migrants in interviews (Halfacree, 1994, 1995) were almost all well-known features of the rural idyll. Similar views and associations have been described elsewhere in Britain (for example, Cloke *et al.*, 1998; Jones, 1993).

Tying social representations of the rural to material practices – discourse as (at least) language *and* action – begins to suggest the potential for reconciliation between the representational approach to defining the rural and the locality approach. This will now be developed further and fills out the way in which we can rethink rurality.

Networks of Rurality

Towards a Fourfold Model of the Rural

Promoting the use of social representations for understanding rurality is not to argue that this is *the* way in which the rural must be defined. Linking social representations to human behavior and to attendant consequences on the ground *may*, in turn, link into the inscription of rural *localities*. Indeed, to some extent social representations can be seen as the resources through which localities will be produced, as is suggested by Moscovici when he attributes to them causal status. We can thus see representational definitions *complementing* rather than replacing locality-based definitions of the rural. This resists dualistic ways of thinking (Sayer, 1991) that argue that we must choose between two alternatives.

Gray (2000) has recently demonstrated how the European Union (EU) has defined the rural in Europe in this way, as both mental construct *and* locality. He argues that the EU has mixed these 'two modes of conceiving rurality' (p.32), as its concept of rurality has evolved in four phases:

- Mental construct. The EU understood the rural as defined by agriculture. Specifically, when the EU thought of rurality, it thought of a specific form of agriculture: small family farms. Such farms were seen as the bedrock of rural society – everything 'rural' was ultimately built upon them. There was also a strong moral element, with these farms' society being based on hard, honest work, etc.
- Locality. The rural then became a locality informed by this idea of what made the rural a meaningful concept. Boundaries were drawn on maps to define rural areas based on agriculturally related characteristics: topography, resources, farm development potential. It was thought that within these localities a distinct set of processes acted. A '*rural* problem' was defined where there were threats to small family *farming* and, thus, to *rural* society.
- Mental construct. Since the 1970s, the mental construct of the rural held by the EU has changed, as a result of the evolving character of rural places. The rural is now regarded as much more autonomous from agriculture. Heterogeneity of spaces and activities, with consumption (e.g. leisure) as well as production, is the new social representation.
- Locality. Drawing boundaries around the rural can now no longer be guided largely by agricultural criteria. Significant processes that inscribe rural

localities are no longer just agriculturally related. For example, in Highland Scotland, effort has been made to recognize localities according to Gaelic culture and crafting (Black and Conway, 1995).

Seeing a dual character to rurality in this way raises a number of interesting questions concerning the relationship between social representations of the rural and any rural locality, no matter how loosely either are defined. Specifically, to what extent are the two terms mutually self-supporting, and to what extent is there dissonance between them? Both are suggested in Gray's study. This issue can be represented simply and diagrammatically in Figure 15.1, and provokes three questions in particular:

- How much is the gap?
- What is the gap (how do the two definitions differ)?
- To what extent is the gap being closed as a result of people 'using' their mental constructs of the rural to change the rural locality so that it fits better ('imagination' → 'reality')? Or is the 'reality' of the rural locality closing the gap in the other direction ('reality' → 'imagination')?

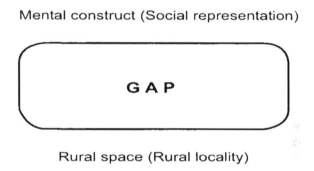

Mental construct (Social representation)

GAP

Rural space (Rural locality)

Figure 15.1   The relationship between social representation and locality definitions of the rural

There is further complexity to be added to Figure 15.1, however. This emerges from recognizing that the idea of social representations of the rural covers a wide range of styles, especially relating to their levels of formality. In my original paper (Halfacree, 1993), social representations were 'lay discourses' associated with the 'accounts' (Thrift, 1986) of the world – in our case, of the rural – held by 'ordinary' people. In contrast, attempting to define the rural as a locality was described as an 'academic discourse', a more distanced and 'theoretical' approach, less informed by the feelings and ideas of ordinary people. However, Jones (1995, pp.37-38) argued that lay discourses must themselves be divided or, rather, that we should see an arena of discourse that lies between the lay and the academic:

In considering the rural, both popular and professional discourses should also be identified and taken into account. ...lay discourses can be separated into externalized discourse, that which is communicated in some way, and personal discourse, the processes of reflection.

Lay discourses are then taken to be *all* the means of intentional and incidental communication which people use and encounter in the processes of their everyday lives, through which meanings of the rural, intentional and incidental, are expressed and constructed. Popular discourses are... those that in some way are produced and disseminated within various cultural structures... Professional discourses are of those whose work is in some way related to the object of discourse [rural locality]... Academic discourses are of those who are studying the object within a discipline, with the aim of understanding and explaining the object of discourse. (my emphasis)

A useful way of building on Jones's critique is through linking it with Henri Lefebvre's (1991, p.33, 38-39)[4] threefold understanding of spatiality, his 'conceptual triad'. First, there are *spatial practices*. These are the actions - flows, transfers, interactions - that 'secrete' society's space, facilitating socioeconomic reproduction. Practices are linked to everyday perceptions of space, and to the rules and norms that bind society together. Secondly, there are *representations of space*. These are formal conceptions of space, articulated by planners, scientists and academics. Representations of space are conceived and abstract but expressed directly in such things as monuments, the workplace and bureaucratic rules. Thirdly, there are *spaces of representation*. These diverse and often incoherent images and symbols are associated with space as directly lived. It is space symbolically appropriated by its users; the 'social imaginary' (Shields, 1999, p.164).

Combining Lefebvre's ideas with those of Jones provides us with a more complex model of rurality, outlined in Figure 15.2. It has four components:

- A *rural locality* inscribed through relatively distinctive spatial practices, and interrogated and described in abstraction largely through academic discourses.
- *Formal representations of the rural* similar to professional discourses but also incorporating a Baudrillardian hyper-real rurality, with the 'map' completely adrift from the territory.
- *Everyday lives of the rural*, incoherent but saturated with popular discourses when the 'culture' underpinning the latter is understood as broadly as possible.
- The *lay discourses of the rural* – the social representations – that Jones saw as '*all* the means of intentional and incidental communication which people use and encounter in the processes of their everyday lives'. These are rooted in but can never be reduced to the aforementioned individual elements relating to Lefebvre's spatial triad.

All four of these elements comprise a *network of rurality* whereby, in echoes of 'actor-network theory', social links acquire 'shape and consistency and therefore some degree of longevity and size' (Thrift, 1996, p.24).

---

[4] For descriptions and elaborations of this schema, see also Harvey (1987, pp.265-9); Merrifield (1993, pp.522-7); Shields (1999, pp.160-170).

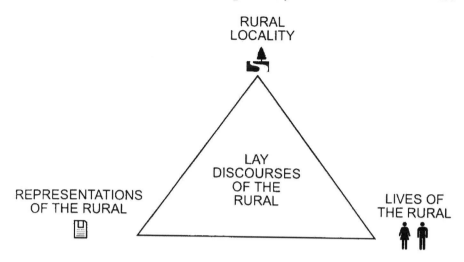

Figure 15.2   A fourfold model of a rural network

A network of rurality only comes about through everyday geographical practises, since these are what bring all the elements together. Moreover, this practising of the rural through the network shows rurality fundamentally to be a site of representational struggle, as suggested by Mormont, Pratt and others (see Cloke and Little, 1997; Milbourne, 1997). This can be illustrated in the British context in terms of the future geography of the so-called postproductivist countryside as we tentatively (see Wilson, 2001) leave the productivist era.

Practising the Rural: The Case of the Postproductivist Countryside

Productivism characterized first and foremost the experience of British agriculture from its post-1945 reconstruction until at least the late 1970s. Defined as positioning agriculture as a production maximizer – a progressive and expanding food-production-orientated industry (Marsden *et al.*, 1993) – it was encapsulated within successive government statements and policies. For example, eager to overcome the boom-bust cycles to which the agricultural industry had been exceptionally prone and to reward the efforts of the farming community during wartime, the 1947 Agricultural Act and the European Community's Common Agricultural Policy subsequently sought to provide a secure base for farmers from a wide range of directions (*ibid.*, pp.59-61). The entrenchment of a corporatist relationship between the agricultural industry and the state (Winter, 1996) reinforced this security.

Crucially, productivism was not just experienced by the farming community but filtered into every corner of British rural life, and beyond. For example, nonagricultural institutions concerned with rural issues tended both to acknowledge and to accept a leadership role for agriculture. Productivism was the glue that consolidated a 'structured coherence' (Harvey, 1985, p.139*ff*.) for rural Britain, holding economy, state and civil society together in a *relatively* stable fashion at the local level (Cloke and Goodwin, 1992).

Thus, a relatively distinct rural network emerged in the post-1945 era centered on a model of 'productivist' agriculture. Each element of this network was, however, diluted and compromised by the remnants of earlier networks. First, a *rural locality* was inscribed through the predominance of particular agricultural practices, plus multifaceted support services, focusing on increasingly industrialized models of food production. This industrial model was tempered, though, by the persistence of less capitalistically rational models, often buttressed by state welfare payments. Second, *formal representations of the rural*, as outlined in official government publications such as the Scott Report of 1943, the 1947 Agriculture Act and the Common Agricultural Policy, nominated, normalized and nurtured the countryside as a food production resource. However, this was overlain by a communitarian and ideological moral stewardship role for farmers and landowners. Third, not least through the connections of productivist agriculture to the wider civil society, *everyday lives of the rural* were lived through this productivist vision. Again, though, this was uneven, as remnants of less productivist agriculture and other economic activities remained. Overall, then, whilst *lay discourses of the rural* were essentially productivist, this was a mediated and heavily moderated capitalist spatiality.

The (pseudo-)productivist rural network today is under immense and probably fatal strain. British agriculture is in 'structural crisis' and cannot be cured by the kinds of 'technical fixes' that rescued it from previous more minor 'conjunctural crises' (Drummond *et al.*, 2000). All the elements of the productivist rural network have been undermined (Halfacree, 1997, 1999). Consequently, Ilbery and Bowler (1998) talk of a 'postproductivist transition', characterized by reduced food production and state support, and the internationalization of the food industry within a more free market global economy, albeit with increased environmental regulation placed on the industry (*cf.* Wilson, 2001). Three 'bipolar dimensions of change are recognized' (Ilbery and Bowler, 1998, p.70): intensification to extensification, concentration to dispersion, and specialization to diversification. However, Ilbery and Bowler are careful not to homogenize and essentialize these trends, recognizing diverse 'pathways of farm business development' emerging (Bowler, 1992).

Once again and crucially, the crisis of productivism and its response in the postproductivist transition is not just of significance to agriculture. Both have a knock-on impact on the everyday contours and textures of rural life. This sense of interconnectedness was revealed during the British Foot and Mouth crisis of 2001-02, which brought the twin but related issues of rural crisis and an increasingly uncertain rural future to the fore. As almost never before, it resonated with an overwhelming sense of confusion about things rural. Established 'certainties' – even as far as it producing food at all – were sharply questioned. For example, government sources suggested that around a quarter of British farms – almost all of them small – will have closed or merged by 2005, with 50,000 people forced to leave the industry (*Guardian*, 2001). Thus, as the productivist network's local structured coherence fails, questions of what is to take its place increasingly come to the fore. What are the new networks of the rural likely to be comprised of?

One response to this question is to imagine alternative maps of potential rural networks in the postproductivist countryside (Halfacree, 1997, 1998, 1999). Each of these differs in some respect from that which was hegemonic in the productivist era. Let us consider four:

- *Super-productivism.* This vision restates productivism, but this time in a much less moderated form. It is apparent within the activities of agribusiness, in the genetic modification of plants and animals, and in biotechnology generally (Goodman and Redclift, 1991). Here, formal representations are shorn of any 'moral' stewardship element, with land treated solely as a productive resource linked to profit maximization. Indeed, such is the physical impact of agribusiness that everyday lived rural spaces have little scope to diverge from the representations that guide and legitimate the industry; the capitalist rationality (*sic.*) of agribusiness impinges in all directions. Overall, a strong rural network is likely to be apparent, exhibiting a relatively high degree of coherence.
- *Rural idyll.* This vision directly opposes – in terms of rural spatiality – super-productivism. It takes and develops the erstwhile 'moderating' element of the old productivist rural representation, namely its moral angle as expressed most strongly through ideas of 'community'. Here, a rural locality may have agriculture as a backdrop but its key spatial practices are consumption-orientated, notably leisure, residence and attendant migration (counterurbanization). The formal representation underlying these practices tends to be the *rural idyll*, upheld by rules such as those of the planning system. However, unlike super-productivism, the extent to which everyday lives in the rural conform to this spatial imagination vary, reflected in battles within the planning arena over what are seen as suitable developments in rural locations (Ambrose, 1992). Whilst in some places there may well be an 'ascendance of certain aesthetic representations of the countryside over previous economic ones' (Murdoch and Marsden, 1994, pp.215-16), this is highly uneven. Overall, therefore, this rural network might be expected to be quite fraught and unstable.
- *Effaced rurality.* As noted earlier, many academics have sought to emphasize just how lacking in distinctiveness the categories 'rural' and 'urban' really are. The rural has, in effect, been effaced by the geographical development of late capitalism. Thus, (formerly) rural areas might be pulled into networks where the rural-urban distinction is ignored, leaving the rural only as a ghostly presence, experienced through folk memory, nostalgia, hearsay, etc. Any talk of networks of rurality thus becomes problematic.
- *Radical visions.* Radical visions have been neglected to date in discussions of the topography of the postproductivist countryside (Halfacree, 1999; see also Bové and Dufour, 2001). By 'radical' I am referring to networks of the rural that have a much more anticapitalist underpinning than those outlined above, since, of course, many of the latter are also radical. More specifically, these networks are likely to involve a rejection of the sanctity of private property, represented in land ownership, and to express a move towards less capitalistic life styles and everyday priorities. They may also represent a transgression of the urban-rural division as conventionally understood. For these radical visions, the rural locality

is likely to revolve around decentralized and relatively self-sufficient liv
patterns. Attendant representations involve the countryside as a diverse ho
accessible to all. Consequently, the lives of the rural should involve a celebrat
of the local and the individually meaningful.

The Importance of the Rural Imagination

[T]he opposition between city and countryside remains, and may now take on r
social significances depending on the ideological or cultural frame of reference
which the agents refer: natural universe vs. urban artefact; a world of sociability vs.
abstraction of large organizations; a world of skills vs. the alienation of industrial wo
Different versions of the urban-rural opposition are currently being construct
(Mormont, 1990, p.41)

The previous section has suggested that both imaginative and more mater
expressions of rurality are experienced through networks of rurality. The content
what I started this chapter by describing as social representations of the rural
such as the English rural idyll – thus needs to be 'taken seriously' for a
appreciation and understanding of the ever-evolving geography of a particu
place. The spatial imagination never remains solely virtual, floating in t
cognitive ether, but is always, if unevenly, grounded. This in itself is a cruc
realization, and is in many respects the spirit behind Whatmore's opening quote.

However, the importance of the rural spatial imagination goes beyond
materialization when we start to ask *why* certain forms of that imagination, such
the rural idyll, tend both to predominate and to be linked to major social tren
such as counterurbanization. Both this content and the related actions can be se
to 'speak' of deeper currents within our everyday lives and experiences (Halfacr
1997). Such expressions can either sit neatly and unproblematically within the
experiences, or they may be read as some form of critique of them. For examp
evidence from counterurbanization can be used to begin to look more deeply
what we might term the 'lure of the countryside' – a lure that appears widespre
within many aspects of contemporary society – and its attendant soc
representations of the rural. There are at least four different ways in which we c
understand this appeal (Halfacree, 1992).

First, we can explore the commonly held belief that the content of c
imaginative rurality and the associated lure of the countryside are reflexive o
'natural' human desire for life within such a milieu. It reflects, in other words
'craving for the expression of man's [*sic.*] unity with nature' (Treble, 1989, p.5
In support of this, one could cite the apparently ubiquitous appeal of rural
throughout history, across space and between classes (for example, Marx, 19(
Hadden and Barton, 1973; Williams, 1973).

There are numerous problems with this deceptively simple explanati
however, not least when we note that everyone does not have such an imaginati
or desire to get 'back' to the rural. However, its major inadequacy is that it fails
consider just what 'human nature' is and how it is shaped by everyday soc

experience. To summarize, naturalization of the lure of the countryside fails to recognize the extent to which:

> actual needs are not spontaneously self-generating in the breasts of individuals, but always exist in the context of a particular social-economic organization and series of institutions and are thus necessarily subject to and moulded by the 'coercion' and 'constraints' of the theory of needs which underlies that organization. (Soper, 1981, p.9)

Thus, Urry (1995a), in his parallel criticism of any naturalization of rural tourism, argues that it takes the cultivation of a *'cultural* desire' for this practice to develop.

This takes us to a second explanation, diametrically opposed to the first, which reduces the content of our rural representations and the corresponding lure of the countryside to the capitalist marketplace. Again, at first sight, this simple explanation seems quite plausible, as there is no doubt that considerable profit is to be made by commodifying rurality. For example, both advertising and marketing have an increasingly 'green' tinge (Thrift, 1989). Hence, we have:

> Commodities wearing vernacular dress – food, paints, furniture, even 'Russell Hobbs' adorn their electric kettle, 'country style', with wheatsheafs; even alarm clocks are marketed ... with flowered borders ... (Samuel, 1989, p.liv)

The 'vagueness' (Wright, 1985) of the content of our rural imaginations is particularly significant here, as has been noted for many years:

> To advertise the purity of brown bread or the natural origins of some prepackaged frozen food, this insubstantial picture can prove extraordinarily potent. A ragged boy clambering up the cobbles of Clovelly bears a wholemeal loaf, the services of a bank are sold via the charms of a small country village where everyone is known to everyone else ...; the advertising men have taken their cue from minor Victorian watercolourists and novelists and are feeding us the same diet. (Darley, 1975, p.236)

Yet, whilst Whatmore (1993) notes how the commoditization of space, including rural space, has extended from physical use values to symbolic value systems (see also Cloke, 1993) – (rural) space as cultural capital (Bourdieu, 1984) – this explanation is also far from adequate. First, it may assume that people are dupes of the market and are unable to exert significant agency. Second, even if such determinism is not assumed then the explanation still falls down as it once again naturalizes the appeal of the rural that has been tapped into by the marketplace. Thirdly, a more postmodern understanding, whereby consumers knowingly and playfully engage with a simulated rurality, still fails to explore just why this simulation takes such a form. Clearly, the market has a role to play in channeling engagement with the rural and in moulding the rural networks themselves, but greater attention still has to be given to why rurality is such a fertile source of profit and can be sold to all kinds of consumers.

A similar criticism can be made of a third explanation for the content of our social representations of the rural and their link with the lure of the countryside.

This refers to the association frequently made between rurality and class. Most specifically, it refers to the association between the reproduction of service class identity and rurality (for example, Urry, 1995b).[5] As Thrift (1987) and others have noted, in order to affirm their position in society – their 'distinction' (Bourdieu, 1984) – the service class engage with specific consumption (as well as production) practices. Rurality features strongly here as the service class encapsulate the idea of an intellectual and subtly distinctive 'cultural nobility' (Thrift, 1987), engaged with a 'symbolic rehabilitation project' (Bourdieu, 1984) for the rural:

> Appropriating 'nature' – birds, flowers, landscapes – presupposes a culture, the privilege of those who have ancient roots. Owning a chateau, a manor house or grange is not only a question of money; one must also appropriate it ... appropriate ... the art of living of the aristocrat, or country gentleman, indifferent to the passage of time and rooted in things which last. (*ibid.*, p.281)

Once again, this explanation has some worth. The service class do seem well set up intellectually to appropriate that elusive thing called rurality, and they certainly have the economic resources to do so. However, there are problems. First, the service class is far from comprehensively 'ruralized' (Hoggart, 1997), being also associated with gentrification, for example. Moreover, service class people are not the only ones that adopt particular understandings of rurality and related practices. In Britain, for example, 'idyllic' ideas of rurality amongst rural residents are not strongly class based (Halfacree, 1992). Of prime significance once again, however, is the failure of this explanation to explore just *what* is it that service class people both invest in and get out of the rural.

The final explanation for the content of our rural representations and their link to actions such as counterurbanization focuses directly on the issue that is skirted over in the other explanations, namely *why* people sign up to representations such as the 'rural idyll'. Specifically, I argue that it reflects a concern with the character and direction taken by contemporary society (Halfacree, 1997), especially what Harvey (1989a) termed the 'postmodern condition'. This condition is characterized by:

> space-time compression, driven by the relentless demands of the capitalist system for capital acceleration and innovation. Temporally, 'flexible accumulation' relies upon accelerating the turnover time of capital, accentuating the volatility and ephemerality of fashions, commodities, the production process, norms, and even values and ideas ... Spatially, communications systems, especially, symbolize the increasing 'annihilation of space through time' ... The power to ascribe the social meaning of space is removed from the people in those spaces and acquired by 'distant forces'. (Halfacree, 1997, p.77)

In order to deal with the 'fierce spatiotemporalities of daily life' (Harvey, 2000, p.237) of this postmodern condition people adopt a wide range of different

---

[5] We could also consider here the links between rurality and an aspiration to aristocratic life styles.

and even contradictory strategies. Some go with the flow and embrace it, whilst others try to deny it point blank. Still others seek a firmer footing from which to deal with its threats and capitalize on its opportunities. Such a strategy requires a range of resources to be deployed in the search for a degree of stability within the dizzy vortex of the postmodern condition. As Harvey (1989a, p.292) puts it:

> The plunge into the maelstrom of ephemerality has provoked an explosion of opposed sentiments and tendencies ... Deeper *questions of meaning and interpretation* ... arise. The greater the ephemerality, the more pressing the need to discover or manufacture some kind of *eternal truth* that may lie therein... The revival of interest in *basic institutions* (such as the family and community), and the search for *historical roots* are all signs of a search for more *secure moorings* and *longer-lasting values* in a shifting world... The *home* becomes a *private museum* to guard against the ravages of time-space compression. (my emphases)

The expressions emphasized in this quote all resonate with both the imaginative content of the rural idyll and with the rationales often given for counterurbanization.

As with ideas of 'nature' (Harvey, 1996), which intersect with those of rurality – as Kundera's Tereza said above, 'Nature is the same as it always was' – the rural is set up as an alternative universe to that of our capitalist postmodern world. It represents a utopianism of both 'spatial form' and 'social process' (Harvey, 2000). As Mormont (1987, p.11) observes:

> The countryside is often considered to be a place where it is possible to put into practice another way of life (from an individualistic perspective), or another model of social and economic organization (in a social world view).

Moreover, as he suggests, just how this imagination gets put into practice varies. The rural may be set up in a reactionary way, used to look back to a supposedly better time and place, as is suggested by many of the impulses behind the heritage industry, for example (Hewison, 1987; Wright, 1985). Or it may be something to retreat into in the present for those with the necessary monetary and other resources. This is suggested in the NIMBY (Not In My Back Yard) 'pulling up the drawbridge' character of much rural 'community' politics, for example (Murdoch and Marsden, 1994; Murdoch and Day, 1998). Or the rural may be envisioned in a more truly utopian way, suggestive of the contours of an alternative existence – urban as well as rural. In this respect, consider again Mormont's 1987 description of rurality as 'a prominent aspiration' (p.19) that can assume a radical edge:

> Rurality is claimed not only as a space to be appropriated ... but as a way of life, or a model of an alternative society inspiring a social project that challenges contemporary social and economic ill. ... Peasant autarky, village community and ancient technique are no longer relics, but images which legitimize *this social project of a society which would be ruralized* ... The aim is not to recreate a past way of life but to develop forms of social and economic life different from those prevailing at present ... (Mormont, 1987, p.18, my emphasis)

The contemporary political significance of rurality appears abundantly clear from this quote, wherever it ultimately leads. The rural must be an essential component, therefore, of any new conceptualization of human settlement.

Conclusion: The Rural within Conceptualizations of Human Settlement Systems

Where, in the light of the discussion given in this chapter, does 'the rural' fit within any new conceptualization of human settlement systems? In answering this question by way of a conclusion, I shall not even begin to suggest any new conceptualization. This is beyond the scope of this chapter and is clearly a major undertaking in its own right, as is readily apparent from this whole book. Instead, I wish to draw attention to a number of points that we need to be aware of and be highly sensitive to when seeking to produce such a classification.

First, and most fundamentally, the rural – or something synonymous – must be a part of any new classification system. Reports of the death of the rural in the West, as Whatmore made clear in the opening quote, have been greatly exaggerated. My placing of the rural in the context of fluid and mutable networks (Figure 15.2) opens up its scope considerably and takes us away from seeing it as an absolute state or 'thing', a reified object that may or may not still be found clinging to life out there on the margins. Instead, the rural is much more alive, mobile and versatile, manifesting itself to varying degrees in any time-space as representation, practice and experience.

Second, and emerging from the first point, the residualization of the rural that characterizes almost all conceptualizations of human settlement systems to date needs to be resisted. None of the sections of this chapter suggest the rural to be in any way residual. This depicts a teleology, albeit often mostly by implication, of the rural place as something that will 'eventually' – the precise date is typically left open – be transformed into a component of a modern urban place. There are parallels here with our understanding of the relationship between so-called less developed and developed countries. Typically, this relationship is seen in terms of 'stage models' of development, which assume we can 'use an historical interpretation of how rich countries became rich as a futuristic speculation of how poor countries can become rich in their turn' (Taylor, 1993, p.9). Taylor calls this assumption the 'error of developmentalism' because it 'completely misses out the overall context in which development occurs' (*ibid.*), failing to see the integral coexistent connection between the two categories, less developed and developed. In our case, the crucial linkage missed is between rural and urban when the rural is seen 'merely' as a residual.

The brief illustration in this chapter of the significance of social representations of the rural in contemporary Europe suggests that at the societal level rurality is far from insignificant with respect to or disconnected from the urban. Indeed, the final section above proposed rurality as presenting a critical commentary on the urban. Moreover, as the discussion of the postproductivist countryside exemplified, rural representations may be integral to the production of a new set of rural spaces. The extent to which these new spaces are distinctive and prominent with respect to the urban environment will depend on the strength and distinctiveness of their rural networks. In this respect, there is clearly much scope for hybrid urban-rural or 'intermediate' spaces – *urbs in rure* or *rus in urbe*, for example. Again, the extent to which these should be regarded as somehow 'transitional' is questioned, since such an interpretation relies on a developmentalist assumption. In-betweenness needs to be seen in a nonevolutionary sense.

Third, and coming out of the network model of rurality – although it is also implicated in the earlier locality and social representation definitions – we must be wary of seeking some kind of 'essence' to the rural as a category. Whilst borders and limits will have to be decided upon for any workable classification of the settlement system, it is important that this is kept as open and fuzzy as possible. Rurality, in other words, may be very different in different places and at different times – it is irredeemably contextual. Moreover, what we try to capture through any classification of settlement systems may well be associated more with one or more element of the network than with the other elements. In principle, the multidimensionality of the rural cannot be distilled into one measure, as Cloke's index attempted, since different ruralities may have little or nothing in common with each other.

Fourth, the rejection of a linear model of the human settlement system with the rural as residual opens up our appreciation of the creative potential of humans for producing the rural. This is most apparent in so-called developed countries and is reflected in Murdoch and Pratt's (1993) concept of the 'postrural'. Thus, one might conclude that today we have the outlines of a 'new' rurality – illustrated in this chapter in terms of conflicts over the production of a postproductivist countryside. Nonetheless, we need to be cautious of such a periodization since, certainly in the modern age, the rural has always been produced in a way that is congruent with the dominant mode of production (Smith, 1984). Thus, for example, in the developing world, the rural may appear, once again, as a residual category but its production and reproduction, and its links with both the urban parts of the same countries and the rest of the world, is increasingly acknowledged (Taylor, 1993). This is, of course, the essence of the uneven geography of capitalist globalization (Harvey, 2000; Tabb, 2001).

Fifth, as the locality perspective on defining the rural most explicitly recognizes, seeing the rural as an important and meaningful category within the conceptualization of human settlement systems is not to suggest some even relatively independent rural world. This is also clear from the previous observation. If urbanization is taken as a reasonable synonym for capitalism in its most highly evolved guise, then rural places are as 'urban' as the city. Hence, the continued appeal of urban systems and other functional approaches to defining and positioning places, as implicated in metropolitan/nonmetropolitan classifications, for example. Such classifications are also useful for exposing any gaps between the 'reality' of rural places and their 'imagined' characteristics, as expressed in Figure 15.1.

In summary, the rural and urban have traditionally been represented as two poles of a dichotomy within otherwise quite divergent conceptualizations of human settlement systems. Although there may be good reasons for keeping this setup for pragmatic purposes in some classificatory systems (for example, see various chapters in this collection), the analysis of the rural today presented in this chapter calls for us to reject, *in principle*, this dichotomous structure. This is the single most important lesson that the UN, national statistics agencies and all bodies concerned with measuring human settlement systems must come to terms with. The urban and the rural as lived networks are not – and never were – *a priori* mirror images of one another. This legacy of our dualistic way of thinking (Sayer, 1991) is hard to transcend, but it is a vital challenge if the importance of the rural today is to be acknowledged fully. More pragmatically, when defining the rural within any classificatory system, we must engage with the intrinsic contextuality of all definitional schemas. Specifically, recognizing Whatmore's tenacious dynamism of the rural, the definition that we deploy must, first and foremost, be (shown to be) appropriate to the task to hand. There must be no more spurious searches for that 'one size fits all'. This clearly undermines excessively complicated classifications in favor of simple, accessible and, for the task to hand, explicitly meaningful formulations.

PART V
MOVING FROM THE CONCEPTUAL TO THE OPERATIONAL

Chapter 16

Multiple Dimensions of Settlement Systems: Coping with Complexity

Mike Coombes[1]

Attention now moves towards the more operational aspects involved in translating conceptual ideas about identifying and distinguishing settlements into specific measures. This chapter considers some opportunities for improving definitions of urban and rural areas. The complex processes which, particularly in more developed countries (MDCs), are reshaping settlement patterns and creating urban systems make the simple framework of the traditional binary urban-rural divide seem inadequate. A mix of academic literature – but predominantly British and recent – is drawn upon here, together with some major crossnational statistical and policy-related documents (e.g. Decand, 2000).

Chapter 1 has made it very clear that the first task is to reconsider what the terms urban and rural now mean in the complex reality of MDCs' settlement patterns. Here the term 'settlement' will refer to a single separable builtup area, whether it be a village or a conurbation, defined using United Nations (UN) principles for identifying urban areas (OPCS, 1984). Three main *dimensions* of settlements are shown to provide a framework for evaluating different ways to demarcate rural from urban areas, leading on to the development of some more appropriate measures. Attention centers first on options for bricks-and-mortar definitions, then moves on to deal with measures which take account of the wider context within which any rural or urban area is set. Finally, the paper considers some implementation issues posed by the newer methods of definition, and especially the question of how several indicators can be combined in a multidimensional approach to representing settlement patterns.

What are the Fundamental Issues?

There is now a wide consensus that in MDCs there is an increasingly 'fuzzy' distinction between urban and rural areas. This trend led to a commonly accepted model in which the two categories were seen as the two parts of a continuum, with the difficulty then lying in choosing the best point at which to draw a line between

---

[1] The author is very grateful to his colleague Simon Raybould for implementing the analysis described in Annex 3 and providing the three maps.

one set of areas and the other. The current position is more complex. Indeed, the distinction between urban and rural areas is arguably now so 'fuzzy' that the two categories need to be understood as each grouping a set of areas with a family resemblance among themselves *but* with no single 'litmus test' distinction between them and the areas in the other group. From this view, the continuum model is too restrictive because it assumes that one single dimension can provide an adequate basis for distinguishing the two categories satisfactorily.

A key reason for the increasing problems of definition is, of course, that modern ways of living and working have systematically challenged longstanding contrasts between the rural and urban domains. Much of the countryside now shares many of the characteristics previously seen to typify urban areas. A major driving force has been the growing preference for living away from larger cities: this preference might be dated back to royal relocations such as to Versailles, but of course it is now an option open to many. The fundamental process here is of increasing mobility, and a crucial consequence is the growth in flows and linkages between cities, towns and countryside. At a high level of abstraction, cities have been 'deconstructed' and portrayed as purely nodes in a 'space of flows' (Castells, 1989). This conceptual development has been followed in the policy context by the European Spatial Development Perspective's suggestion that to counterpose the urban and the rural is now an 'outdated dualism' (Nordregio, 1999). Less contentiously, this critique reinforces the emphasis on flows and linkages between cities and countryside.

Yet there is plenty of evidence that key urban-rural differences such as settlement size do indeed influence people's life chances (e.g. Denham and White, 1998). More generally, Chase-Dunn and Jorgenson (2002) argue that the 'settlement size distribution – the relative population sizes of the settlements within a region – is an important and easily ascertained aspect of all sedentary social systems' (p.1). The distinctions between urban and rural areas are less clear-cut than once they were, but they are still real (Carter, 1990). For example, there are rural policy concerns, particularly about accessibility and the natural environment, which do not take the same form or intensity in urban areas.

Reviewing an emerging consensus nearly 30 years ago, Rees (1970, p.276) summarized the importance of settlement patterns with the following, slightly cautious, claim: 'The densities at which people live have profound effects on their lives, or so the classical urban ecologist or experimental psychologist would have us believe'. In the USA at the time especially, city size and central area density were seen to be positively correlated, whilst the density of urban neighborhoods was negatively correlated with distance from the city center. These two regularities were reinforced by cities' average densities being positively correlated with their overall size. In addition, people's behavior was frequently argued to be shaped by the population density levels where they live.

Many changes to the patterns Rees summarized have led to today's more complex picture. For instance, many larger cities have lost population, particularly in their inner areas. Urban decline has brought social problems, so that lowered population densities are now often associated with those social and economic problems previously associated with high densities. Also, substantial new towns and other developments often include large populations living at low densities.

The continuing restructuring of advanced economies has also led to an ever-increasing emphasis on flows, both in the form of physical mobility and as electronically-mediated interaction such as teleworking. Increasing flows are often a corollary of increased specialization, and within regions specialization tends to reinforce polynuclear structures where each town and rural area has a mutually interdependent role (Batty, 2001). At the same time, the increasing flows of people across a modern region also results from lifestyle trends, and dual-career households in particular. It is especially important to note that both these micro and macro level trends have transformed patterns of flow *spatially* as well as greatly increasing their number. The traditional model, with a simple pattern of flows centered on a single city, is often now replaced by a reality featuring linear or 'edge' cities and networks of towns, all of which are linked by multidirectional flows including 'reverse' commuting and chained school-shop-work trips. As a result, the crucial step of recognizing the importance of flows in MDCs has to be followed through with a far wider perspective than a simple 1950s-based analysis of commuting into cities. Unless flows in all directions are considered – along with flows other than commuting if possible – then the analysis still presumes that all areas conform to a mid-century model of a monocentric urban system.

This emphasis on flows is part of situating any rural or urban area in its context. Taking the context into account is, as argued elsewhere (Coombes and Raybould, 2001), crucial to recognizing that there are several dimensions to the differences between modern urban and rural areas. Examination of recent trends in MDCs suggested three main *dimensions* which were not substitutable one for another:

- the intensity or concentration of settlements;
- the size of settlements; and
- accessibility to services and other facilities.

The last of these most clearly relates to the earlier discussion of flows, although urban intensity can also be seen to be increasingly characterized by a multitude and diversity of movement. Most pressingly for this paper, these different dimensions will produce different rankings from rural to urban, because place X can be more 'rural' than Y in terms of one dimensions while it is more 'urban' than it in one or both of the other two respects.

The Challenge of Definitions

As stated above, the categories rural and urban are seen here to represent 'family resemblance' across a variety of characteristics. All rural areas will share *most* of the characteristics which make up a stereotypical contrast with urban areas but, taking any one characteristic separately, a few predominantly rural areas will be more like a typical urban area whilst a minority of urban areas will be more like a typical rural area. This 'fuzzy' distinction implies that methods to delimit urban and rural areas on the ground:

- will be unlikely to generate a clear 'definitive answer' with a single criterion,
- may better meet different purposes by drawing on different approaches, stressing different characteristics, and
- can, at best, provide a set of definitions suitable for a majority of the very broad range of users of standard urban-rural definitions.

Before discussing alternative ways of identifying rural and urban areas, it is important to establish the criteria against which the alternatives should be assessed. The discussion so far has aimed to sketch out what the categories urban and rural mean in MDCs, and clearly the principal criterion here is that the definition should represent the most important contrasts between the two categories. In practice, this leads back to the three key dimensions of intensity of settlement, settlement size and accessibility.

A first practical issue concerns the 'building block' units of analysis for the definitions. Two of the three key dimensions focus on settlements, but there is an issue over whether these can be taken as given. While it is a valuable starting-point if all settlements have been identified according to the UN builtup-area principles, questions do remain to be answered before these definitions can be considered ideal. The first question is whether the detail of UN definitions can be reviewed to keep separate substantial towns which are only tangentially linked by development, thereby avoiding the problem faced by Pumain *et al.* (1991) whose definitions could link together a 'necklace' of towns across distances nearing 100 kilometers. Another is whether settlements should be subdivided so as to help distinguish areas with different levels of access to certain facilities. A parallel question is whether there is an optimal way to partition the area between towns to create the 'building blocks' needed for accessibility analyses. In general, smaller units of analysis lead to superior definitions, but the more important imperative here is avoiding units which group (parts of) larger settlements together with substantial areas of countryside.

As for the definitions which are to be created, it is important that they produce boundaries which separate one area from the next. This may seem obvious, but it was *not* the criterion for the methods in Coombes and Raybould (2001) where the outputs were needed at the level of larger administrative areas (i.e. to measure the *degree* to which such a broad and inevitably mixed area is urban rather than rural). Setting aside several specific policy concerns of that research, two key guidelines identified there are also still relevant in the present context:

- robustness – the data used as input must be consistently reliable across all areas, and
- plausibility – the output values should be largely in accord with prior expectations.

Asking 'where is the most urbanized part of Britain?' provides a suitable rehearsal of these issues. The most plausible candidate is the City of London (the 'square mile' which centers on St. Paul's Cathedral), with its intensive land use and its centrality within Britain's major metropolis. Yet both the two most commonly used measures of urban character – population size and population density – suggest that the City of London is far from the most urban area of the country. The simple reason for the population size measure's result is that the City includes only one square mile (approximately) out of the vastly wider London conurbation and so its total population is inevitably quite low. This can be resolved by shifting analysis to the wider settlement level, although of course then the City as such is no longer identifiable. The density measure's counter-intuitive result highlights three other important issues to be faced when trying to devise better urban-rural measures:

- The units of analysis are critical. Analyzing the City of London in isolation from its neighbors radically underestimates the number of people for whom the City is a core part of their daily lives. An immediate response could be to include incommuters and measure the 'daytime population' (cf. Goodchild and Janelle, 1984), but a more fundamental response is to recognize that adopting 'off-the-shelf' administrative areas was never likely to provide a satisfactory geostatistical unit of analysis (Coombes, 2000).
- Land area is problematic. Density measures treat the measure of land area (the denominator) as of equal importance to the measure of population (the numerator). Yet the land area includes not only residential areas but also industrial and undevelopable areas which are irrelevant to the population. This effect varies markedly between one area and another, due to idiosyncratic boundaries, so land area measures fail the robustness criterion set out above.
- Context cannot sensibly be ignored. A population density measure for the area purely within a boundary such as the City of London's produces a result which would implausibly remain the same if the City was relocated to a remote offshore location! Less obviously, the same principle could be said to apply the whole London conurbation, even though its significance too substantially depends on its place within a wider region.

Population density is all too often used as a 'proxy of first resort' for a more robust measure (cf. Dorling and Atkins, 1995). Its familiarity has led it to being used in an extraordinarily wide range of ways, but it is probably most plausible as an indicator of the intensity of settlement. Its simplicity has encouraged the sort of overinterpretation in which low density areas are often described as 'leafy' or 'peripheral' and high density areas as 'overcrowded' even though the measure does not indicate any of these features either directly or consistently.

This critique of the population density measure does at least draw attention to users' evident preference for easily understood measures and definitions. In particular, highly technical 'black box' methods would have to produce hugely superior results before their benefits would be seen by many to outweigh the disadvantage of their obscurity. This is a problem for the Urbanization Index proposed in Coombes and Raybould (2001) as an indicator of settlement intensity in preference to population density. On the same basis, the fact that the urban-rural dualism has been utilized for so long by so many surely shows that users prize simpler categorizations. Thus any new approach to definition needs to err more towards few – inevitably broader – categories, rather than producing a larger number of categories so as to provide finer distinctions between types of area.

What can be Built upon?

The three key dimensions distinguishing rural from urban areas mentioned above – intensity of settlement, settlement size, and accessibility – pose a problem when simplicity of definitions is a priority. Either one or two of the dimensions must be overlooked or there is increased complexity. Whilst each of the dimensions is fundamentally distinct, intensity and size are both bricks-and-mortar aspects of urbanization, whereas accessibility prioritizes flows and linkages. The need for simplicity here prompts an assessment of the relative merits of settlement size and intensity measures, plus any alternatives, to provide a basic bricks-and-mortar distinction of urban from rural settlements.

Table 16.1 summarizes the most frequently used urban-rural characteristics, within a framework derived from a 30-year-old international overview of urban area definitions (UN, 1969). Table 16.1 also shows that this framework comfortably embraces criteria which were identified in a more recent crossnational comparison of *rural* area definitions (OECD, 1994). The two studies also found that the different criteria were used with similar frequency, which tend to support the earlier view that conventionally they are two sides of the same coin.

Table 16.1   Conventional criteria for defining urban and rural areas

UN (1969) classification of urban criteria listed in descending order of frequency of use	OECD (1994) classification of rural criteria showing the number of countries using them (out of 24)
	Settlement population size *14*
Population size (of administrative area or settlement)	Administrative area population outwith conurbations *6*
	Administrative area population size *8*
Population (or housing) density	Population density *7*
Economic activity	Agricultural share of workforce *3*
	In/out commuting ratio *4*
Other urban characteristic(s)	Centrality or service levels *2*
Administrative status	Administrative status *5*

Leaving the more heavily-used criteria at the top of the table to the last, this review can quickly dispose of the *administrative status* of areas as not only an unscientific criterion but also an increasingly irrelevant one as many countries reorganize their administrative geographies to group rural and urban areas together. Next comes something of a 'catch all' category, with the right-hand side of the table giving the example of *service levels* which, of course, is here more related to accessibility measures than to the bricks-and-mortar approach. The *economic activity* criteria also include an access-related factor in the form of the commuting measure. Agriculture has for some time been far from fundamental to rural areas in many modern economies so it is an unsuitable indicator of urban-rural profile, just as manufacturing used to be concentrated in larger urban areas but now is more often a feature of small towns.

The paper's earlier critique of the *population density* measure guarantees that the recommendations here will accord with a British official review (ODPM, 2003) which came out against density as a basis for defining rural areas. Looking at the settlement intensity dimension more broadly, the question remains whether there is in any case a level of population concentration which is either necessary or sufficient to distinguish rural from urban areas. Whereas a highly compact form is highly evocative of Victorian urban areas, its absence does not make a spaciously planned new town like the UK's Milton Keynes into a rural area, just as a small clustered village's compact form is not enough to make it urban. In other words, intensity *is* a distinct dimension along which high or low values tend to be characteristic of urban and rural areas respectively, but in modern developed countries it is not a robust basis for defining the urban-rural boundary.

This leaves *population size* – which for present purposes can be taken as settlement size as opposed to administrative area size – as the remaining candidate for the bricks-and-mortar urban-rural discriminator. Certainly in everyday terms, towns and cities are urban, whereas villages and countryside are rural. A settlement size criterion, like any solo discriminator, can only be a 'blunt instrument' *even if* the anomalies it produces are relatively few. The specific size threshold above which settlements are deemed to be urban is inevitably a difficult choice. There are only a few countries like Scotland whose urban hierarchy is sparse enough for certain thresholds to readily distinguish towns which are widely accepted as comparable. By contrast, in England a threshold such as 10,000 cuts through the settlement size ranking at a point where there are many very different types of town. A very recent statement by the British government (DTLR, 2002) illustrated this by being unable to choose between 10,000 and 20,000 as the threshold for urban policies.

The difficulty of achieving broad agreement on a settlement size threshold means that the aim becomes choosing the 'least worst' value where the urban-rural boundary should be positioned. The next section of the paper asks whether there are new alternative approaches – emerging since the work underlying the review in Table 16.1 – which might side-step the need to accept such a difficult compromise solution.

How can Definitions be Improved?

Just as the challenge of urban-rural definitions has become more complex, so the potential for more complex forms of analysis has greatly increased (e.g. Burrough 1996). It is no longer reasonable to continue following the earlier approach of relying almost exclusively on a single 'handy' indicator like population density, as if no other indicators were available. Where much more information is available, the challenge becomes choosing from options which can range from satellite imagery to lifestyle datasets. Moreover, data can be accessed at increasingly fine scales, often due to postcoding, whilst Geographic Information System (GIS) techniques make all this information much more manipulable. Nevertheless, new possibilities do not *necessarily* imply that entirely new approaches are needed: there are advantages of continuity in retaining or building upon existing approaches. Given that no single 'definitive' solution will meet all needs in any case, starting entirely afresh would need careful justification.

Before getting too optimistic about new possibilities arising, it is important to keep in mind that the growth in the availability of small area data has not occurred everywhere. This means that the task of finding methods that are both appropriate and applicable crossnationally may actually have been made more difficult rather than less. That said, there now *is* satellite imagery available for most countries, so this can provide an important additional data source for urban-rural definitions. Keeping the focus here on establishing basic urban-rural boundaries, rooted in the bricks-and-mortar tradition, the question then becomes how to synthesize information of different types so as to output improved boundary definitions.

Although there can be no doubt that GIS has greatly eased the process of bringing together diverse datasets, it is equally well known that there has been far less progress in providing new ways of analyzing the collated datasets. In particular, there has been no breakthrough which provides a simple synthesis of multidimensional datasets. The more complex methods which present themselves as alternatives here include:

* a *cluster analysis* of the multivariate database,
* a *synthetic index* approach, and
* certain *rules-based* definitions which work through a series of steps.

There are relatively few examples of cluster analyses used to distinguish urban from rural areas. The most familiar versions are the private sector's geodemographic classifications, but their use of terms like 'suburban' are often unscientific because they have no consistent basis in the data analysis. A particularly good example of a synthetic index is a new British 'town centeredness' index (Hall and Thurstain-Goodwin, 2000), but its sequence of complex geostatistical modeling is not readily explainable to users. A particularly relevant recent example of a rules-based approach is the new Bamford *et al.* (1999) classification of rural areas according to their travel time from larger settlements, although this may not be very transferable to countries where a set of key urban size thresholds will not find ready agreement.

A disadvantage of both the index and the cluster analysis approaches is that they are less able to implement 'everyday' ideas of how to combine criteria. For example, it might be thought that all settlements above a certain size should be recognized as urban areas, together with smaller ones which have other characteristics (e.g. local facilities) similar to those of larger settlements. Combining criteria in this way can be relatively straightforward within a set of rules – as illustrated by the 'conurbation' definitions of Taylor *et al.* (2000) – but a cluster analysis or index would need the size criterion to be applied as a separate initial step. In effect, they then become part of a rules-based approach in order to create a method in which users can understand the interplay of the various factors.

Another way in which rules-based approaches tend to be different is that they use relatively few criteria, and ones that are recognizable by users. For example, a new Scottish rural area classification (Hope *et al.*, 2001), while needing high-level GIS techniques and substantial datasets, uses key criteria of settlement size and drive times that are readily understood. By contrast, cluster analyses and indexing methods frequently synthesize numerous indicators, some of which may be heavily preprocessed or in some other way less recognizable to lay users, and then depend on a combinatory technique that is widely seen as 'black box' to a greater or lesser extent. Whereas users can directly engage in a debate about appropriate drive times, it is hard for them to choose one particular form of principal component analysis, for example, even though these decisions may shape the final outcome more profoundly than the choice of input indicators (Coombes and Wong, 1994). As a result, the previous emphasis on simplicity or transparency of methods leads to a preference for a rules-based approach if several input variables are needed to produce appropriate definitions.

For the present, it seems likely that settlement size is the only plausible candidate to be a sole discriminator between the two types of area, although this will be a 'blunt instrument' and there is unlikely to be ready agreement when it comes to choosing the population size threshold above which settlements will be deemed to be urban. As yet, there is even less consensus likely on the way in which other data, such as satellite imagery, can be brought to bear on the problem so as to reduce the number of anomalous results. This lack of a clear prospect for improvement means that there is insufficient reason to accept the disadvantage of using a less readily understandable procedure. The conclusion here in favor of a rules-based procedure in preference to a cluster analysis or synthetic index approach will become relevant again as attention turns from the bricks-and-mortar aspect of rural-urban distinctions towards issues related to accessibility.

Recognizing the Context

Two types of conclusion can be drawn to summarize the discussion up to this point. The first conclusion is a straightforward reaffirmation of settlement size as the primary – and possibly sole – indicator of urban areas in the bricks-and-mortar sense. At the same time, accessibility-related analyses are needed to consider the broader context of each settlement. No substantial role is seen for a measure of settlement intensity in rural-urban definitions (except that a minimum level of settlement intensity helps to identify the outer boundaries of individual settlements whose population size can then be measured).

The second type of conclusion is the set of guidelines which have emerged progressively through the discussion to establish types of measurement which will be more 'fit for purpose' in urban-rural definitions. It is timely to restate those which are relevant when seeking a suitable way of mapping the context of each settlement through analyzing the flows which portray the interaction between settlements and their context:

- flows are multidirectional and not necessarily focused on a central city,
- commuting is just one of several types of flow which could be relevant here,
- the boundaries should not cut through settlements, and
- more understandable methods of analysis are much preferred by users of the definitions.

Identifying clusters of flows and interactions around settlements calls for an approach which is at least partly rooted in the models of time geography (Pred, 1984). These models portray local boundaries, if produced by analyzing flows, as a summary of the local population's daily pattern of movements. Producing genuinely multi-dimensional definitions calls for analyses which reflect numerous types of interaction, such as migration or access to services, but unfortunately interaction datasets tend to be scarce, and indeed not every advanced country even has information on commuting patterns.

The most well-established approach to defining the surrounding areas with which cities are linked is typified by Metropolitan Statistical Areas (MSAs) in the United States (Spotila, 1999). Viewed from the criteria set out here, this approach has the crucial flaws of presuming an urban-centered structure, and ignoring all flows other than those to the urban core from elsewhere. There are rather fewer examples of boundary definitions which do *not* rigidly presume a certain pattern of flows; most of these are to be found among the range of approaches to labor market area (LMA) definition. A recent review (OECD, 2001) showed that many countries now have LMA definitions, most of which – though not all – rely exclusively on commuting patterns as flow data. The remainder of the discussion here will briefly review ways in which the wider context of urban areas can be identified in three different types of circumstances:

1. Where commuting patterns are the only available flow data.
2. Where additional flow data can be drawn upon.
3. Where no flow data at all is available.

In the first type of circumstance, the need is for a set of consistently-defined LMAs, which are largely *self-contained* in commuting terms with few journeys crossing their boundaries (Goodman, 1970). It is notable that most countries have devised their own LMA definition procedures, although the British software was adopted by ISTAT for their definition of Italian local labor market areas (Sforzi *et al.*, 1997) and has also been used in Spain (Casado-Diáz, 1996). A comparative research program applied alternative methods to several countries' commuting datasets, concluding that the British method was the 'best practice' model on which was based Eurostat guidelines for LMA definitions (Eurostat, 1992). Annex 1 (see the end of this chapter) summarizes this procedure, which has been termed the European Regionalization Algorithm (ERA). It can be seen there that flows in all directions – not just those into cities from elsewhere – are equally important to producing optimal LMA boundaries. Figure 16.1 illustrates LMA boundaries produced by ONS and Coombes (1998) using the British software: the center of Bristol has a large catchment area, but the smaller city of Bath nearby emerges from the analysis with a sufficiently self-contained LMA centered on itself.

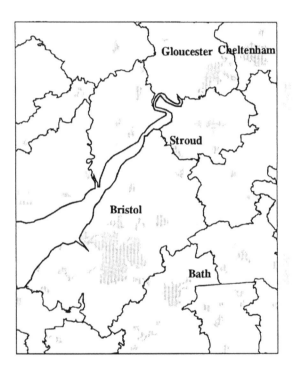

Figure 16.1 Labor Market Areas in the West of England

Turning to the second type of circumstance, it has already been stressed that commuting patterns alone can be no more than a very partial proxy for a more rounded assessment of flows between rural and urban areas. For example, Claval (1987) called for an examination of numerous facets to the linkages which make up a region in practice. Until recently there has been no method for combining the evidence which data on different types of flow can provide towards such a multidimensional portrayal of regions. Coombes (2000) has now demonstrated an appropriate method, the starting point for which is viewing each strand of evidence as part of a 'fuzzy' picture, with the task of the analysis then being to distil the recurring patterns which make up the underlying structure.

The methodological innovation hinges on splitting the procedure into two phases. Phase 1 involves compiling numerous analyses from numerous datasets, and then in phase 2 the results from these analyses are collated within a synthesizing analysis. Phase 2 centers on creating *synthetic data* for which the input is the range of evidence provided by phase 1. Described in more detail in Annex 2, the approach involves layering sets of phase 1 boundaries on top of each other and counting the number of layers in which there is no boundary between each pair of areas. This provides an assessment of the 'strength of evidence' that two areas should be grouped together, on the grounds that they are linked together by most of the input classifications. Figure 16.2 shows – for the West of England again – those adjacent areas which are linked by high synthetic data values. This synthetic dataset is then analogous to a flow matrix, because it represents the level of connectedness of pairs of areas, and so can be analyzed with ERA to produce boundaries grouping rural and urban areas which have a high level of connectedness. The separation of Bath from Bristol – which was seen in Figure 16.1 – is clearly shown here.

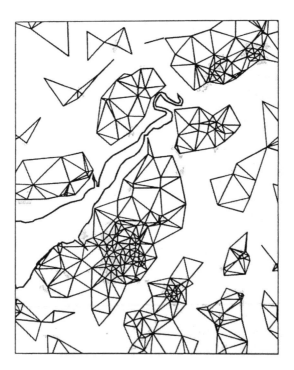

Figure 16.2   The pattern of high synthetic data values in the West of England

In the third type of circumstance – where there are *no* flow datasets to draw upon – how could patterns of linkages be represented in the form of boundaries? There are, in fact, a number of analysis techniques which can estimate patterns of linkages given some relevant information. A fairly simple, yet plausible, example can be outlined by estimating the pattern of commuting where no flow data is available *but* the distribution of both jobs and employed residents is known. For some time now, spatial interaction models have been able to estimate the pattern of flows which could link where the workforce live with where the jobs are located. Such a matrix of estimated flows can then be analyzed by a method such as ERA to create hypothetical LMAs. A critical problem of spatial interaction models is that they are undeniably 'black box' in their workings. Glover and Openshaw (1995) had to devise a much simplified 'front end' so as to encourage use of a web-based version. Although this ameliorates the problem slightly, a more fundamental issue remains in that few users can understand how the results were produced. A much simpler option has been devised here, providing a very straightforward method which should be readily understood and easily transferable (Annex 3). Requiring only the distributions of jobs and residents as input, plus a centroid for each 'building block' area, the method produces very plausible estimates of commuting patterns. Figure 16.3 covers the West of England again, and shows that the method estimates the main cities' catchment areas well. Indeed, it even hypothesizes a flow across the Severn estuary at exactly the point where the Severn bridge exists, despite the analysis having no input data on transport infrastructure.

The form of analysis outlined in Annex 3 – and more complex spatial interaction models – do still need additional data to that which is available from nearly every national census. For example, the estimated commuting patterns here required the location of jobs to be known. A different form of modeling is needed if the data available at a detailed scale relates to the resident population only. In such situations there is even greater reliance on the key principle underlying spatial interaction modeling, *viz.* that the probability of interaction declines with increased distance.

In the GIS era, this distance-deterrence principle may be made more sensitive to local circumstances than by simply assuming that a straight-line distance between two places represents the difficulty of traveling between them (Higgs and White, 2000). There are often now measures of actual route lengths, allowing the estimation of point-to-point travel times based on the speed of travel along different route segments. All this sophistication, of course, depends on the availability of these extra datasets, and here the interest in such models is restricted to circumstances where data availability is extremely limited.

Mention has already been made of analyses such as that of Bamford *et al.* (1999) that measure the relative ease with which more rural areas can access urban centers. These analyses could readily identify the areas with a level of accessibility above a predefined level. Given the necessary route data, this approach can draw boundaries round areas within X minutes' drive-time of major centers. Without this information, the approach is limited to simpler analyses which would then appear on maps as purely hypothetical (e.g. circular catchment areas). In both forms of application, the crucial first step in the process is choosing the centers whose catchment areas are to be modeled. This step draws attention to the model's presumption of monocentric regions, in which the only interaction considered radiate to/from the center. A final point to remember is that, as with most models, key thresholds need to be chosen and these will often not command consensus due to their great influence on the resulting boundaries. This problem is likely to be substantial for these models because users may expect differentiated results – e.g. 'within X minutes of a city of size 1 *or* Y minutes of a city of size 2 *or* ...' – which would require a consensus on the choice of numerous thresholds of both city size and accessibility level.

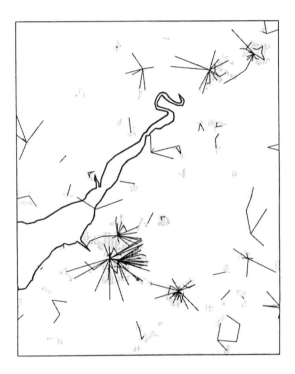

Figure 16.3   A pattern of hypothetical commuting flows in the West of England

Implementation Issues

Drawing clear conclusions here is scarcely possible because many different alternatives have been discussed, not least to cope with the great variation in data availability that has to be faced when considering new international standard definitions. Emphasis has been placed repeatedly on flows and linkages, but data coverage of these patterns is acutely limited. All the same, the single main conclusion here has been the need to combine a bricks-and-mortar definition of each urban settlement with a mapping of its context. One implementation issue which is provoked by that conclusion is how these two elements of rural-urban definitions should be combined. In tackling this issue, the earlier discussion about users' preference for simplicity becomes an important constraint. A simpler approach will have the additional advantage that it may reduce the number of thresholds to be agreed, thereby avoiding the type of problem noted above in which several different hinterland sizes had to be chosen to match several different size bands of city.

Table 16.2 compares ways of implementing the call here for urban area definitions to be set in context by representing their interactions with nearby areas. The rows in the table each reflect a different level of data availability. Thus the top row is the 'ideal' situation in which there is more than one dataset about patterns of relevant interaction, whereas the second row considers options if there is just one such dataset (e.g. commuting patterns, or just possibly migration flows). The third row provides a reminder that certain types of network – whether fixed networks like roads or service networks like bus services – may be useful as a 'proxy' for data on actual patterns of interaction. The last row assumes that no relevant data is available, apart from the map locations needed to estimate distances between places.

Table 16.2 Representing the linkage of urban and rural areas: analyzing interaction, with differing levels of data availability

| | Linkage to: | |
Linkage shown by	Foci which the relevant data show to attract flows	Foci which are predetermined in the analysis
Multiple forms of interaction	Localities in Britain (Coombes, 2000)	City Regions in Britain (Coombes, 2000)
Commuting (or possibly migration)	Local labor markets (e.g. Casado-Díaz, 1996)	Metropolitan areas (e.g. Spolita, 1999)
Potential for, or signs of, interaction (as shown by networks)	Bus service hinterlands (Green, 1950)	Road network accessibility (e.g. Bamford *et al.*, 1999)
Probability of interaction (as shown by relative distances)	See Annex 3	Basic 'gravity' models (Glover and Openshaw, 1995)

Table 16.2 also has two columns. The left-hand one covers those analyses in which there is no preselection of key 'foci' for the pattern of flows (nb. this type of analysis may also handle the multidirectional flows which are increasingly common in polynuclear regions). The right-hand column is for analyses that predetermine which places are the foci for the interaction patterns analyzed (e.g. the central cities which are identified first in most metropolitan area definitions, so that the only commuting flows then examined are those of noncore residents who work in these cores).

The principles outlined in this chapter indicate that the analyses in the left-hand column of Table 16.2 are preferable to those in the right-hand one and, data permitting, those higher in the table are to be preferred to those lower down. This interpretation is consistent with the approach of Frey and Speare (1995) who argued that methods of labor market analyses in Europe (e.g. Casado-Diáz, 1996) were preferable to metropolitan area definitions in the USA (not least because the latter only analyze the small subset of commuters who travel to work in the predefined urban cores). Although the table is mainly a summary of material already discussed earlier in this chapter, there has been no previous reference to Green (1950). The latter's use of bus service information to identify urban centers and their hinterlands may be highly relevant in some less developed countries where public transport still dominates mobility patterns, and where census information may well not include information on the pattern of commuting. Another useful reminder provided by the table is that patterns of local migration can offer a valuable alternative, or supplement, to the more familiar commuting data for these types of analysis. Finally the table draws attention to the role of spatial interaction models in situations where very little data can be found.

In most cases, the wider area within which the urban area is set will be defined as a labor market area (LMA). The next step is to differentiate these LMAs in an appropriate way. Population size is the one universally-available measure, but the categorization could be in terms of either the whole LMA or its largest urban area. The development of polynuclear regions suggests that the whole LMA is the more appropriate metric. This leads to the outline recommendation that the eventual classification uses two population-size measures:

- settlement size (with a threshold providing the simplest rural-urban categorization), and
- size of the labor market area (or nearest-equivalent available boundary) in which the settlement is situated.

Implementing such a classification doubtless faces numerous hurdles, ranging from huge issues such as uneven data availability to the more technocratic (e.g. coping with settlements straddling more than one LMA). These issues lie in the future, beyond the major uncertainty of establishing whether this approach is reasonably acceptable to rural and urban experts, data supplies and users.

Concluding Comment

Underlying this chapter is the thesis that the new GIS-facilitated flexibility *increases* the importance of the decision over which areas are best used for data analysis, because the rapid reduction in barriers to data availability increases the need to identify the most 'fit for purpose' urban and rural definitions. The identification of these areas is far from deterministic: a number of alternatives can be shown to meet different requirements. There are few easy answers when trying to respond to new opportunities and user needs whilst adhering to the ethos summarized by the Fundamental Principles of Official Statistics (agreed in 1991/2 by the United Nations Economic Commission for Europe). In practice, these Principles involve a tension between trying to satisfy all the citizens' entitlement to public information, and aiming to retain the public's trust in official statistics by discouraging inappropriate usage of statistics. An additional tension for this paper has been between the drive to innovate – stimulated by acute dissatisfaction with existing classifications and aided by new geostatistical opportunities – and the belief that users prefer simple and stable classifications. In this light, it may be less surprising that a chapter opening with a discussion of new multidimensional complexity ends recommending a two-dimensional classification which builds on the longstanding measure of settlement size.

Annex 1   Defining a city's wider context using commuting data

The basic features of the ERA method derive from an earlier algorithm (ONS and Coombes, 1998), but that multi-step method has been simplified to remove numerous steps which required the setting of applications-specific parameters (e.g. 'select areas which score higher then x on parameter y'). Reducing the procedure to its basic components makes it much more transferable, in that it can be used with different datasets and for different purposes without long periods of experimentation to determine optimal settings for each parameter. At its simplest, the ERA procedure can be described in two steps:

1. consider all input areas to be potential output regions, calculating their values on the statistical objective function set (e.g. level of self-containment), and ranking them accordingly
2. disassemble regions which fail to meet the statistical criteria (starting with the region furthest from meeting the objective function) and group the constituent areas individually with whichever other region they share most flows.

In technical terms, ERA's tendency to be self-optimizing is due to three factors:

1. The software can cope with very different sized 'building block' because in its early stages it groups them into localized clusters without trying to get 'everywhere right first time' and then it gradually iterates towards a more optimal solution.
2. The groupings are not constrained by contiguity, which allows the procedure to always choose the linkage that maximizes the grouping's integration; in fact very few areas are eventually grouped non-contiguously because the groupings are driven by patterns which reflect people's reluctance to travel longer distances than are essential.
3. The procedure is not rigidly hierarchical, in that two areas which are grouped together at an early stage may then later be disassembled and grouped into separate areas.

Annex 2   Defining a city's wider context using multiple datasets

The essential basis for the synthetic data is the understanding that a set of boundaries is a classification in which each building block area is allocated to one and only one region. Thus each analysis undertaken in phase 1 produces a classification of all the N building block areas from which the final definition are to be composed. For each area, the initial information in each classification is the region number to which that area is assigned. The crucial step (phase 2) is then to transform this information into binary data in a matrix by taking each pair of areas and identifying whether they are ('1'), or are not ('0'), classified into the same region. In this way, each classification becomes expressed as a binary matrix of N*N cells (although the matrix is in fact symmetrical, so only half of it is needed). For example, if area B was in the same region as area C but in a different region to area D then the cell BC would take the value 1 while cell BD would be 0 (and cell CD would also take the value 0).

The crucial benefit from re-expressing each separate classification in this binary form is that these matrices can then be cumulated to produce the synthetic data needed. For example, if the results from the phase 1 analyses were collated in this way, the value in each cell of the final synthetic data matrix would vary from 0 (the value for any pair of areas which were not in the same region according to any of those analyses) up to 3 (the value for any pair of areas which all three analyses had put in the same region).

Annex 3   Defining a city's wider context in the absence of flow data

The method can be seen as a development from the analysis by Small and Song (1992), who attempted to identify 'wasteful' commuting by creating a hypothetical pattern of commuting which would minimize work trip travelling. The analysis here makes a similar assumption in favour of filling each job with the nearest people available to work. The method can be briefly outlined in a few steps, as below:

1. Assume jobs in each building block area are filled by the same area's residents so far as possible.
2. Identify each area as having surplus jobs *or* surplus work-seeking residents *or* a balance with all its jobs and residents allocated.
3. Proceed through all area-area pairs, in ascending order of distance apart:
    3.1   STOP if the next pair are too far apart for commuting to be plausible, otherwise proceed.
    3.2   *if* one of that pair of wards has surplus jobs and the other has surplus residents, fill as many of the surplus jobs as possible with these residents and update the areas' surplus/balance status; whether or not such a change has been made, return to step 3.1.

Chapter 17

Using Remote Sensing and Geographic Information Systems to Identify the Underlying Properties of Urban Environments

John R. Weeks

Urban places represent built environments that are physically distinguishable from the natural environment, and are thus potentially identifiable through the use of remotely-sensed sources such as satellite images. The urban environment can be defined by classifying images and then combining that information with census data to create a quantitative index of the urban-rural continuum. This is based on the premise that variability in the built environment is associated with variability in human behavior, and that this variability captures the nature of urbanness in human societies. The chapter begins with a justification of the use of the built environment as a signature of urban places. It continues with an overview of how satellite images can be used to distill information about the urban environment, and of the role that geographic information systems (GIS) play in the analysis. It then illustrates this approach to understanding the urbanness of places using data from Egypt. Variables derived from satellite images are combined with census data to improve our understanding of the spatial variability in human behavior in the context of the urban-rural continuum. Finally, ways in which this type of analysis could be used to measure and understand phenomena such as urban sprawl and multinucleation of metropolitan areas are suggested.

The Built Environment as a Signature of Urban Places

Urban places are typically defined by demographers according to criteria of population size and density. To be urban requires that a sufficient quantum of people are living in sufficiently close proximity to one another so that life is demonstrably different from that in rural areas. That difference is often expressed in terms of economic activities. In particular, urban places are routinely defined as concentrations of people who are engaged in nonagricultural activities (Weeks, 2002). Definitions based on size, density, and economic activity all imply a dichotomy between urban and rural, and that notion is almost certainly accurate from an historical perspective. Until only a few hundred years ago, most cities were bounded by protective walls, which offered a clear distinction between the city and the noncity population. In the US in the nineteenth and early twentieth centuries, rural turned into urban when you reached streets laid out in a grid. Today, such clearly defined transitions are rare. As discussed in the opening essay of this volume, researchers have complained for decades about the arbitrariness of drawing a line between places that are rural on one side and urban on the other.

Over time the distinction between town and country may have become blurred, but in general the distinction illustrates an important point: urban places are identifiable by their infrastructure. As Smailes (1966, p.33) suggested, 'the geographer must regard as urban a particular manmade type of landscape.' Urban places – be they towns, cities, or megalopolises – have, at a minimum, buildings and roads that make them different from the rural countryside. Historically, that difference has always been present, but modern urban places have vastly more complex infrastructure, including electricity lines, gas pipelines, water storage and treatment facilities and water transport pipes, sewers and waste treatment plants, landfills and other refuse facilities, bridges, tunnels, and various aspects of mass transportation and telecommunication.

Yet, many of these aspects of infrastructure, especially the communications-related ones, have now reached into what had previously been thought of as rural places, changing the lives of those residents in the process. This reminds us that important aspects of urbanization are ideational. There is an explicit recognition that urban people order their lives differently from rural people; they perceive the world differently and behave differently. At the same time, living in a rural area in most industrialized societies does not necessarily preclude participation in urban life. The flexibility of the automobile, combined with the power of telecommunications, can put most people in touch with urban life. Even in remote areas of developing countries, radio and satellite-relayed TV broadcasts played on sets powered by a portable generator can make rural villagers knowledgeable about urban life, even if they have never seen it in person (Critchfield, 1994). In the process, people in rural places are becoming more urban, and this serves to change the character of the places where they live.

The increasing connections between the urban and the rural have the effect of urbanizing rural places, helping to explain why the world is on a trajectory toward being predominantly urban: not only are people moving to cities; the cities are moving to people. The direction of movement is important to consider. Rural places tend to be characterized by what they lack – electricity, running water, sewage, paved roads, schools, health clinics and hospitals, diversity of employment opportunities, not to mention the lack of amusements and amenities such as sports teams, theatres, and restaurants. While there are always some people who do not want such things, the evidence suggests that most rural residents prefer to have more, rather than less, of these improvements. For most of human history, a person had to migrate to a city to participate in urban life, but it is now possible for governments to extend many of the characteristics of urban life into rural areas, permanently changing the nature of those places, both ideationally and physically.

As important as ideas are, the signature of an urban place is the built environment, represented most obviously by buildings, roads and sidewalks. A person raised in the city will still be urban even when isolated in the countryside, just as a farmer who moves to the city may never fully adapt to the kind of life demanded by the city. But both people readily recognize the gradations of urbanness or ruralness by the differences in the built environment. Furthermore, differences in the built environment can be shown to be related to differences in the human behavior taking place in those environments. Byrne (2001, p.149) reminds us that, in fact, 'the built environment for urban residents is the locus of the social.' As Winston Churchill once said, 'We shape our buildings, and afterwards our buildings shape us' (Churchill, 1943). This is another way of saying that the context *in* which we live influences *how* we live. Duncan made the classic statement of this:

> A concrete human population exists not in limbo but in an environment. Moreover, to continue to exist, it must cope with the problems posed by an environment which is indifferent to its survival but offering (in varying degree) resources potentially useful for the maintenance of life. By mere occupancy of an environment, as well as by the exploitation of its resources, a human population modifies its environment to a greater or lesser degree, introducing environmental changes additional to those produced by other organisms, geological processes, and the like. Thus, in the language of bioecology, one may say that not only does the environment 'act' upon the population but also the human population 'reacts' upon its environment...The 'adjustment of a population to its environment, therefore, is not a state of being or static equilibrium but a continuing, dynamic process. (Duncan, 1959, pp.681-82)

When Duncan uses the word 'environment', he is referring to the natural environment, in the way that human ecologists have tended to do, but a substitution of 'built environment' for 'environment' keeps the meaning while applying it specifically to human life as organized in cities, towns and villages. The term 'local context', or 'local environment', means the complex of social activities that are taking place within a given built environment.

Social scientists tend to focus on the population and social organizational parts of this system, and spend less time thinking about the environment in which these parts are embedded. In particular, sociologists and demographers tend to be vague, if not dismissive, of the built environment of the buildings, parks, roads, bridges, and the associated infrastructure that humans create out of the natural environment and which become the places where everyday life takes place. Yet, the built environment is, in fact, the actual environment in which a large fraction of humans spend their entire lives. The natural environment is so transformed by urbanization that the majority of urban residents spend little time touching soil and interacting with flora and fauna.

To understand what an urban area is, we can begin with the idea that the local environment of social structures and institutions is the context within which individual lives are understood, and then add to that the notion that the outward manifestation of the social environment is the built environment (the buildings, streets, and infrastructure) created by the people living in those places. This is important both theoretically and methodologically because the social world exists only in our heads, whereas the built environment is the physical representation of the social activities of humans and it may be more measurable than are the attitudes and behaviors of the people themselves. Furthermore, as Bronfenbrenner (1995) notes, individuals and their environments are in constant, reciprocal interplay: there is a dynamic relationship between the built and social environments. Urban places are microcosms of a larger society, shaped by the interaction of demographic processes, social processes, and the built environment in which these processes are being played out. In other words, the urban transition is not just a national pattern that can be laid over a society and understood in those general terms. It is a process that occurs place by place over time, and if we can understand how the demographic and social processes intersect with and interact with the built environment, then we should have a greater understanding of the underlying source of dynamism in urban morphology.

This implies that, if we can quantify aspects of the underlying properties of the built environment, then we can produce an index that measures the degree and/or type of urban place, helping us thereby to move beyond a dichotomy of urban-rural into a genuine continuum of urbanness that encompasses properties of the built environment as well as the behavior (or at least the characteristics of) the people in those environments. Seen in this way, the built environment is not just a proxy for an urban place, but rather it represents an important stage upon which urban life is played out, and different stages demand and/or permit different kinds of human activities. The physical and social worlds are thus highly interconnected. A built environment left unattended by humans slowly reverts to nature, just as a human population living outside a built environment lives within the primitive world of nature, with all of the attendant tasks and risks associated with that life.

I argue that, precisely because the built environment is a signature of urban places, we may be able to use the technology of remotely-sensed imagery and GIS to assist in the development of new approaches to the measurement and quantification of the rural-urban continuum.

Using Remote Sensing to Capture the Urban Scene

Remotely-sensed images range from high-resolution aerial photographs and digital imagery to low-resolution satellite images (which may be either photographs or digital imagery), with most imagery used in social science falling in the middle of that range. Resolution refers to the size of the scene on the ground captured by the smallest pixel (picture element) in the image. Thus, a 1-meter image means that the smallest amount of detail in the image is 1 meter by 1 meter in size on the ground. Images also vary according to the bandwidth of light captured by the sensor (camera or other recording device), ranging from panchromatic (gray scale) to multispectral (visible red, green, blue, and near-infrared bands, as well as other bands that are not visible to the naked eye). A basic premise of remote sensing is that the earth's features and landscapes can be discriminated, categorized, and mapped according to their spectral characteristics. The nuclear reactions of the sun produce electromagnetic energy, and this energy is propagated by electromagnetic radiation at the speed of light through space, reaching the earth's atmosphere practically unchanged. Part of it is absorbed as it passes through the atmosphere, and the remainder continues on to the earth's surface. The part that continues is then either reflected or absorbed by objects on the earth's surface and reradiated as thermal energy. 'Passive' remote-sensing systems operate by measuring the energy which is reradiated or reflected from the object of interest back to the remote sensor. The sensors are most often optical (measuring light reflectance), but they may also be thermal (measuring heat reflection), or something else, depending upon the wavelength of the specific kind of energy that the sensor is designed to measure (Lillesand and Kiefer, 2000).

In order to appreciate the value of remotely-sensed images for demographic analysis, it is crucial to understand exactly what it is that can be extracted from an optical satellite image. The image itself is composed of a mosaic of individual pixels representing information that has been captured for an area on the ground that is equal to the resolution of the image. The information recorded for each image depends upon the particular sensor. For a panchromatic image, information is recorded for only one band of reflectance, based on the brightness of the pixel in the visible range of wavelengths between approximately 0.4 and 0.7 micrometers (μm). We typically call this a black-and-white image, although really it is mainly shades of gray, with black and white representing the two extremes. Technically, it is brightness at the satellite that is recorded, but through a series of adjusting techniques, we are able to estimate what the brightness is on the ground at that place shown on the image. For a multispectral image, information is recorded for two or more bands of reflectance. The Indian Remote Sensing multispectral image (IRS-IC LISS-III) which is used in the research reported here records three bands in the visible and near infrared (VNIR) range, including green (0.520-0.590 μm), red (0.620-0.690 μm), and near infrared (0.770-0.860 μm) at 24-meter spatial resolution, and one band in short wave infrared (SWIR – 1.50-1.70 μm) at 71-meter spatial resolution.

For work at the equivalent of the census-tract level of analysis, the ideal image is a relatively high-resolution multispectral image. In the research that is discussed here, only commercially available satellite images are employed. The 'highest-end' commercial options include 1-meter resolution IKONOS panchromatic images and 4-meter IKONOS multispectral images, which can be merged with the 1-meter pan image to create representative 1-meter color imagery. There is currently no archive of these images for the study site in Egypt, and so they would have to be specially ordered at considerable cost. Existing archived images for the late 1990s and early 2000s include Landsat (US) Thematic Mapper 30-meter multispectral images, SPOT (French) 20-meter multispectral images, IRS (Indian Remote Sensing) 5-meter panchromatic images and 24-meter multispectral images, and SPIN-2 (Russian) 2-meter panchromatic images.

Analysis of Images for Urban Areas

Images were first derived from aerial photography in 1858, and they have been useful in the analysis of urban areas for several decades. The Australian Bureau of Statistics notes that an urban center with a population between 1,000 and 19,999 is to be delimited 'subjectively by the inspection of aerial photographs, by field inspection and/or by consideration of any other information that is available' (Australian Bureau of Statistics, 2001). In the 1960s, Noin (1970) derived estimates of the rural population of Morocco by examining aerial photographs to determine the number of housing units in rural areas and then applying a household-size multiplier to dwellings to estimate the population. One of the more sophisticated applications of aerial photography was that by Green (1955), who developed a method to analyze the social structure of urban areas based on a set of surrogates obtained from the aerial photos. Using black-and-white photos, he created social indices based on characteristics such as neighborhood location, single-family homes, and density of housing. In most applications of aerial photography, however, information is obtained from an image captured on film by means of *interpretation* of the image by humans. This requires a great deal of training and practice and it is not necessarily replicable from place to place and time to time. When images are captured digitally, it is possible not only to interpret them, but also to *classify* the information using a mathematically-derived algorithm that is replicable and which can produce a statistically quantifiable result.

The classification of digital images has represented a breakthrough for analyzing the earth's surface because the process can be automated on the computer and repeated for images representing different times and places. Image classification is the process whereby all pixels in the image are categorized into a land-cover class or theme (Lillesand and Kiefer, 2000). As we (or, more accurately, computers using an algorithm that we have developed) look at each pixel, the question is: Does this pixel represent vegetation (and perhaps a specific type of vegetation), or bare soil, water, shade, or an impervious surface (such as asphalt or cement)? These are the basic building blocks of the natural and built environment and each type of land cover is associated with a particular 'spectral signature,' which represents a combination of wavelength values shared by one class of surface (such as vegetation), but not by the others. The higher the resolution (i.e. the smaller the pixel size), the more accurately we are able to classify a pixel because it is more likely that the pixel will include only one type of land cover. On the other hand, for lower-resolution images, the more likely it is that the pixel will represent a mixture of different land covers, forcing us to make decisions about how appropriately to classify the image. Once we have classified the image according to land cover (the physical property as seen from the air), we are in a position to use information from other sources to make inferences about the way in which the land is being used (which is a socially derived category). From this process we are able to create variables describing the environmental context of a specific place.

The panchromatic image is not capable of classification into land-cover types, but it can be used to derive information about brightness and about the texture at the earth's surface. The variability from one pixel to another in the amount of brightness (the gray scale) at night, for example, can be used as an index of economic wellbeing. In a rural village light at night may indicate electrification, indicating a higher level of economic development than a similarly situated village emitting less light. Weeks (2003) used the brightness of night-time lights to assess the relationship between lighting and crime in an urban center in California. Light reflected from the surface during the day is more varied, of course, than at night and this provides a way of measuring texture at the earth's surface. Very little variability from pixel to pixel in the amount of brightness would indicate a homogeneous surface, and a great deal of variability would indicate a heterogeneous surface. In a city such as Cairo, Egypt, the older parts of the city, characterized by low-rise buildings and narrow streets – what Rodenbeck (1999, p.224) has called the 'higgledy-piggledy burrows' of Cairo's popular quarters – will exhibit a relatively homogeneous texture, whereas newer 'irregular' suburbs may be expected to exhibit considerable heterogeneity in texture. By combining the land-cover classification with the texture classification, we are in a position to describe the physical nature of the urban scene captured by the remote-sensing device.

In general, vegetation is easier to classify than are humanbuilt structures, and so the classification of remotely-sensed urban imagery is much more cutting-edge than is the classification of rural areas (Jensen and Cowen, 1999). Most of the literature on the classification of remotely-sensed images has emphasized the creation of variables describing the natural environment of plants, soil, and water. Urban environments include a complex mix of buildings, streets and other infrastructure, as well as vegetation, soil, and water, often interwoven with one another. Methods for dealing with urban images are still evolving, but at an increasingly rapid pace (Gruen *et al.*, 1995; Cowen and Jensen, 1998; Rindfuss and Stern, 1998; Jensen and Cowen, 1999). Seeking social meaning in imagery holds the promise of providing information that speaks to the core research issues of the social sciences (Geoghegan *et al.*, 1998). The interface between remote sensing and social science depends on the kind of features that can be detected such as landuse/landcover, buildings, infrastructures, roads network, and also on how often and to what detail such data can be obtained (i.e. spatial and temporal resolution).

Numerous studies have documented the ability to extract population information either directly from remotely-sensed data, or indirectly by analyzing information derived from the imagery (Lo, 1995; Elvidge *et al.*, 1997; Lo *et al.*, 1997; Mesev, 1998; Ryznar, 1998; Tanaka *et al.*, 1999; Weeks *et al.*, 2000; Rashed *et al.*, 2001). Lo (1995) was able to derive population estimates from remote imagery by testing a number of regression models that link spectral radiance obtained from a multispectral SPOT image with high and low population densities in some metropolitan areas in Hong Kong. Elvidge *et al.* (1997) and Doll *et al.* (2000) have used satellite images to identify the relation between population, gross domestic product and electric power consumption in 21 countries. They concluded that VNIR (visible and near infrared) emissions of nighttime lights could be successfully used to define and update the spatial distribution of human populations, particularly in urban areas. Other studies have shown that community-level demographic characteristics such as income and education are strongly correlated with variables extracted from high-resolution remote imagery (Jensen and Cowen, 1999). Examples of such variables extracted include building sizes and densities, parking lots, existence of water tables, street widths, and health of landscaping.

These studies have demonstrated the potential value of remotely-sensed images, but we are just now in the process of developing algorithms for the automatic extraction of information from images in ways that allow us to use these data in a quantitative spatial analysis of the urban scene over a fairly large area.

Using GIS to Analyze the Urban Scene

Although we may have successfully classified the pixels in the image according to land-cover type and texture, the analysis of that information requires that we link the location of each pixel to other information about what is happening on the ground at that location. For the purpose of analyzing the urban structure, we are particularly interested in joining the data from the remotely-sensed image to information gathered at the local level in the census, such as the equivalent of the US census tract. We accomplish this in a GIS environment, which allows us to match data from one layer (such as the image analysis) for a specific geographic area (such as a census tract) with data from another layer (such as the census data) for that same geographic area. It is not an exaggeration to say that the remotely-sensed data would be useless to us unless they were incorporated into a GIS.

A GIS is a computer-based system that allows us to combine maps with data that refer to particular places on those maps and then to analyze those data and display the results as thematic maps or in some other graphic format. The computer allows us to transform a map into a set of areas (such as a county or a state or a census tract), lines (such as streets or highways or rivers), and points (such as a house or a school or a health clinic). Our demographic data must then be georeferenced (associated with some geographic identification such as an address, ZIP code, census tract, county, state, or country) so that the computer will link them to the correct area, line, or point. If the computer knows that a particular set of latitude and longitude coordinates represents the map of Egypt, then our data for Egypt must be 'georeferenced' to that particular location. Or, if we have a survey of households, then each variable for the household would be georeferenced to the specific address (the point) of that household.

The 'revolutionary' aspect of GIS is that the georeferencing of data to places on the map means that we can combine different types of data (such as census and remotely-sensed data) for the same place, and we can do it for more than one time (such as data for the 1986 and 1996 censuses of Egypt and imagery for those two dates). This greatly increases our ability to visualize and analyze the kinds of demographic changes that are taking place over time and space. Since the census data will be aggregated at a specific geographic scale, with variables such as proportion of adult women who are literate, the data from the image must also be aggregated up to that same level of geography. Thus, we must create summary indices of land-cover types and texture information that match the geography of the census data. It is easiest to explain this process through an illustration.

An Illustration from Greater Cairo and Menoufia, Egypt

The urban area of Greater Cairo comprises the governorate of Cairo on the east side of the Nile River as it travels through the metropolitan region, together with the portion of the governorate of Giza that is along the west bank of the Nile River within the metropolitan region, plus the southern tip of the governorate of Qalyubia – which currently represents the northernmost reach of Greater Cairo. The area's location is shown in Figure 17.1. Nearly one in five Egyptians lives in the Greater Cairo region and for centuries it has been a quintessentially primate city, dominating the social, economic, and political life of the region. Its location is a geographically strategic crossroads (Palmer-Moloney, 2001). 'When you see Cairo in its full setting, the whole city suddenly makes sense. Look south and you can see the long flat river coming out of Africa; look west and you can see the first veins of the rich Delta; look north and the river is heading determinedly for the Mediterranean and for Europe ... Cairo itself is built at the meeting place of Africa, Egypt, Europe, Arabia, and Asia' (Aldridge, 1969, p.5). The United Nations Population Division (UN, 2003) lists the population of Cairo to be 9.5 million as of 2000 (the twentieth most populous city in the world), with a projected population of 11.5 million in 2015 (when it would be the nineteenth most populous).

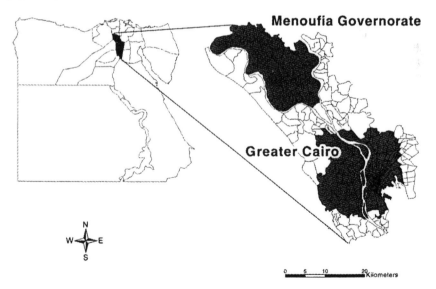

Figure 17.1 Greater Cairo and the governorate of Menoufia in context

Menoufia is a predominantly rural governorate (the equivalent of a state in the US or a county in the UK), just to the northwest of Greater Cairo. By government definitions of urban (which are based on administrative criteria), 80 per cent of Menoufia's population of about 3 million people resides in rural places. We have satellite imagery of Cairo and Menoufia acquired for 1996 (multispectral) and 1998 (panchromatic) and the census tract ('shiakha', literally the area controlled by a sheikh) boundaries from the 1996 census. Our goal is to combine data extracted from the image for each shiakha with data from the 1996 census for that area, in order to test the idea that data from the images will improve our ability to quantify the nature of human settlements.

Extracting Variables from Remotely-Sensed Imagery

As discussed above, there are two different types of variables that can be obtained from the image: (1) land-cover classification; and (2) texture. If the resolution of the image is sufficiently high (e.g. 1m or less), it may be that each pixel will represent only one type of land cover, and so we can make an accurate 'hard' classification. However, for lower-resolution images – the kind that are more readily available for different places and for different times – each pixel is likely to represent a mix of different land-cover types. As a result, a hard classification will probably represent only a part of the pixel and will inaccurately describe the remainder of the area covered by that pixel. A variety of techniques, including maximum likelihood classifiers, have been employed to try to increase the accuracy of the overall (or 'hard') classification of each pixel (see, for example, Curran *et al.*, 2000; Mesev *et al.*, 2001).

Our approach to classifying the urban scene is to employ a 'soft' approach, called spectral mixture analysis (SMA). Since our multispectral image has a resolution of 24 meters, we know that the probability is very low that any single land-cover classification will accurately represent a particular pixel. In the 'soft classification' approach, each pixel is assigned a class membership probability for each land-cover type. Fuzzy classification and SMA are two families of techniques designed to provide a 'soft' classification of mixed pixels. The basic difference between them is that SMA is based on a physical model of the mixture of discrete spectral response patterns (Roberts *et al.*, 1998), thus providing a deterministic method for addressing the spectral mixing problem rather than a statistical method as in the case of the fuzzy approach (Mather, 1999). SMA allows us to decompose each pixel into the percentage of the pixel that is represented by the major land-cover classifications that can be derived from the image. In this way, we create a profile for each pixel of its constituent parts, and by aggregating those values over the entire shiakha, we are able to define the land cover of the shiakha in terms of the percentage of the earth's surface that is covered by particular types of cover. We have thus far favored the use of SMA over fuzzy-classification techniques, because it serves our purpose of deriving standardized and comparable RS measures that can be utilized with census data in a GIS to study demographic dynamics in urban areas.

SMA was developed initially for use in classifying the natural environment, but we have shown that it also makes sense for urban environments. The details of the procedure are found elsewhere (Rashed *et al.*, 2001), but here it is worthwhile discussing some of the underlying assumptions. Spectral mixing occurs when the spectrum measured by a sensor is a mixture of the spectral response of more than one component within the scene (Adams *et al.*, 1993). That is, various materials with different spectral properties are represented by a single pixel on an image. A spectral mixture model is a physically-based model in which a mixed spectrum is modeled as a combination of 'pure' spectra, called *endmembers* (Adams *et al.*, 1993; Roberts *et al.*, 1998). Linear SMA is the process of solving for endmember fractions, assuming that the spectrum in each pixel on the image represents a linear combination of endmember spectra that corresponds to the physical mixture of some components on the surface, weighted by surface abundance (Tompkins *et al.*, 1997).

The conceptual model selected to extract image endmembers from the RS data is Ridd's VIS model (Ridd, 1995). The VIS model represents the composition of an urban environment as a linear combination of three types of land cover, namely green Vegetation, Impervious surfaces, and bare Soil. Just as soils may be described in terms of their proportions of salt, silt, and clay using the traditional triangular diagram, so various subdivisions of urban areas may be described in terms of proportions of vegetation, soil, and impervious surface. Ridd's VIS model offers an intuitively appealing link to the spectral-mixing problem, because the spectral contribution of its three main components can be resolved at the subpixel level using the SMA technique. The model was originally applied to American cities, but it has also been tested with data from Australia (Ward *et al.*, 2000) and Thailand (Madhavan *et al.*, 2001). The results show that the model is robust outside the United States, although the model may require an additional component (e.g. water/shade) to achieve an accurate characterization of the morphology of non-US cities.

Successful SMA application relies on the accuracy of endmember selection. If the endmembers are incorrect in the physical sense, then the fractional abundances are also incorrect and the results of SMA become meaningless. The selection of endmembers can generally be done in two ways, either by deriving them directly from the image (referred to as *image endmembers*), or from field or laboratory spectra of known materials (referred to as *reference endmembers*). Since we do not have reference endmembers collected from the study sites, we use image endmembers in the SMA stage. Several methods of identifying image endmembers have been described in the literature (Milton, 1999). The approach we have adopted to select image endmembers is a compromise between manual and automated ('Purity Pixel Index') approaches (see Boardman, Kruse, and Green, 1995).

The result, then, is a set of endmember fractions representing the percentage of each of four different types of land cover, classified from the modified VIS model as vegetation, impervious surface, bare soil, and our addition to the model of water and/or shade. Texture transform analyses are then used to add additional information to the classification results, to quantify the degree of variability in the image from one pixel to the next. After classifying the imagery data and obtaining quantifiable indicators of variations in the physical environment, they are linked to census-derived variables within the GIS environment. To do so, the several classes, as well as the texture bands, are converted into raster grids such that each class is represented by a single grid. The next step is to convert each grid into a polygon coverage, with each pixel being converted into a polygon with an area matching that of the remotely-sensed image's resolution (in this case 24 square meters). Resultant coverages are then clipped to the outside borders of the study area. Next, a coverage of the shiakhas is laid over each class's coverage and a unique code is assigned to each pixel (represented by a polygon) according to which census tract (shiakha or locality) it was located within. For each class coverage, the area of all pixels belonging to the same census tract is summed, and the total area is then normalized by calculating the ratio of the resultant area to the census tract's area. The end product is a normalized value (ranging from 0 to 100) for each of the classes assigned to each census tract, indicating the percentage of pixels in that census tract that are classified into this particular class. Each class represents a different aspect of the urban environment (such as areas dominated by tall buildings, areas dominated by low-rise buildings, areas with vegetation, areas with water or shade, areas that are highly variable in texture, and areas that are homogenous with respect to texture). These classes thus represent the variables derived from the images, and the percentage of pixels in each class is the measurement of that variable within that geographic area.

Predicting an Urban Gradient from the Remotely-Sensed Image

The quantification of human settlements in terms of their degree of urbanness should permit us to move from thinking in terms of a rural-urban dichotomy to the notion of an urban gradient. This builds on the concepts inherent in Christaller's (1966) central-place theory, but with the modification that, while most theories of urbanization take each city as the unit of analysis, our approach is to look at each city as a dynamic region undergoing constant change with respect to its urban characteristics. In the gradient conceptualization, we would expect that the geographic center of a city would be most urban, and that urbanness would decline with distance from the center. Over time, of course, we would expect to find that diffusion of urbanness, and various processes such as counterurbanization (Champion, 1989), would contribute to an increase in the urban characteristics at increasing distances from the center. Indeed, this might lead to mutations into multiple nuclei as some areas away from the center take on the characteristics that at an earlier time were solely the properties of the center.

Population density and nonagricultural activities are the most often used indices of an urban place, and have been used as the initial measures of an urban gradient in Cairo and Menoufia, with the task then of evaluating how well they are predicted solely on the basis of the information derived from the remotely-sensed images. In other words, how closely associated are the usual definitions of urban places with the characterization of the environment context that includes information obtained from a satellite image?

We have data for 300 shiakha in Greater Cairo, and an additional 314 shiahka in the neighboring rural governorate of Menoufia. Population density in Cairo ranges from a low of 618 person per square kilometer (in an area on the edge of the Mukatim desert to the southeast of downtown Cairo) to a high of 359,000 (in the parts of south central Cairo dominated by high rise apartment buildings), with a mean for the city of 45,000. The distribution is highly skewed, with the highest densities being found in a relatively few neighborhoods and with most neighborhoods having more moderate levels of density. Density is generally lowest in the outer ring of Cairo, but there are some clearly defined spatial patterns that can be seen in Figure 17.2, which maps population density in Greater Cairo and Menoufia.

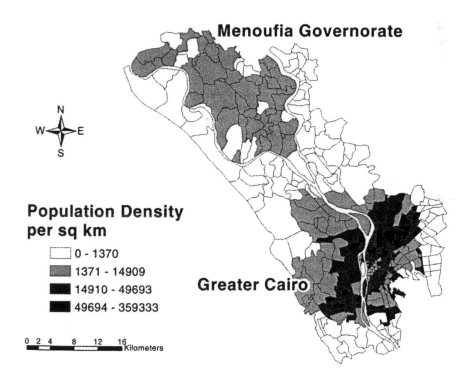

**Figure 17.2   The urban gradient in Cairo and Menoufia measured by
        population density**

In Menoufia the highest density is 14,138 per square kilometer, with a low of 145 and an average density of 1,851. That density definition includes the entire area of each shiakha. If we look only at the built area of each shiakha, we find that the rural villagers are actually living much more densely. The built area of the average village comprises 10 per cent of the land area of the shiakha. If we assume that all people in the shiakha reside within the boundaries of observable built area, then the average population density in Menoufia is just under 45,000 persons per square kilometer, almost exactly the same as in the Greater Cairo area. However, to be consistent with the general principle of defining density as population per total area, I have used the standard definition of population density in this analysis.

Economic activity was measured in terms of the percentage of economically active males aged 15 and older who were employed in any sector other than farming, fishing, hunting, or mining. Within Greater Cairo, this ranged from a high of 100 per cent (largely in the center of the city) to a low of 41 per cent in the northernmost shiakha, which is actually in the governorate of Qalyubia, with an average of 97 per cent of each shiakha's labor force being employed outside of agriculture. Even in Menoufia, despite its characterization as a rural governorate, the average village had 65 per cent of males employed outside of agriculture, with a range from a high of 94 per cent in a northeastern shiakha to a low of 29 per cent in the southern tip of the governorate. The bivariate relationship between these two traditional measures of urbanness is only 0.580, indicating an interaction between the two variables, but also some relative independence: they are not measuring exactly the same phenomena. However, the spatial pattern is very similar for the two variables with the exception that density varies quite a bit in the center of Cairo, whereas the percentage of males employed outside of agriculture does not.

A combined index was created by averaging the z-scores calculated for each of the two variables. This implicitly weights each variable equally and additively. The spatial distribution of this variable is shown in three dimensions in Figure 17.3. The general pattern is for the center of the city to be more 'urban' with the index of urbanness dropping off especially in the northwestern direction, and to be low in Menoufia.

Now the question is whether these measures of urbanness within Greater Cairo and Menoufia can be predicted by variables extracted from the remotely-sensed image. We have five predictor variables from the image: (1) the percentage of an area that is classified as vegetated (VEG); (2) the percentage of an area that is classified as representing soil or materials (such as bricks) made from local soil (SOIL), (3) the percentage classified as impervious surfaces (such as concrete or asphalt roofs or roads) (IMP); (4) the percentage classified as water or shade (which will largely be derived from the shadows of buildings) (SHD); and (5) a texture measure that indexes contrast from one pixel to another (CON).

Table 17.1 shows the results of the bivariate correlation coefficients between each of these variables. Overall it can be seen that the correlations are quite high between each of the measures from the satellite imagery and both of the urban definition variables. The contrasting-texture variable emerges as the single best predictor of both population density and the percentage of males in nonagricultural activities, but the soil and vegetation variables are close behind. The closeness of fit of all of these variables suggests two things: (1) the remotely-sensed data can provide a good proxy for the census-based variables; and (2) by combining the variables we may be in a position to better model the urban-rural continuum.

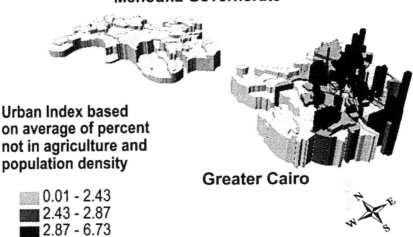

Menoufia Governorate

Urban Index based on average of percent not in agriculture and population density

- 0.01 - 2.43
- 2.43 - 2.87
- 2.87 - 6.73

Greater Cairo

Figure 17.3 The urban gradient in Cairo and Menoufia measured by a variable combining population density and proportion of males not in agriculture

Table 17.1 Bivariate correlation coefficients between census-based measures of urban-rural and the variables derived from the remotely-sensed imagery

	Population per sq km	% males not in agriculture	Contrast in texture (CON)	Vegetation (VEG)	Shade/ water (SHD)	Impervious surface (IMP)	Soil (SOIL)
Population per sq km	---	0.580	-0.617	-0.613	-0.472	0.614	0.578
% males not in agric		---	-0.827	-0.803	-0.596	0.699	0.834
CON			---	0.786	0.670	-0.702	-0.881
VEG				---	0.530	-0.739	-0.835
SHD					---	-0.754	-0.678
IMP						---	0.781
SOIL							---

The ability of the remotely-sensed data to predict the census-based variables was tested through step-wise ordinary least-squares regression, and the results are shown in Table 17.2. With population density as the dependent variable, four of the five variables from the image emerge as statistically significant predictors of population density. The most important of these, as measured by the standardized beta coefficient, is the amount of contrasting texture (variability at the earth's surface). Texture is negatively associated with density, reflecting the greater variability of the surface in those places where density is low. Table 17.2 shows that the higher the fraction of impervious surface, the higher the population density, as one might expect. Also in the expected direction are the associations of more vegetation and more soil with lower densities. Overall, these variables combine to explain 46 per cent of the variation from shiakha to shiakha in Greater Cairo and Menoufia governorate with respect to population density. The z-normalized Moran's I is calculated as a measure of spatial autocorrelation in the residuals, and the value of 1.50 suggests that there is not a statistically significant amount of spatial autocorrelation.

Table 17.2 Predicting census-based urban indices from the remotely-sensed images

Predictor variable	Dependent variable: Population density		% males not in agriculture		Z-normalized average of density and % males not in agriculture	
	Beta	Significance	Beta	Significance	Beta	Significance
VEG	-0.273	0.000	-0.292	0.000	-0.312	0.000
SOIL	-0.268	0.001	0.283	0.000		ns
IMP	0.344	0.000		ns	0.212	0.000
SHD		ns		ns		ns
CON	-0.392	0.000	-0.348	0.000	-0.415	0.000
Adjusted R^2	0.463		0.757		0.742	
Moran's I (z-normal)	1.50		1.14		1.42	

See Table 17.1 for key to predictor variables. Blank cells indicate that the beta coefficients were not statistically significant at or beyond p=0.05.

Using the proportion of the male labor force in nonagricultural jobs leads to slightly better results (Table 17.2, middle columns). Once again, the contrasting texture is negatively associated with nonagricultural economic activity and, not surprisingly, the percentage of an area classified as vegetated is strongly and negatively predictive of this aspect of being an urban place. The percentage classified as soil is positively associated, probably indicating that more builtup areas, even when made of brick, are less likely to be associated with agricultural occupations. Overall, these variables combine to explain 76 per cent of the variability in the nonagricultural percentage, and Moran's I once again indicates that there is no statistically significant pattern of spatial autocorrelation in the residuals.

The index that combines population density and the proportion of the labor force in nonagricultural jobs (Table 17.2, right-hand columns) also picks up the most significant predictor variable from each of those two dependent variables. Thus, for the combined index, the most important predictor among the remotely-sensed image variables is the contrasting texture, followed by the amount of vegetation, which is negatively associated with urbanness. The other statistically significant predictor variable is the amount of impervious surface, which is positively associated. The other variables drop out of the model. Overall, these three variables extracted from the remotely-sensed image combine to explain 74 per cent of the variation in this combined density/nonagricultural index.

Using the Remotely-Sensed Variables to Create an Index of Urbanness

Given the high correlations between the usual indices of urban and the data extracted from the remotely-sensed image, I have created an index of urbanness that represents the combination of data from the remotely-sensed image and the more traditional measures of urban place. Principal components analysis (PCA) was used to combine the five variables derived from the remotely-sensed images with population density and the proportion of the male labor force engaged in other than agriculture. The PCA provides a convenient way to weight each of the variables and combines them using the statistically-derived component score coefficients. All seven variables loaded into one component, with nearly equal component score coefficients. In essence, the resulting index amounts to having added up the z-scores for each variable, in much the same way as for the index that combined only population density and the proportion of the male labor force engaged in nonagricultural activity.

This composite index was normalized so that the lowest score was zero and the highest score was 1. This produced a mean of 0.35 with a standard deviation of 0.24. Because of the high intercorrelation among variables, the spatial distribution of urbanness is similar to that which was generated by other measures, as can be seen in Figure 17.4, in which a three-dimensional map is used so as to better visualize the urban gradient. That map shows a gradient of urbanness from the center, particularly on the east side of the Nile (the older part of Cairo) spreading out especially to the newer urban areas on the west side of the Nile (in Giza governorate) and as Cairo stretches into the farmland of the delta, with less urbanization in Menoufia. Nonetheless, there is variability in urbanness, even in predominantly rural areas.

It was hypothesized that these differences in urban characteristics, representing aspects of both the built and social environments, would be associated with differences in the demographic characteristics of places. Although we do not have a great deal of detailed data with which to test this, Table 17.3 shows the average values for three different types of population characteristics drawn from the 1996 census of population, for each of ten decile categories of the urban-gradient index, where 1 indicates the lowest level of urbanness among the 614 shiahka in Greater Cairo and Menoufia and 10 represents the highest level. The table also shows the average values for each variable that is a component of the urban-gradient index.

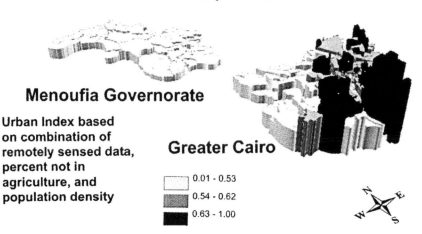

Menoufia Governorate

**Urban Index based
on combination of
remotely sensed data,
percent not in
agriculture, and
population density**

Greater Cairo

☐ 0.01 - 0.53
▨ 0.54 - 0.62
■ 0.63 - 1.00

Figure 17.4   **Urban-gradient index combining population density, proportion
of males not in agriculture and remotely-sensed variables**

Table 17.3   **Three demographic characteristics and urban-gradient index
variables for areas of Greater Cairo and Menoufia, according to
level on the urban-gradient index**

Urban grad-ient in deciles	% popu-lation aged under 15	% women 15+ never married	% women 15+ with more than primary education	% males not in agri-culture	Popu-lation per sq km	CON	VEG	SHD	IMP	SOIL
							Variables in the urban-gradient index			
Least urban										
1	37.6	21.7	28.5	55.9	1,375	194.8	30.7	41.1	28.2	2.2
2	37.4	22.2	28.7	61.6	1,700	195.2	27.3	40.1	31.9	2.3
3	36.9	21.8	31.9	65.3	1,531	194.5	25.9	39.2	34.3	2.4
4	37.5	22.1	31.8	66.3	1,771	192.1	23.8	38.8	37.6	1.9
5	37.1	21.8	34.9	73.3	2,067	190.6	22.1	37.1	39.7	2.3
6	31.9	21.9	41.7	89.0	12,266	105.8	16.6	31.9	45.1	12.1
7	23.2	27.5	52.0	98.7	35,139	58.9	2.9	29.2	50.3	20.7
8	23.2	28.2	52.1	99.0	43,782	54.5	1.7	25.8	54.6	22.0
9	26.8	27.5	49.5	99.0	51,632	54.7	1.2	21.0	59.9	22.7
10	28.6	25.6	43.1	99.0	79,990	58.0	0.9	15.4	68.5	25.6
Most urban										

See Table 17.1 for key to remotely-sensed variables.

These data show that the percentage of the population that is under 15 years
old is considerably higher in the lower few deciles, and then drops after the fifth
decile is reached. Conversely, the percentage of women aged 15 and older who are
never married (a proxy for age at marriage) is lowest in the least urban deciles and
then rises beginning in the seventh decile. There is a more striking trend with
respect to the education of women. In the lowest decile of urbanness only 28 per

cent of women aged 15 and older have more than a primary level of education, whereas that nearly doubles to 52 per cent by the seventh decile. It can be noted that the most urban decile (10) is not the one with the most urban demographic characteristics in terms of age structure, marriage patterns, and educational levels. This is partly a scale problem, and partly a problem with the influence of population density on the index of urbanness.

The scale issue relates to the fact that, as can be seen especially in Figure 17.2, the shiakha in our data set are not of uniform size. In particular, the shiakhas in the older part of Cairo are considerably smaller in area than those in the suburbs. As a consequence, these smaller geographic areas can be more influenced by one or two large high-rise apartment buildings. In Cairo, as in many cities of developing countries, these high-rise blocks tend to house younger couples with their children, thus raising the population density (a decidedly urban characteristic) without all of the other attendant urban characteristics being present. Indeed, the central part of Cairo, near Tahrir Square, would be thought of by most people as the most urban part of the city, but its density is lower than in those newer areas with high-rise apartments, and it has more vegetation and more textural variability than the newer concrete blocks of apartments.

These data are suggestive of the idea that, even within the boundaries of a large city, an index of urbanness allows us to distinguish among differences in social behavior, and even to rethink what 'urban' means. The less urban portions of Cairo are characterized by demographic features associated with more traditional attitudes toward women and families – less value placed on education of women, and more value placed on younger age at marriage and family-building activity – but so are the most urban places, with the more traditionally urban places between those extremes. This sort of variability is instructive because it forces us to recognize the spatial variability that exists within an area that is normally thought of as monotonically urban.

Discussion and Conclusion

Our interest is in developing an urban-gradient index for all inhabited areas (excluding wilderness and desert regions), and we have taken some preliminary steps to do this in Egypt, using data for Greater Cairo along with data for the largely rural governorate of Menoufia. Agricultural areas are usually defined almost automatically as rural because of the low population densities that are obviously associated with places in which a large fraction of the land is devoted to growing crops. However, in many developing countries like Egypt, the population in these places resides primarily in villages, rather than being dispersed across the countryside, and in fact the population density may be quite high in these places. In Menoufia, more than 80 per cent of the shiakha have at least 2,500 people, and nearly one-third have population densities within the built area of the village that are as high as, or higher than, those found in the suburbs of Cairo. Furthermore,

more than 90 per cent of these villages have more than 50 per cent of adult males involved in a nonagricultural economic activity.

Our analysis shows very clearly that, by every measure of urban, an urban gradient exists in the study area, going from least urban in Menoufia to most urban in the edges of central Cairo. The percentage of males not in agriculture rises slightly within the urban deciles in Menoufia, and then jumps to higher levels in the suburbs of Cairo. The percentage of females with more than a primary level of education shows a general tendency to rise slightly as the urban gradient increases in Menoufia and then it rises steeply with the increase in urbanness in Cairo. The percentage of women never married is consistently low in Menoufia and is similar to the suburbs of Cairo, but then it rises within Cairo. Conversely, the percentage of the population that is younger than age 15 is consistently high in Menoufia and similar to the levels in the lower urban deciles of Cairo, and then it drops as the urban deciles increase within Cairo.

This chapter has thus demonstrated the way in which the extraction of data from remotely-sensed images can increase our quantitative understanding of the nature of urban settlements. Nevertheless, there are many limitations to the analysis. The analysis is particularly limited by the modifiable areal unit problem (MAUP) defined well, for example, by Fotheringham and Wong (1991). The areas for which we have census data are those places that are defined for us by CAPMAS (Central Agency for Population Mobilisation and Statistics). The areal extent of newly defined places in the suburbs is very different from the older places in the center of the city and this complicates the analysis. On the other hand, if we are confident that the data from the remotely-sensed image can provide us with proxy data for human settlements, then we are in a position to define regular grids on the image and conduct an analysis that would dramatically reduce the impact of the MAUP because it would not be dependent on the census data and thus on the census boundaries. Our results suggest that the remotely-sensed imagery offers a very promising set of possibilities in that regard.

How to create a quantitative index of the urban gradient that best combines the variables is also open for discussion. We have employed an essentially reductionist approach in this analysis by using principal components analysis, but there are numerous other ways that indices could be constructed. The task ahead will be to see if several different methods yield similar results, increasing our confidence in the robustness of the use of remotely-sensed images. Although we cannot yet claim that the same classification scheme will lead to the same interpretation of urban areas in every area of the world, a fuzzy-set approach may allow us to make similar kinds of distinctions about urbanness in geographically very different places. Thus, we were able to show that Menoufia is predictably less urban than Cairo, but that the more urban places in Menoufia are similar to the less urban places in Cairo.

Data from the remotely-sensed images can be used in the analysis of urban processes such as urban sprawl (including exurbanization and an assessment of periurban areas), counterurbanization, and multinucleation. We require data from two or more dates in order to conduct such analyses, because it is change in places over time that we must measure. These analyses essentially measure the impact of human settlement by quantifying the change in the environment occasioned by human activity. This serves as a remotely-derived proxy for human behavior taking place on the ground that we might not otherwise be in a position to measure. Thus, the analysis of remotely-sensed images, and their inclusion in a geographic information system, offers us an additional set of data and insights by which to understand the nature and dynamic processes of human settlements and may offer ways of detecting change in the urban environment that is not measurable by other means.

Chapter 18

Reflections on the Review of Metropolitan Area Standards in the United States, 1990-2000

James D. Fitzsimmons and Michael R. Ratcliffe

This chapter describes and reflects upon the latest review of metropolitan area standards undertaken in the United States, the fifth since these standards were introduced for presenting the results of its decennial census of 1950. The Office of Management and Budget (OMB, 2000b) announced its 'Standards for Defining Metropolitan and Micropolitan Statistical Areas' in 2000. This marked the culmination of ten years' work which, in addition to OMB's efforts, included research and testing by the US Census Bureau, extensive review and guidance by a federal advisory committee that included representatives of six statistical agencies, and abundant comment from the public, government agencies and Congress.

Under the revised standards, the concept of a metropolitan statistical area or a micropolitan statistical area is that of a geographic entity containing a significant population nucleus, plus adjacent communities that have a high degree of integration with that nucleus. This general concept has remained essentially the same since the US federal government first defined metropolitan areas more than 50 years ago, and those familiar with the 1990 standards will see much that is familiar in this latest version. At the same time, however, none of the provisions of the 1990 standards remained unchanged, and some of the changes – such as the expansion of the classification to address areas with smaller population centers even as the standards themselves grew shorter and simpler – are substantial. Combined statistical areas receive explicit recognition for the first time, and the new standards introduce some changes in terminology.

The chapter begins by sketching the governmental context of the standards review. It then outlines the many steps the review involved, starting in 1990 and continuing through a period of intense activity at the end of the 1990s. It goes on to describe the main features of the revised standards and the rationale for the decisions behind them. This is followed by a glimpse into the workings of the new standards as afforded by a test with 1990 data, giving an indication of the changes likely when the 2000-based areas are announced in 2003. Finally, an account of unresolved issues emphasizes that the review forms part of a larger, continuing effort to improve depiction of the nation's settlement and activity patterns. In this way, we stress the major effort that can be involved in the task of identifying human settlement structures, as well as provide insights into the types of conceptual and practical issues that have to be addressed in such an exercise.

Historical and Contemporary Governmental Context

By the 1940s, it was evident that the value of metropolitan-scale data produced by US federal statistical agencies would be greatly enhanced if those agencies used a single set of geographic definitions for the nation's largest centers of population and activity. Growth patterns around major centers had become more expansive due to technology and rising incomes, and the patterns varied across the country. Recognizing the extent and variability of this expansion in the first half of the twentieth century, federal agencies individually had taken steps to address the issues in data reporting. A result of these agency efforts was a variety of statistical areas at the metropolitan level (including 'metropolitan districts', 'industrial areas', 'labor market areas', and 'metropolitan counties') that used different criteria applied to different geographic units. Variations in methodologies and the resulting inconsistencies in area definitions meant that one program's reported statistics were not directly comparable with another's statistics for any given area.

Starting in 1947, OMB's predecessor, the Bureau of the Budget, led a group of agencies that included the Census Bureau, the Bureau of Labor Statistics, and seven others in charting a course for providing standard statistical areas that would accommodate the changing settlement patterns of the day (Klove, 1952). That effort yielded what were then called 'standard metropolitan areas' in time for their use in 1950 census reports. Various agencies of the federal government continue to define a wide range of geographic areas for both statistical and nonstatistical purposes, but since the appearance of the standard metropolitan areas, comparable data products for metropolitan areas have been available.

Neither the metropolitan areas nor the standards used in defining them, however, remained static. OMB has reviewed and revised the standards in the years preceding their application to new decennial census data. The Metropolitan Area Standards Review Project (MASRP) that produced the 2000 standards was the fifth such effort. These reviews in general have been concerned with the standards' responsiveness to changes in the settlement and activity patterns of the country and the evolving expectations of data users. In the course of revisions based on the earlier reviews, however, the standards grew in length and complexity, and by the late 1980s and early 1990s OMB was concerned about the standards' conceptual clarity. Others, too, had complained that the standards were overly complex and burdened with *ad hoc* criteria. MASRP addressed, as its first priority, issues regarding conceptual and operational complexity.

Other reasons for a particularly thorough review of the 1990 standards included the fact that the settlement and activity patterns of the country had seen significant changes in recent decades. Developments in transportation and communications technologies, increased disposable incomes, and evolving preferences had translated into lower-density residential patterns, changes in commercial and industrial location practices, and new relationships between work and home, all of which had potential implications for statistical areas. How successfully had previous revisions of the standards taken account of these changes on the ground? Also, computer-related advances in data collection, storage, and analysis, and particularly enhanced capacities for data geocoding, made it feasible to consider working with smaller geographic units in constructing statistical areas. Using smaller units would offer the possibility of defining areas with greater geographic precision. Reflecting the range of concerns, OMB made clear that all aspects of the 1990 standards were subject to review and possible revision.

The Process of Reviewing the Standards

MASRP comprised many elements. Key review steps included publication of four *Federal Register* notices between 1998 and 2000.[1] OMB and the Census Bureau organized two conferences in Washington, both of these attracting individuals from around the country to discuss previously identified issues or raise new ones. A Congressional hearing in 1997 focused on specific issues in the definition of metropolitan areas. Also, as shown below, a number of publications came out of the review, including journal articles and a working paper that presented work completed in a set of external studies.

Led by OMB, MASRP depended heavily on the participation of others from both inside and outside of the federal government. The Census Bureau was responsible for much of the research effort throughout the decade, including both investigation of individual issues and framing entire definition approaches to facilitate and focus discussion. In 1998, OMB formed an interagency standards review committee, which provided detailed advice and recommendations during the period of intense activity leading up to the publication of the revised standards. Finally, a wide range of individuals and groups from outside federal government took advantage of opportunities to comment on proposed approaches in numerous forums provided by the review.

Early Years - Defining the Project

The Census Bureau established parallel research contracts with four universities in early 1991. These contracts were modest in size and intended to provide a reading of the nature and scale of the review task ahead. The charge to the investigators was broad: to suggest alternative approaches to representing the country's settlement structure. Drawing on discussions with two external working groups that it convened the previous year, the Census Bureau identified a number of issues on which the studies should focus:

- Geographic units or building blocks to be used in defining statistical areas.
- Criteria by which geographic building blocks would be grouped.

---

[1] The *Federal Register* is a US Government publication that provides notice of efforts underway or actions taken.

- Availability of data for statistical area definition.
- Relationships among statistical areas that would be defined.
- Recognition of geographic entities within statistical areas.
- Possible role for local opinion in area definitions.
- Frequency of updating.

Some of these issues had been part of discussions surrounding metropolitan areas since the 1940s (though they may have taken on new meaning in recent years). All of them remained prominent throughout MASRP.

An additional charge to the contracts' investigators was to evaluate the feasibility of addressing in a single approach the full range of the nation's settlement pattern, from the least intensely settled areas to the most intensely settled. Behind this request was a view coming out of the working group discussions that it would be useful to account for the entirety of the country in a new approach to statistical areas.

The results of the studies demonstrated dissatisfaction with the 1990 standards and suggested a range of possible alternative approaches. Two of the studies' reports, in particular, found the existing standards wanting in terms of their conceptual foundations, and all four studies suggested approaches that differed in key respects from the 1990 standards. The studies' conclusions also differed substantially from one another, notably in terms of type of geographic building blocks and the criteria for aggregating them.

In 1995, the Census Bureau assembled the reports from the four studies as a working paper (Dahmann and Fitzsimmons, 1995), which formed a centerpiece of discussion at the review's first conference. A key question presented there was whether government-defined metropolitan areas had retained their importance in an era when agencies and individuals were increasingly able to take advantage of powerful geographic information system software and define areas for their own purposes. The response of those assembled was that there remained a clear role for government-defined metropolitan (and nonmetropolitan) areas: namely, the role that met the original purpose of providing standard areas for which numerous agencies produce data. The conference also featured general agreement that the statistical areas defined should account for the entire territory of the United States and that the areas should be defined using the same set of rules throughout the country.

During the ensuing months, staff from the Census Bureau and OMB made presentations before various professional, data user, and local groups, describing major issues under discussion, steps taken, and plans for future steps. In 1997, concerns in several locales regarding qualification of metropolitan areas precipitated a Congressional hearing. Eleven Members of Congress sought information on the basis of existing (1990) minimum qualification requirements for metropolitan areas, with some being concerned that possible changes in standards might make that status more difficult to obtain. Information presented by OMB at the hearing stressed that all aspects of the 1990 standards were subject to review, presented a timetable for the review, and assured the Members of Congress that the review would be open and take into account local concerns.

Period of Intense Activity, 1998 to 2000

In 1998, OMB chartered the Metropolitan Area Standards Review Committee, charging it with examining the 1990 standards in view of the work completed earlier in the decade and providing recommendations for possible changes to those standards. The review committee included representatives from the Census Bureau (which chaired the committee), Bureau of Economic Analysis, Bureau of Labor Statistics, Bureau of Transportation Statistics, Economic Research Service, National Center for Health Statistics, and, *ex officio*, OMB. This committee would become one of the three principal sources of advice to OMB on the review, along with Census Bureau staff and the full range of public comment.

Earlier in 1998 staff at the Census Bureau had begun to evaluate systematically the information gathered in the preceding years and identified areas needing additional work. The goal of this effort was to prepare a detailed set of alternative area definition approaches that could feed into the work of the review committee, along with reaction from data users. In the end, the Census Bureau delineated and presented to OMB four separate approaches. This work was published as the first of the MASRP's four *Federal Register* notices (OMB, 1998), along with a request for comment, and was discussed at the review's second conference that took place in January 1999.

These discussions, together with 40 written comments prompted by the notice, formed the starting point for a series of meetings of the review committee, 38 in all. The review committee delivered to OMB a draft of the proposed standards, which was published in a second notice (OMB, 1999), again with a request for comment. This time, with more tangible issues at hand and clear movement towards final standards, the notice drew 673 comment letters, a large share of which (over 470) was the product of a campaign mounted in one area to find a way in which it would qualify for metropolitan recognition.

Following further meetings, the committee forwarded proposed (revised) standards to OMB, which published them in an August 2000 *Federal Register* notice with a request for comment (OMB, 2000a). This third request for comment yielded 1,672 letters, but, as before, many of these (more than 1,450) were the products of campaigns. After several more meetings, the committee then conveyed its final proposed standards to OMB, with the final Standards for Defining Metropolitan and Micropolitan Statistical Areas being published in a notice on December 27, 2000 (OMB, 2000b). This notice included OMB's comments on the review committee's recommendations and made some minor changes in the committee's work.

Outcomes of the Review

This section briefly describes the major features of the revised standards as published by OMB (2000b) and then goes on to present the rationale for the decisions made on key issues. The standards use densely-settled urban cores of at least 10,000 population as the basis for identifying statistical areas that portray the extent of social and economic ties between communities, as measured using journey-to-work data. The standards yield two principal types of core based statistical areas (CBSAs): 'metropolitan statistical areas', defined around at least one Census-Bureau-defined urbanized area of 50,000 or more population, and 'micropolitan statistical areas', defined around at least one urban cluster of at least 10,000 and less than 50,000 population (Figure 18.1).

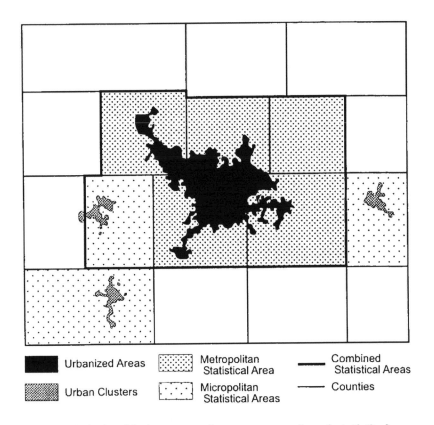

■ Urbanized Areas   ▦ Metropolitan Statistical Area   ── Combined Statistical Areas

▦ Urban Clusters   Micropolitan Statistical Areas   ── Counties

Figure 18.1   Relationship between urban cores, core-based statistical areas and combined statistical areas

Counties and equivalent entities are used as geographic building blocks throughout the United States and Puerto Rico (also, a conceptually similar set of areas will be defined in the six New England states using minor civil divisions). The largest metropolitan statistical areas are subdivided into smaller groupings of counties, or 'metropolitan divisions', based around locally important employment centers. Adjacent CBSAs that have sufficient employment interchange are grouped together to form larger statistical regions, referred to as 'combined statistical areas'.

Geographic Building Blocks

The choice of geographic unit to be used as a building block had implications both for the nature of the areas to be defined and for the ability of federal agencies to provide statistical data for the resultant areas and their components. When reaching a decision on the building block issue, the review committee and OMB considered the following points, all of which supported the use of counties and equivalent entities when defining metropolitan and micropolitan statistical areas throughout the United States:

- Consistency. Use of consistent geographic building blocks makes possible meaningful comparisons of statistical areas nationwide.
- Data availability and utility. Data for a geographic building block must be available from a variety of sources if the resulting statistical areas are to serve as a standard across agencies and be useful to a wide range of users.
- Stability of boundaries. The effort required to maintain data bases that include statistical areas and the usefulness of portraying demographic and economic change in areas over time put a premium on use of building blocks with stable boundaries.
- Familiarity. The building block should be meaningful and recognizable to data users.

Whether to use counties or county subdivisions as building blocks had been a central issue in the late 1940s during development of the standard metropolitan areas. Resolution of the issue at that time in favor of counties gave precedence to greater availability of data at the expense of superior geographic precision. The concerns raised in the 1940s about geographic precision, however, have never gone away and were central issues again for MASRP. Counties vary considerably in land area and generally are larger in western states than in eastern states. The smallest county located within a current metropolitan area is New York County, New York (which comprises Manhattan), with a land area of 59 square kilometers. At the other extreme, the largest metropolitan area county is San Bernardino County, California, which covers 51,842 square kilometers (a larger area than Switzerland). The typical county (1,596 square kilometers) is considerably smaller than San Bernardino, but small counties also can contain territory and population that are not integrated with any core of at least 10,000 population.

Alternatives to counties, however, each lost out on more than one criterion. The key disadvantage of all of the smaller entities considered was the relative paucity of the federal statistical data (especially economic data) that is collected, tabulated and published for them. In addition, census tracts have less stable boundaries than counties and are less recognizable to data users, while county subdivisions pose difficulties of consistency across the country and are not well known in some states and, finally, ZIP Codes (i.e. postal codes) lack specified boundaries and are inconsistently represented as geographic areas.

The appeal of greater precision will keep the discussion of appropriate geographic building blocks alive for future reviews, when perhaps geographic bases for statistical programs will have been enhanced. Research will continue to evaluate the use of census tracts, in particular, for special-purpose areas or complementary classifications. Already, the Economic Research Service, in conjunction with the federal Office of Rural Health Policy and the University of Washington, has developed a nationwide rural-urban commuting area classification using census tracts (Morrill *et al.*, 1999).

Form and Function in Statistical Area Definitions

The methodology incorporated in both the 2000 standards and their predecessors has taken into account the observation that large centers have both form and function. That methodology uses such variables as population size and density to measure the form, or the structural component, of the centers; i.e., what we see in the landscape. Settlement form is the basis of the Census Bureau's urbanized areas and urban clusters (which are collectively termed 'urban areas'). The functional component – interactions of people and activities among places as measured by commuting flows – is key to determining the centers' daily reach. Throughout the review of the standards, substantial agreement existed that population density and daily commuting continue to be the best means for defining areas consistently nationwide. At the same time, however, many observers concur that both the structural and functional components of cities and their surroundings have changed significantly since metropolitan areas were first defined (Adams, 1995; Berry, 1995; Dear and Flusty, 1998; Adams *et al.*, 1999).

Metropolitan and micropolitan statistical areas are representations of the process of urbanization, which after Harvey (1989b) can be characterized as the 'flow of capital across space'. This process is evident in both the built environment (buildings, houses, and roads) that represents the structural component of urbanization and in the functional interactions between urban cores and outlying communities. How to measure and account for both the structural and functional elements in the urbanization process was central to the review of the standards. MASRP discussion considered whether evolving satellite imagery or geographic information systems technologies had yielded gains that could be captured for at least some aspects – the structural component more than the functional aspects – of statistical area definition. The review concluded, however, that in large part the measures of the structural and functional aspects of the urbanization process used in previous standards are still relevant.

The structural aspect of the urbanization process is represented statistically by the Census Bureau's urban areas, which are based on population density. The functional aspect of the urbanization process is manifested in the less tangible (but more or less measurable) flows of people, goods and services, ideas, and other cultural phenomena streaming into, out of, and around the core of a metropolitan or micropolitan statistical area. The extent of a metropolitan or micropolitan statistical area reflects a field of influence surrounding an urban core, but it does not imply that the entire area is urban in a structural sense. Instead, these are 'daily urban influence areas' in which urban, suburban, and rural areas are associated with an urban center (or centers) through a set of functional relationships as measured and demonstrated by journey-to-work patterns.

The role of individual places and population nuclei in the CBSA concept figured prominently in the review. For the 1950 census, a city of 50,000 or more residents was necessary for designation of a metropolitan area. Over time, it became possible also to designate a metropolitan area on the basis of 'twin cities' totaling 50,000 or more population and, eventually, designation came to be based on the presence of either a city or an urbanized area of 50,000 or more population. The single central city of 50,000 or more persons retained conceptual primacy, however, since designation on the basis of an urbanized area required in addition that the entire area have a total population of at least 100,000.

Observable changes in the urban landscape suggest that individual places are becoming less important than the network structure itself, and places become nodes in a complex system of social and economic linkages created and organized under constantly shifting economic and political circumstances. These developments point to the growing interdependence of places in general and some blurring of individual place identities, underscoring the need to identify metropolitan and micropolitan statistical areas as a means of describing socially and economically integrated regions. Also, in combination with continuing decentralization of population and urban functions away from historical central cities, the changing balance of network and node suggests that individual central cities of 50,000 or more population are no longer appropriate starting points for identifying daily influence areas. Instead, urban cores, comprising both the historical central city as well as other decentralized economic centers that have developed in the inner and outer suburbs, are the organizing entities that dominate and influence their surrounding regions.

Central Counties and Statistical Area Categories

The locations of urban areas with at least 10,000 population are used to identify the 'central counties' of metropolitan and micropolitan statistical areas – the counties to and from which journey-to-work volumes will be measured in order to determine the extent of the CBSAs. The review committee advised that categorization as metropolitan or micropolitan statistical areas should be based on the population in the urban areas they contain, reasoning that the range of services and functions provided largely derive from the sizes of the cores. The specific issue of whether to categorize a CBSA based on the population of the most populous core or on the total population of all cores within it received considerable attention throughout the review. In the end, the review committee and OMB concluded that a single core of 50,000 or more population provides a wider variety of functions and services than does a group of smaller cores, even when such a group may have a collective population greater than 50,000. A single core of at least 50,000 population, therefore, is required for qualification as a metropolitan statistical area. Although the available literature on urban form and function supports this decision, such research has tended to focus on individual cities rather than urban agglomerations. The lack of scholarship focusing on the social and economic influence of urban agglomerations frustrated the review, requiring it to draw inferences from the existing body of work (e.g., Esparza and Krmenec, 1996). Subsequent reviews of the standards will need to reconsider this issue as urban forms and structures continue to evolve in the United States.

Functional Integration and Outlying County Qualification

The metropolitan area concept has relied heavily on functional integration in determining geographic extent. In the original definitional work on standard metropolitan areas, functional integration was measured in terms of telephone-call patterns. The 1960 census was the first to collect commuting data, which the metropolitan area program then began using as its measure of functional interaction between central counties and potential 'outlying counties'. The functional measure specified in the 2000 Metropolitan and Micropolitan Statistical Area Standards is again the journey to work.

The 1990 and earlier standards required the use of other, structural measures of 'metropolitan character', in addition to the functional measure of journey to work, for determining the qualification of outlying counties and thus the geographic extent of a metropolitan area. Potential outlying counties had to satisfy these additional measures, which included specified levels of population density and percentage of the population in urban settlements, to qualify for inclusion in a metropolitan area, regardless of the intensity of the journey-to-work patterns. Whether the combination of functional and structural measures was still appropriate was a central issue in the review of the standards. During MASRP, Census Bureau staff and the review committee reached the conclusion that the relationship between the metropolitan character measures and social and economic linkages had become less meaningful over time in the United States. OMB supported this conclusion, so the structural measures of metropolitan character have no role in outlying county qualification in the revised standards.

MASRP considered functional integration measures other than commuting, focusing in particular on whether new or improved data sets that could meet definition needs had become available from other agencies or the private sector. For instance, some commenters suggested using data on shopping and recreational activity patterns in addition to commuting patterns. Others, recognizing that social and economic linkages now take place on the 'information superhighway', suggested internet usage patterns for determining the extent of areas. A solid body of research about the relationships between telecommunications networks, internet usage, and urban form and function was lacking at the time of the review, though there are indications that this situation is changing (e.g., Moss and Townsend, 1998; Wheeler *et al.*, 2000; Walcott and Wheeler, 2001). Such information, as well as other data sets pertaining to daily and weekly activity patterns of households and media penetration, would result in a more comprehensive definition of interactions between central and outlying counties, but currently there are no nationally consistent and easily accessible data sets for any of these variables.

Three priorities guided the decision to continue using journey-to-work data as a measure of integration between central and outlying counties: data used to measure functional connections among counties should describe those connections in a straightforward and intuitive manner, be collected using consistent procedures nationwide, and be readily available to the public. These priorities pointed to the use of data gathered by federal agencies and, more particularly, to journey-to-work data from the Census Bureau. The percentage of a county's employed residents who commute to the central county or counties of a CBSA is seen as an unambiguous measure of whether a potential outlying county should qualify for inclusion. The percentage of employment in the potential outlying county accounted for by workers who reside in the central county or counties is a similarly straightforward measure of ties, and including both criteria in the revised standards addressed the conventional, and the less common reverse, commuting flows. Census data are the product of nationally uniform collection procedures and are available to all.

Changes in daily mobility patterns and increased interaction between communities are indicated by increases in intercounty commuting over the past 40 years (Pisarski, 1996), and the 2000 standards require a higher minimum level of commuting to work than was previously required. The percentage of workers in the United States who commute to places of work outside their counties of residence increased from approximately 15 per cent in 1960 to nearly 25 per cent in 1990. Raising the commuting percentage required for qualification of outlying counties from the 15 per cent minimum of the 1990 standards to 25 per cent was appropriate against this background of increased overall intercounty commuting coupled with the removal of all metropolitan character requirements from the outlying county criteria. In other words, since out-of-county commuting has become more commonplace, a higher percentage of commuting is necessary to demonstrate ties comparable to those indicated by a lower commuting rate in 1960.

MASRP also considered the 'multiplier effect' that commuters would have on the economies of the counties in which they live. The size of the multiplier effect depends on the size of a region's economy and the employment base, but a review of the literature indicated that a multiplier of two or three generally is accepted for most areas. Applying a multiplier of two or three with the 25 per cent minimum commuting requirement means that the incomes of at least half of the workers residing in the qualifying outlying county are connected either directly (through commuting to jobs located in the central county) or indirectly (by providing services to local residents whose jobs are in the central county) to the economy of the central county or counties of the metropolitan or micropolitan statistical area.

Merging Contiguous Areas

In addition to considering the ties between central counties and outlying counties, the 2000 standards recognize that close social and economic ties, as measured by commuting, exist between some contiguous areas based on separate cores. When strong ties exist between the central counties of two contiguous areas, similar to the ties between an outlying county and a central county or counties, the standards call for merging the two areas to form a single metropolitan or micropolitan statistical area. Areas that merge could feature two or more cores of similar size in close proximity or a large core with smaller 'satellite' cores distributed around it hinterland.

Principal Cities

Because Census Bureau-defined urban areas are the cores of CBSAs, the identification of 'central cities' as required by previous metropolitan area standards is no longer necessary for qualifying and defining areas. Nevertheless, specific cities within individual metropolitan and micropolitan statistical areas are important for analytical purposes as centers of employment, trade, entertainment, and other social and economic activities. The 2000 standards use a variety of population and employment data to identify these 'principal cities' within each area. These should capture many previous central cities as principal cities, but the new standards also recognize recent changes in the US urban landscape by making it possible for some newer outlying employment centers to be identified as principal cities, too.

Metropolitan Divisions

In the largest metropolitan statistical areas (those with cores of 2.5 million or more population), patterns of interaction between counties can be complex and often focused around multiple economic and social centers. Some of these centers and groupings of counties around them would constitute sizable areas on their own in different settings. The usefulness of subdividing a number of the largest metropolitan statistical areas was evident when analyzing interactions within (1) expansive urban regions that contain multiple historical employment centers, (2) areas containing two or more large cities, each with its own observable sphere of influence, and (3) areas in which suburban counties have emerged as important employment centers.

The 2000 standards identify two types of counties around which to define metropolitan divisions. In the first category are those counties that are strongly self contained, as demonstrated by high percentages (65.0 per cent or greater) of employed residents who remain in the county to work and by high ratios of jobs to resident workers (0.75 or greater). These 'main counties' can stand alone as social and economic units within the larger metropolitan statistical area or provide the social and economic center around which a group of counties is organized.

A second category of counties consists of those with high ratios of jobs to resident workers but a lower percentage of employed residents working within the county (50 per cent to 64.9 per cent). These 'secondary counties' can be identified as social and economic centers, but they also connect strongly with one or more adjacent counties through commuting ties. Such counties are only moderately self contained and can provide the organizing basis for a metropolitan division only when paired with one or more counties of similar or greater economic strength. As such, they must combine with another secondary county or with a main county when forming the basis for a metropolitan division. The standards assign the remaining counties of the metropolitan statistical area to the main and secondary counties with which they have the highest rates of commuting interchange. Metropolitan divisions, if present in a metropolitan statistical area, will account for all of its territory.

Combining Statistical Areas to Form Larger Regions

The standards recognize ties between contiguous areas that are less intense than those captured by mergers, but still significant, by combining those areas. Because a combination thus defined represents a relationship of moderate strength between areas, the combining areas will retain their identities as separate metropolitan or micropolitan statistical areas within the combination. Combined statistical areas will offer an additional level of statistical geography for those seeking to analyze demographic and economic data for larger regions.

The professional literature sheds little light on the conceptual issue of when two neighboring metropolitan statistical areas may be considered a single area. Local experience frequently offers contradictory advice, as some neighboring areas with modest commuting interchange regard themselves as unified, even as other pairs with much stronger commuting ties see themselves as having little in common. The standards addressed this range of uncertainty by requiring combinations if the level of employment interchange is substantial, but by calling for local opinion in cases when the interchange meets minimum requirements but is weaker.

Frequency of Updating

The frequency with which new metropolitan and micropolitan statistical areas are designated and existing areas updated was of considerable interest to data producers and users throughout the review. In the past, many observers argued for minimizing changes in area definitions during the course of an intercensal decade to ensure that databases could be maintained consistently and economically. The counterargument was that definitions should be updated to reflect changed conditions as rapidly as the data permit. The frequency of updating depends in part on decisions concerning geographic building blocks, criteria for aggregation and, ultimately, data availability. Recent practice has been to review data for potential qualification of new areas annually, using Census Bureau population estimates and special censuses. During MASRP, a consensus emerged around annual updates for designations (i.e. defining new areas) and five-year reviews of the definitions of existing areas, the latter being frequent enough to capture changing conditions but without impairing database stability. OMB's current plan is that, following announcement of the new statistical areas in 2003, there should be annual designation of new areas based again on Census Bureau population estimates or special censuses as in the past. Meanwhile, updates of existing areas will await the availability of commuting data for all counties from the Census Bureau's American Community Survey in 2008.

Application of the Metropolitan and Micropolitan Statistical Area Standards with 1990 Census Data

When the new standards were tested with 1990 census data, it was found that metropolitan and micropolitan statistical areas accounted for approximately 195.9 million people, or 90 per cent of the US population (Table 18.1).[2] By comparison, actual metropolitan areas defined using the 1990 standards applied to 1990 census data accounted for 80 per cent of the US population. The metropolitan and micropolitan statistical areas defined under the new standards with 1990 data covered approximately 3.6 million square kilometers, accounting for nearly two-fifths of the nation's territory – nearly twice the 20 per cent covered by the actual 1990-based metropolitan areas. Table 18.1 also shows the number of metropolitan and micropolitan statistical areas that were defined when the new standards were applied to 1990 census data as well as the number of counties accounted for by these areas.

Table 18.1   Proportion of US population and area, and number of counties, resulting from the test application of the 2000 standards with 1990 census data

CBSA type (and number of areas)	% 1990 US population	% US total area	Number of counties
Metropolitan Statistical Areas (331)	78.8	20.9	891
Micropolitan Statistical Areas (496)	10.7	17.7	581
Outside CBSAs	10.5	61.4	1,669
United States	100.0	100.0	3,141

The new standards look likely to result in a change of status for many counties. Certainly, when they were applied with 1990 census data, there were many cases of counties departing from their status under the 1990 standards. In particular, over 600 counties changed from nonmetropolitan status to being components of test metropolitan or micropolitan statistical areas. Second, 61 counties switched from metropolitan to micropolitan area status. These were generally on the fringes of actual metropolitan areas, contained locally important employment centers, and thus had relatively small outcommuting to the metropolitan area's central counties. In addition, 17 counties changed from metropolitan area status to being outside CBSA status. These, also typically located on the fringes of metropolitan areas, had generally qualified under a criterion in the 1990 standards that summed commuting to the central counties of individual 'preliminary' metropolitan areas that had then been merged to form a single area.

[2] Definition of these hypothetical areas required assumptions that will not be appropriate or necessary in the 2000 standards' first official application with 2000 data.

Future Directions: Statistical Area Research Projects

Development of an approach to statistical areas that would account for all of the territory of the country was an early concern of the review. Although tremendous variation in settlement and activity patterns characterizes the country, the 1990 standards defined individual metropolitan areas and left all territory outside metropolitan areas simply as 'nonmetropolitan'. Some comments argued that this dichotomous approach was inadequate. It quickly became apparent, however, that an approach relying only on functional criteria would have difficulty accounting for all settlement, and attempts to fashion a hybrid approach based on functional and structural measures that would account for all settlement were unsuccessful. OMB and the Census Bureau will continue work in this area.

A related issue was the classification of types of locales, such as inner city, suburban, exurban, and rural, and whether such types should be identified within larger statistical areas. Formal definitions of such settlement types would be of use to the federal statistical system as well as to researchers, analysts, and other users of federal data. MASRP decided, however, that these types would more appropriately fall within a separate classification that focuses exclusively on describing settlement patterns and land uses. Increased public interest in issues such as urban sprawl and the spread of urban development in the United States argues for continued research into new and better ways of defining different settlement pattern categories. OMB and the Census Bureau are committed to continuing this line of research, too.

Use of areas for nonstatistical purposes is another line of continuing inquiry. Comments received throughout the review indicated a need to distinguish more clearly between, on the one hand, using metropolitan and micropolitan statistical areas to collect, tabulate, and publish economic and social statistics to inform public policy and, on the other, the use of the area definitions as a framework to determine eligibility and allocate funds under nonstatistical programs. The revised standards retain the long-term statistical purpose, but the review committee and OMB recognized the need to begin a collaborative, interagency process that could result in the development of geographic area definitions that are appropriate for the administration of nonstatistical programs. Such a process could result in the identification of existing geographic area classifications and modifications to those that are already in use by agencies (for instance, there are at least six definitions of 'urban' or 'urban place' currently in use by US federal agencies). It could also lead to the development of guidelines that explain appropriate use of specific area definitions in various circumstances. A longer-term goal of such an effort could be the development of one or more geographic area classifications designed specifically for use in the administration of nonstatistical federal programs or of guidance for agencies that need to define geographic areas appropriate for use with specific programs.

Also needed is further attention to other unresolved or only partially resolved issues. The current desirability of counties rather than smaller units as geographic building blocks, for example, in part reflects the nature of geographic bases of statistical programs, and it may be appropriate to upgrade those bases. Questions remain about the appropriate determination of statistical area size thresholds, reflecting uncertainties associated with monocentric versus polycentric structure. Also unresolved are issues about size classes for data presentation purposes.

In conclusion, OMB's new standards offer an approach to defining metropolitan and micropolitan statistical areas that is both conceptually consistent and easy to understand and implement. The new standards employ publicly available data for nationally consistent geographic units in order to define a set of functional areas for which a wide variety of statistical data will be available. More research, however, is needed to develop additional geographic classifications that identify a continuum of settlement types as well as the social and economic linkages between those areas. As such, therefore, the new 'Standards for Defining Metropolitan and Micropolitan Statistical Areas' represent an important stage in a process that is ongoing.

PART VI
THE WAY FORWARD

Chapter 19

Conclusions and Recommendations

Graeme Hugo and Tony Champion

Two fundamental conclusions can be drawn from the papers in this book. Firstly, 'place' is of central significance in the analysis of population change and demographic behavior. Where people are born, live and work *do* exert an influence on population processes and patterns. This makes it imperative that the nature of the settlement system is faithfully and accurately reflected in demographic research and data collection. Secondly, it is apparent that the conceptual and definitional bases which are currently used in most countries and for international monitoring are becoming increasingly inadequate reflections of the contemporary and evolving settlement systems. It was the unanimous view of the meeting on which this book is based that there is a pressing need to review the settlement classifications used both in statistical agency data collections as well as demographic research. Publication of this volume is intended as a first step in this process.

This final chapter starts by presenting our views on some key issues in future settlement classification, in particular justifying the emphasis on classifying places rather than people and reiterating the case for a fundamental change in ways of doing this. The central thrust is towards greater flexibility so as to cope with the ever-changing nature of settlement systems and to cater for the increasing diversity of user needs. The chapter goes on to argue that the best way of achieving this flexibility is through the use of the smallest possible building-block areas. These can be aggregated together to form more accurate representations of individual settlements than is normally possible with the administrative areas that are still widely used for this purpose. The greater flexibility and choice afforded by this approach raises the issue of what are now the most appropriate ways of depicting the settlement system and its component parts, so the next section puts forward suggestions as to the key elements. At the lowest level, we retain the concept of builtup urban places, though stressing the need for their more detailed and accurate delineation, and then move on to broader concepts of settlement entities that exist primarily through functional linkages. The latter are shown to necessitate a rethink of the way in which 'rurality' is conceived, as well as demanding that much more attention should given to measures of accessibility as opposed to landcover criteria. The next part of the chapter addresses the question of how these changes might be operationalized, notably by taking advantage of developments in geographical information systems (GIS) and also in remote sensing. Given that what is proposed may involve considerable effort in switching from current practice, the book concludes by justifying this investment in terms of the benefits for users and by making some recommendations on the way forward, addressed to data collectors and providers as well as to the main user communities.

Key Issues in Future Settlement Classification

Logically, the first and most fundamental issue relates to whether settlement classifications should seek to categorize people or places. In practice, this issue has not received a great deal of attention from the contributors to this volume. This can be attributed to the primary aim and starting-off point of research in this area. While terms such as 'urban population' suggest that we are classifying people, our approach is determined by our central hypothesis that where people reside is a significant factor in explaining their demographic characteristics and behavior. Settlement type is, in effect, an 'independent' variable to be taken into account alongside other factors such as socioeconomic position, labor force participation, ethnicity and status of women. It is for this reason that the classifications should apply to places on the earth's surface, with people being categorized according to the type of settlement that they inhabit.

Given the emphasis on classifying places, the second most important issue that has been addressed by this book concerns the extent to which new thinking is needed on the best ways of doing this. Again, there would seem to be no contest. Chapters in this book have dealt with urbanization in each of the major regions of the world and a number of individual countries with a wide variation in economic development, socioeconomic structure and cultural context. In all areas there is dissatisfaction with the established international standards for settlement classification. While it is apparent that the amount of relevant information, research knowledge and technical expertise available to inform and facilitate revision of settlement classifications is usually much greater in more developed countries, the need to reform classification has been seen to be just as pressing in less developed contexts. Indeed, any suggestion to rethink settlement classifications needs to be accompanied by a strategy to allow it to occur in less developed, as well as more developed, contexts.

Reinforcing this point is the breadth of interest in, and support for, a review of current practice in the international monitoring of urbanization and settlement-system change. While the current book has overwhelmingly been the work of academic researchers, it is recognized that any attempt to fundamentally change the portrayal of settlement systems needs to involve other important stakeholders. Certainly it is researchers who have demonstrated the inadequacy of existing settlement classifications in accurately depicting the reality of how people are currently sorting out the contexts in which they live and work. However, it is also important to involve policy makers and planners in these deliberations since it is crucial that they be able to access information which relates to meaningful communities and groupings of people if they are to plan effectively for them. Another group, which is also represented in the authorship of this volume, comprises the national statistical agencies that have the prime responsibility for undertaking data collection.

A further general point that emerges particularly strongly from the later chapters in this collection is that an individual settlement can be classified in more than one way. It can be a large or small place on one type of classification and a remote or accessible place on another. A settlement can be seen either purely in terms of its own internal make-up or in relation to other settlements and the wider geographical context. In a world of increasing spatial interaction, the emphasis is switching to the latter or to a combination of the two, such as the now fairly common practice of identifying physically-defined agglomerations within wider functionally-defined regions. Indeed, the basic vocabulary of settlement concepts has broadened substantially over recent decades, and not just in the more obvious direction. At the same time as ever more extensive population 'containers' such as extended metropolitan areas and polynuclear urban regions have been identified, so also has more attention been given to distinguishing different types of settlement areas within urban agglomerations. A similar case has been made for recognizing the diversity that exists within the other part of the simple urban-rural dichotomy.

Some may question the argument made here for a sweeping change in settlement classification on the basis that it will negate the intercensal comparison of trends and patterns. There are two ways of countering this reaction. Firstly, it is clear that the current conceptual basis of classification is highly inadequate and becoming more so with each census round. Perhaps most telling is the large proportion of countries for which the simple urban-rural dichotomy will be almost completely irrelevant by 2030, according to the UN's projections. A break away from this system has to be made and the sooner the better. There is also increasing disquiet – as evidenced by several chapters in this book – about the way in which the world's largest cities are defined. Secondly, census authorities have become highly skilled in backcasting, which allows new classifications to be introduced and applied not only to the present census but previous censuses as well. Similarly, they have developed concordances or lookup tables that allow old definitions to be applied to the current censuses at the same time as new definitions are introduced. These developments have been greatly facilitated by recent advances in computing technology.

Leading on from this, however, is the issue of whether there is still a role for a single all-encompassing settlement classification. For many decades, statistical agencies and analysts have been calling for greater uniformity in the geostatistical units for which data are made available, with some success as evidenced by the progress made towards statistical harmonization in the European Union and by the achievements of CELADE in Latin America. In fact, however, this need no longer be the case. For one thing, there is now a wider range of different users and applications than ever before, each with their own preferences for geographical frameworks. Also, there is now the technology available for meeting this variety of user needs, based on the principle of starting with small building-block areas or even individually-geocoded households and aggregating these into user-defined settlement classifications. Instead, therefore, what we should be pressing for is much greater flexibility in the presentation of data for settlement systems so as to allow data to be presented for a myriad of different types of areas. Rather than undermining the task of international comparison, this should provide extra choice in ways of doing this in a meaningful way. The next two sections present more detail on the methods of achieving this greater flexibility through starting with data for small building blocks and then on the wide range of settlement concepts which provide the basis for defining the spatial entities to which the building-block data will be aggregated.

Small Building Blocks for Greater Flexibility in Settlement Definition

Given the ever-changing nature of settlement systems and the diversity of user needs, it is imperative to inject a considerable degree of flexibility into settlement classification schemes. This will readily allow comparisons to be made over time despite reclassification of particular settlement entities as well as make it possible for demographic data to be presented for customized areas that are relevant for particular applications. As just mentioned, a principal key to the development of flexible classification schemes is the use of building-block spatial units which are as small as possible. The smaller those units, the greater the freedom to take into account changes in the settlement system while allowing accurate intercensal comparisons. Also, the smaller the unit, the greater the chance that it will be relatively homogeneous with respect to its physical characteristics and its resident population, thus aiding ecological analysis and policy targeting among other things. Modern developments in census taking and in the technology and methodology available for the analysis of census data make the use of small building blocks hugely more feasible now than three decades ago when the Goldstein Committee made its recommendations (see Chapters 1 and 2).

Currently censuses vary between countries in the geographical and population sizes of the smallest spatial units for which they provide data. These units generally, however, were originally conceived in the pre-computer era, when they often equated to the area and population that a single census enumerator could take responsibility for. They are frequently too large to fit accurately within the boundaries of settlement classifications. There are, however, two developments in censuses, mainly in more developed countries, which are promising in this respect.

The first of these relates to the development of smaller basic units than currently used for the dissemination of data. The key issue is that the reporting units should be designed specifically for data output rather than being restricted to those used for the collection of data. In Canada, for example, the 'blockface' is designed to allow, through aggregation, the construction of user-defined query areas for data extraction and tabulation. The blockface refers to one side of a city street, normally between consecutive intersections with streets or physical features such as creeks or railways. Centroids are calculated for each of the over 400,000 blockfaces in Canada, so that GIS can be used to amalgamate these into user-defined spatial units such as settlement classifications. New Zealand uses a similar 'meshblock' system (Statistics New Zealand, 1997, pp.33-5).

The second trend within censuses is the move toward the geocoding of all households so that in effect the household becomes the basic building block of the census. This means that, while census data will not be made available for individuals or households for well founded confidentiality and privacy reasons, data can be supplied for user-defined areas, such as settlement types of a given minimum size. These two developments are part of a movement toward greater flexibility in the spatial units for which census data are provided, facilitating the supply of information for areas meaningfully defined for the user's purpose. In addition to assisting the use of more meaningful settlement classifications and user-defined statistical areas, they also enable census authorities to respond to the increasing pressure in many contexts to provide community-level, small-area data.

While the moves towards meshblocks/blockfaces and geocoding will allow greater flexibility in the geostatistical units of analysis which can be used, it should also be mentioned that important advances are being made in methods for converting data from one geostatistical system to another. Simpson (2002), for example, has shown how 'geographical conversion tables' can be constructed to merge data sets based for a range of small spatial units in the United Kingdom (electoral areas, census units, postal geography, etc.).

Given the increased flexibility provided by small building-block areas, there is now much more incentive than in the past to put effort into identifying the types of areal unit for which one would like to have population data reported. The most commonly accepted approach is through the concept of *locality*.[1] Localities must not be confused with the basic census building-block areas discussed in the previous paragraph. In the 1958 Census Recommendations issued by the United Nations (1958, p.11), locality is defined as 'a distinct and indivisible population cluster (also designated as agglomeration, inhabited place, populated center, settlement etc.) of any size, with or without legal status including fishing hamlets, mining camps, ranches, farms, market towns, communes, villages, towns, cities and many others' – a definition that has remained largely unchanged in subsequent decennial issues of the UN's census guidelines.

Unfortunately, the UN's guidelines do not indicate exactly how the 'locality' concept should be operationalized. On the one hand, the UN notes that localities should not be confused with the smallest administrative divisions of a country, although the two might coincide in some cases. Moreover, the term 'locality' is used in different ways in different censuses. In some, for example, it is the smallest population agglomeration in a hierarchy of urban places. In others it is defined to incorporate the hinterlands of the population agglomeration, providing a set of mutually exclusive spatial units that cover the entire inhabited area of a country. The UN also allows that one physically-defined urban agglomeration can comprise more than one locality. Nevertheless, for present purposes the point is clear: the smaller the building-block area for which population data are released, the greater is the choice that users can exercise in defining whatever 'localities' that they wish to obtain and analyze data for. We return later in the chapter to the issue of operationalizing this concept.

---

[1] This discussion of localities is based largely on a paper entitled *On the Identification of Localities* prepared by Hania Zlotnik of the United Nations Population Division, which was originally part of an earlier draft of her chapter appearing in this volume.

In terms of the types of localities that might be recognized, it is true that the definition still used in most countries refers only to population agglomeration, along the lines stipulated by the UN's guidelines. Yet, in many nations, settlement patterns are dispersed so that not all people live in agglomerations. This focuses attention on how the national space can be meaningfully divided up into spatial units which are mutually exclusive and cover the entire settled area. Traditionally, this has been via the use of administrative areas as basic spatial units. Thus the United Nations (1969) specifies five different basic settlement categories which recognizes the importance of administrative units – localities, minor civil divisions, administrative centers of administrative divisions, small units specified generally as 'cities', 'towns', 'townships', 'villages' etc., and urban units specified by name only. However, there has long been a questioning of the meaningfulness of administrative boundaries for many areas of social, economic and demographic research and the possibility is that there may be spatial units which are more relevant and can facilitate the planning process.

Much of this discussion has centered around the idea of 'social and economic catchments'. These can be defined in generic terms as follows: The territory occupied by a group of households and individuals who are in some form of regular interaction and which the inhabitants identify as 'their' community or region. As articulated by Smailes (2000, p.128), social and economic catchments have three basic dimensions:

- Territoriality (a habitat/place dimension).
- Communion (a shared feeling of belonging).
- Interaction (the local social system).

A key element of such catchments is that they are centered on a particular urban area or central place which is the focus of much of the social and economic interaction occurring in the catchment and is the location of many of the facilities that people living in it need and use. There is a longstanding literature linking central places with the areas and populations that surround them and are functionally linked to them. These have variously referred to 'city-regions', 'hinterlands', 'umlands', 'functional urban areas', 'urban fields', 'service areas', 'trade areas' and 'labor markets'. A second key characteristic of social catchments is that most (but not all) of the people living within them feel a sense of belonging to the community and the social group living within them. While social and economic catchments can be delimited with sharp boundaries, it is likely that in reality their boundaries are somewhat diffuse. An important characteristic of catchments, however, is that the people living within them share a common interest and purpose.

One of the most important issues relating to social catchments relates to scale. In fact, they can occur at several geographical scales. One can feel attached to, and interact within, one's local area, a wider region, a state or province within a country, and the nation as a whole. Of course, one may feel attachment to one or other of those levels with varying degrees of strength. Smailes (2000, p.160), for example, has argued that the primary social allegiance and place identification of most households is highly local in nature. Hence he places considerable emphasis on the significance of localism in social allegiance. Nevertheless, it is important to recognize that individuals can belong to a number of social catchments at different scales and the degrees of attachment people feel to various scales of social catchments is an area of needed research.

One area of argument relates to the extent to which social catchments overlap with hinterlands of centers based on economic activity such as commuting patterns, shopping patterns and business linkages. Again, some interesting work has been done in non-metropolitan South Australia along these lines by Smailes (2000). He argues that, whereas in the past the geographies of social identification on the one hand, and of commercial and business activity on the other, were once closely linked, they are now 'separating and slowly drifting apart' (Smailes, 2000, p.158).

There is a longstanding theoretical basis for consideration of catchments in geography in the form of Central Place Theory. This theory argues that there is a particular pattern of ordering in the location, size, nature and spacing of central places. Of particular significance is the fact that Central Place Theory maintains that the central places in an area, when ranked according to their size, do not form a continuum but instead there is a 'nested hierarchy'. It is possible to recognize a number of levels of the hierarchy, and the central places in a particular level of the hierarchy share a common set of goods and services. Each separate order of the hierarchy is characterized by places with a particular suite of goods and services, a similar amount of functional complexity and a relatively similar population size. Centers in higher orders of the hierarchy contain the functions of lesser-order centers together with those which characterize their own order.

In practice, one of the most commonly used types of catchment has involved the demarcation of 'labor market areas' around central places using journey-to-work data. This involves analyzing the areas from which a central place draws its workforce from outside the central place (and, in the case of activities like town farming, the areas where the central place sends its workforce to work daily). Censuses in most OECD nations include a question on, not only what work people do, but also the location of their place of work. In the USA, for instance, the cross-tabulation of place of work against place of residence has allowed the depiction of 'commuting zones' (Edmondson, 1995). This form of regionalization has proved extremely useful in a range of applications including planning community-based labor market programs and employment generation programs, but they also have meaning for demographic analysis, including studies of stroke mortality, marriage patterns, household structure, income sources, minority concentrations and black migration (Tolbert and Sizer, 1996, pp.4-5).

What Concepts are Needed in Settlement Classifications?

Once one acknowledges the advantages of using the smallest possible building-block units to compose larger areas for the presentation and analysis of population data, then it is worth giving serious consideration to what these larger areas might comprise. The chapters in this volume have thrown up some common themes, thereby suggesting that it is possible to identify a range of settlement categories that are relevant across a wide range of countries. There is universal agreement here that, while the concepts of 'urban' and 'rural' still have meaning in certain contexts, we should be going beyond simple dichotomous classifications, since in the contemporary world many people live and work in a much more differentiated settlement system. In this section, we make a first pass at what a more sophisticated generic classification would probably look like, though stressing that the nomenclature at this stage is still quite tentative.

Builtup Urban Places

This categorization essentially recognizes that builtup urban areas as identified in traditional classifications are still important. They are distinguished by being an identifiable cluster of residential, industrial, commercial and recreational development of a minimum population density and size. The actual definition will necessarily vary between countries. The point to be made here is that accurate delineation of such areas and strict application of the set criteria is now much more possible than in the past because of use of the increasing use of small building-block spatial units in censuses, as well as because of the recent developments in remote sensing, GIS, etc. Nowadays we should be able to delimit such areas much better than we have in the past, so that underbounding and overbounding of builtup areas should be minimized. A key point is that administrative criteria should not be the basis on which urban areas are delineated, but rather it should be the boundaries of the builtup urban fabric.

A good idea of how the delineation of these builtup urban places should be approached was provided many years ago by Vapñarsky (1978). He identified five conditions that a definition of locality must meet to be complete and operational:

- The definition method should guarantee that every point of concentration detected through it should actually appear in the listing resulting from its application, without repetitions or exclusions.
- It should permit an accurate delimitation of the area of concentration, whether it be continuous or discontinuous, associated with each point of concentration identified.
- It should permit the accurate assignment of a population figure to each area of concentration.
- It should not be based on legal or administrative considerations of any kind.
- It should be accompanied by technical rules that make the definition empirically applicable. Given that instruments such as censuses produce only approximations to real entities, it is required that any technical rules proposed to make the definition operational produce a census entity that approximates the real entity as closely as possible.

Vapñarsky (1978) argued that localities be defined in terms of 'elementary objects' and their spatial interrelations. Elementary objects comprise several categories, including volumes (buildings, containers), precincts (green, cultivated), strips (streets, unpaved roads, paved roads, rail, water), a complementary category (trees, screens) and a residual category (vacant lands, bodies of water). By using geographic representations of these objects at increasingly smaller scales and setting rules about the existence of interrelations between objects that determine whether those objects are or are not part of a locality, Vapñarsky illustrated how it is possible to identify localities spatially and determine their boundaries. The rules rely exclusively on spatial attributes. To illustrate this point, the conditions set for the determination of urban localities, which Vapñarsky called 'agglomerations', are reproduced below:

- Every block constituting the agglomeration must possess a minimum number of built-up stretches and streets.
- The distance separating a block from its nearest neighbor already within the same agglomeration must not go beyond a certain maximum.
- A minimum number of blocks meeting the two above conditions must exist in order for an agglomeration to exist.

Vapñarsky's paper was written long before geographical information systems were developed. This methodology and technology greatly facilitate the approach which he advocated.

Catchments of Urban Places

It has long been recognized that areas adjoining urban places are quite different to other rural areas by virtue of this adjacency. Proximity leads to intensive interaction between the catchment zone and the urban center, often including commuting in both directions. Moreover, the land use and functions of these catchment areas are distinctively different from those in rural areas. These are seen as a 'daily activity space' and are defined as the area over which daily interaction such as commuting occurs at a relatively substantial level. Hence, as mentioned in the previous section, they can be identified using journey-to-work data collected in census enumerations (Ghelfi and Parker, 1995; McNiven *et al.*, 2000; Morrill *et al*, 1999). These catchments are more or less coincident with labor markets and housing markets, thus forming a highly relevant unit for economic and social planning. Moreover, while intensive commuting to urban centers over longer distances is sometimes seen as a feature only of more developed nations, the reality, as demonstrated in several of the chapters in this volume, is that such commuting is a significant feature of less developed parts of the world as well.

There is, of course, a potential problem in that not all countries collect data on commuting, perhaps not surprisingly since it is a relatively costly and complex exercise to produce matrices crosstabulating origin (home) and destination (work). There are, however, a number of possible alternative ways of demarcating catchment areas. In past studies, for instance, telephone traffic, newspaper circulation and television audiences have been used. Secondly, in the less developed parts of the world, the vast majority of commuters to urban places utilize some form of public transport, especially bus and train services. There is an established methodology of utilizing bus services for delineating urban hinterlands in early postwar Europe. The classic study was by Green (1950). The use of GIS can greatly facilitate this process. In most LDCs, governments are responsible for the licensing of such transport, so information on the numbers and routes of bus and train services are generally available. Of course, there are also a myriad of informal and small-scale forms of public transport in such contexts, but the data on main services should be sufficient to demarcate such areas.

Beyond this, where no interaction data is available, there are other alternatives. Firstly, there are a number of intensive field-based approaches which can be used to delimit the boundaries of urban centers' areas of influence (Hugo *et al.*, 2001). Secondly, GIS can be used to demarcate hinterlands of urban places if there are road network data available in digital form. GIS allows one to buffer out from a central place along roads to establish road distance zones. Measurements can be modified to take account of availability of public transport, quality of roads etc. There would need to be local knowledge employed as to what meaningful thresholds need to be identified, and maybe this need to be undertaken only for the larger urban places that exercise a significant pull over their surrounding areas. Thirdly, it is possible to develop surrogate measures for delimiting catchments of urban places especially using gravity models (Carrothers, 1956). The example presented by Coombes in Chapter 16 of this book (see his Annex 3 for the detailed methodology) clearly demonstrates that the application of gravity-model principles produces a pattern of areas that is very similar to those derived from the commuting data that is collected in the UK.

Extended Metropolitan Areas

These are substantial urban centers together with their associated hinterlands, i.e. both the builtup urban places and their catchments for the relevant centers. The minimum size of urban center can vary between countries but probably is of at least 100,000 persons. The key point is that we are dealing with centers with a substantial range of high level functions and a regional role.

There is clearly scope within this class of cities for development of a hierarchy of extended metropolitan areas. The US Bureau of Census adopts the following standard classification of urban centers and localities according to their population size:

Level A:	1 million or more population
Level B:	250,000 to 999,999
Level C:	100,000 to 249,999
Level D:	50,000 to 99,999
Level E:	20,000 to 49,999
Level F:	10,000 to 19,999
Level G:	5,000 to 9,999
Level H:	1,000 to 4,999
Level I:	(Rural localities) 200 to 999

In this classification hinterlands could be drawn for levels A, B and C, and perhaps also for D and E.

Megacities and Wider Urban Systems

Megacities is a term which has come into the literature to designate large urban agglomerations with 8-10 million or more inhabitants. A crucial element in megacities is that they are essentially large 'polynuclear extended metropolitan areas'. Here, essentially urban development has linked together a number of originally separate substantial urban centers. This phenomenon was first identified by Gottmann (1961) when he defined the metropolitan centers extending between Boston and Washington and linked by transport arteries as 'megalopolis'. It is crucial that these are defined so as:

- They include all of the polynuclear extended metropolitan areas that are part of the megacity.
- They are not based on outdated administrative regions which have long been overspilled by urban development.

It is clear that the boundaries used by the UN (2003) to define many of the world's megacities utilize administrative boundaries and do not always reflect the organic whole of the megacity as a functioning urban region. Hence methodologies which include not only the originally separate urban centers but also their merged hinterlands and commuting zones need to be adopted.

It could be argued that such cities are not just distinguished in their size but also by a number of other features such as:

- morphology – several urban nuclei linked by ribbons of urban type development along transport routes;
- rapid population growth, especially on the periphery as a result of inmigration from the center as well as outside of the city;
- strong international links, more multicultural than the national population;
- disproportionate concentration of GDP, services, investment, power;
- strong linkages with the entire national space, and possibly beyond.

The last three of these features link to the point, articulated in Chapter 13 by Bourne and Simmons, that there are emerging wider systems of cities which go beyond individual urban regions and include a number of separate cities strongly linked to each other by a plethora of interactions rather than by continuous tracts of physical urban development. These interactions involve financial transactions, trade linkages and linked production processes as well as major communications linkages (Townsend, 2001) and movements of people. These complex systems may cross international boundaries. An element in these classifications may be world cities (Friedman, 1986) and global cities (Sassen, 1991). In establishing a suite of settlement concepts for the reporting of population data, a decision would need to be made as to whether these wider urban systems constitute an appropriate category or are merely used to produce contextual information for individual cities. Whatever approach was taken on this issue, there would need to be a set of guidelines for delineating the 'megacity' entities, and these would need to be based mainly on functional, notably interaction, criteria.

Rural Areas

A key element in the all categories identified above, apart from the builtup urban places, is the fact that they involve linking the urban and the rural together. This raises the question of the place of the rural in any future settlement classification. Traditionally, rural has been differentiated from urban largely as a residual, once urban areas have been defined. Hence, the term has referred to human settlement where there is not a continuous builtup urban fabric with a population, or number of houses, of over some minimum size. Chapters 14 and 15 of this book have, however, argued strongly that rural areas should not be treated as an undifferentiated residual, echoing the 30-year-old plea of Graumann and the Goldstein Committee (see Chapter 2). Halfacree is adamant that 'rurality' exists as a set of ideas in its own right and is not just the mirror image of 'urbanity', while Brown and Cromartie are keen that the multidimensional nature of rurality is fully acknowledged and that this is drawn upon to aid the task of rural-area subdivision.

There does, however, need to be more research into the concept of rurality. Drawing on these new ways of looking at rurality, attention needs to be given firstly to how the lines are drawn between rural and urban territory and to how the diverse rural area may be meaningfully subdivided. Attempts at the latter have generally been of three types:

- The dividing up of areas on the basis of their degree of rurality, which process may or may not incorporate the urban as part of the classification.
- The dividing up of rural or nonmetropolitan areas on some kind of functional criteria.
- The dividing up of these areas according to their adjacency, or level of accessibility, to large metropolitan centers.

In all, there are a number of criteria on which such a classification of rural areas may be considered. As well as degree of adjacency to urban areas of a particular size, these include population density, settlement characteristics, size of local economy, economic base of communities, diversity of local economy, natural environment, institutional arrangements and sociocultural dimensions. It is very likely that the approach would have to vary between different cultural, economic and social contexts, so research would need to explore whether a universally applicable basis for such a division is possible.

Accessibility

Accessibility is a key determinant of the extent to which any place of settlement is likely to form part of a larger entity such as a labor market area or extended metropolitan region, and thus is important in helping to define the latter. At the same time, it is an argument of this book that urban-rural classification captures only one dimension of settlement systems and that it is also desirable to distinguish between individual settlements according to other criteria including their degree of accessibility/remoteness. As argued in Chapter 16, this dimension of settlement is distinctive from categorization on the basis of urbanness-ruralness. Moreover, as shown in Chapter 1, this dimension can and does impinge upon demographic variables. Degree of isolation from services and variations between areas in the potential for local social and economic interaction are important in understanding the demographic, economic and social behavior of individuals and groups.

The concepts of accessibility and remoteness, however, will vary between cultures and nations, depending on a range of elements such as ownership levels of vehicles, access to and provision of public transport, and physical and social barriers to movement. Hence the precise components of a measure of accessibility in a particular country should be decided individually by that country. What can, however, be the subject of a set of general recommendations are, firstly, the elements to be included in a measure of accessibility/remoteness and, secondly, a methodology to convert these elements into a measure of accessibility.

The main elements that should be included in a measure of accessibility are the following:

*   The location of all urban centers.
*   The distribution of the national population.
*   The national road network.
*   In some cases, the availability of public transport.

The methodology can be specified involving the following steps:

*   Establishment of the urban category/ies to which accessibility is to be measured.
*   Development of a measure of accessibility to those centers, e.g. road distance, Euclidian distance.
*   Modification of this distance measurement by availability of public transport, travel times, car ownership, etc.

What comes out of such a procedure is an ability to classify all parts of a nation (whether a specific location or a spatial unit like a locality or urban agglomeration) in terms of its degree of accessibility/remoteness. Hence it would be possible to categorize places on this basis as well as by population size or any other criteria of how urban or rural they are.

Potential Provided by GIS and Remote Sensing

Already at several points in this chapter, and in this collection of papers more generally, there has been reference to the enormous progress made in GIS and remote sensing in recent years. It has been suggested that these two forms of technology and methodology can be helpful both in the operationalization of better methods of classifying settlement systems and in the presentation of demographic information for a greater variety of areas. While it is all too easy to treat these developments as some kind of 'silver bullet' capable of solving all our problems, in reality they offer a great deal of potential.

Firstly, considering GIS, it is important to stress at the outset that it can only be useful if census (and other data) are produced for small areas, or ideally for geocoded points. Only then can the building blocks be aggregated into meaningful spatial categories. If the basic building blocks (the smallest areas for which data are available) are large, the likelihood that meaningful boundaries can be defined for settlements is greatly reduced.

Given that census data are available for small spatial units, what can GIS do? Most simply, it means that the basic units can be readily combined and recombined so as to create alternative depictions of the settlement system. GIS can be used to quickly and efficiently apply any criteria selected for identifying and categorizing spatial units, permitting population, landcover, cadastral, interaction and other variables to be considered simultaneously in deciding upon boundaries and classifications. It allows accessibility to be easily measured. It also provides a very powerful way of storing, retrieving, manipulating and presenting large amounts of spatially-referenced data. Most importantly, GIS injects a substantial element of flexibility since, if data are geocoded or available for very small areas, there is a capacity to adopt user-defined settlement or other spatial classifications. This level of detail, of course, raises important privacy issues, but there are effective ways in which the confidentiality of individuals can be strictly observed.

Turning to remote sensing, the issues are at this stage more limited. The fact that there are a number of sophisticated global imaging programs means that it is possible, at a cost, to obtain high-quality photographic coverages of virtually everywhere on the earth's surface. Most simply, this means that morphologically-based settlement classifications can be facilitated. For example, it is possible to draw boundaries around builtup areas according to a set of rules relating to population or housing density, contiguity, landuse types, etc. It is possible, too, to differentiate areas of varying population density through analysis of dwellings and land use. As demonstrated by Weeks in Chapter 17, more sophisticated attempts have used imaging not only to draw the lines between various settlement categories but also to make important distinctions within settlement categories according to population characteristics, socioeconomic characteristics, economic functions, etc.

Remote sensing may be of particular significance in situations where no other data are available for distinguishing settlement types and estimating their proportions. Sutton *et al.* (2001), for example, have used night-time satellite imagery to estimate population numbers and density. There is now a sufficient run of data to observe areas where lighting conditions have changed, providing the capacity to identify changes in settlement systems (Elvidge *et al.*, 2000). Thus the opportunities for using remote sensing in settlement classification would appear considerable.

The potential of using GIS and remote sensing together to assist in the development of relevant classifications and data bases of human settlements has been demonstrated in the *Global Urban-Rural Data Base* initiative developed by the Center for International Earth Science Information Network (CIESIN). The goal of this project is to develop a global georeferenced population database that merges conventional census data with satellite and other geographic data. The project aims to produce a complete urban-rural grid surface and provide data for all cities in the world with more than 5,000 inhabitants (Balk, 2002).

Nevertheless, it has to be acknowledged that there remain some obstacles to the adoption of GIS and remote sensing for these purposes, especially in less developed situations. With respect to GIS, as has been pointed out, a basic need is for censuses and surveys to include the facility to both collect and disseminate demographic data for small areas. This would represent a substantial change from present practice for many countries and international agencies, for which concepts of the census remain large scale and macro. Set against this, however, this step accords well with the current shift away from centrally planned and controlled government to more decentralized regional and community development models, involving more devolved planning, decision making and revenue raising.

Hence there is a strong argument to be made for relevant international agencies to facilitate and encourage the use of smaller building-block spatial units in censuses and major surveys. This is not just to allow sharper and more meaningful classifications of settlements, but would also be a means of assisting community and regional development. Targeting of initiatives in such areas as education, health, welfare, family planning and agricultural extension can mean that limited resources can be better used to advance the wellbeing of the poorest and most deprived.

There are, of course, important resourcing issues in both GIS and remote sensing which impinge especially hard on less developed nations. The hardware and software demands of both are considerable. The need for skilled human resources to operate the technology and to analyze and interpret its results is also a major challenge, which is aggravated by the 'brain drain' of information-technology workers from less developed to more developed parts of the world. On both accounts, however, there is scope for optimism. The costs of both hardware and software have fallen dramatically in recent years, while developments in GIS software make it much less demanding of advanced computer skills. As a result, there is now more scope than ever before for benefits to flow from coordinated international programs aimed at upgrading the hardware, software and human resources in GIS and remote sensing in less developed nations.

Access to Data

The suggestions made above about using small building-block areas and benefiting from the latest concepts and technology are aimed towards providing population data and other policy-relevant information for the most meaningful geographical areas and groups, thereby allowing more efficiency, effectiveness, transparency and equity in the allocation of resources. Besides the basic resourcing challenges just described, there is also the question of access to the data needed for these purposes. An area of increasing concern is the tendency for some statistical agencies to bring in policies of 'cost recovery' and 'profit' and accordingly make very high charges for data, meaning that many potential users are not able to purchase the data. It is somewhat paradoxical that, just when we have the technology to readily access and analyze small-area information from censuses, many groups are disenfranchised from accessing it because of its high cost. If the technology is to lead to more people-centered social and economic planning, data needs to be made both financially and technically accessible to the full spectrum of the community and not become the preserve of the wealthy and powerful.

An alternative position being taken by a small but increasing number of governments, such as the UK, is that their statistical collections are a public good which citizens have contributed toward by completing census forms, etc., and thus have a right to access. Moreover, such governments have realized that making data freely available in the community contributes to informed decision-making, with consequent improvements across an economy and society. However, in many countries especially less developed ones, census authorities are prevented from taking such a position because of lack of relevant financial and expert support from funding agencies. There can be few more effective contributions to assisting the development process than ensuring the creation and wide accessibility of sound, accurate, timely and relevant information on the populations of communities. This infrastructure is absolutely basic to good decision-making and the allocation of scarce resources which is based purely on degree of need and not influenced by corruption, elite pressure or unsubstantiated rumor. Significant international assistance is likely to be crucial in order for local communities to gain this level of access.

One encouraging development in this respect has been the development by some international agencies of software that is capable of undertaking many of the types of analysis referred to here and is made available to users in less developed countries at no cost. One example of this from the Latin American region is, as mentioned in Chapter 5, the package REDATAM Win R+ (Retrieval of Data for Small Areas by Microcomputers) which was developed by the Latin American Demographic Center (CELADE). As Brennan-Galvin (2002, p.13) points out:

> The software uses a compressed database for micro-data with the registers of individuals, sets of housing, city blocks or any administrative division identified in the census. On the same basis, one or several censuses, surveys and administrative registries may be combined. Any geographical area of interest or combinations of such areas may be defined through a database to create new variables and rapidly show tabulations in graphic windows. The data from different geographic levels may be combined by order of importance to create aggregate variables and the results may be displaced on maps from REDATAM or transferred to a geographic information system.

This system seems to encapsulate many of the features that we have been stressing in this chapter. This being so, an obvious first step in attempting to operationalize the principles outlined above would be to apply this package to the task of redefining the settlement systems of a sample of countries in Latin America. From this exercise, there would hopefully emerge a clearer picture of the most important types of data that should in the future be supplied to the UN and other international agencies for dissemination to users.

Why are These Changes Needed?

The changes outlined above will require a significant investment and the question must be asked as to whether this is justified. The first level of justification is in terms of the rapid pace of urban growth in many parts of the world and the continued evolution of settlement systems both there and in countries that have already reached a high level of urbanization. Increasingly, planning for people means planning for urban areas, and especially for large cities and their wider regions. Relevant and timely population information for these areas is of crucial and basic significance.

From the perspective of policy makers and planners, there are two areas that need to be considered in particular. Firstly, it is of critical importance that the spatial units which are utilized in developing strategic planning decisions are meaningful units which have an integrity based not only on physical adjacency but on patterns of economic, social and cultural interaction, recognized shared needs and goals, common problems and issues and a degree of community interest. Such units are not provided by most current administrative unit systems and settlement classifications. Secondly, it must be possible to provide relevant information required by policy makers and planners on the basis of those classifications. The resources available to national, regional and local governments are constrained, so it is crucial that these resources are utilized in the most equitable and efficient manner. Only when appropriate data are made available for appropriate planning areas can effective targeting of intervention and service provision be possible.

In both more developed and especially less developed nations, there has been a growing recognition that there are major benefits to be achieved through the decentralization of much decision-making in development planning, service provision and other areas of governance. This has meant that there has been a tremendous increase in the responsibilities placed on regional and local authorities. However, in many countries these bodies lack both the expertise and the relevant information bases to allow them to effectively take on those responsibilities. In an era of increasing democratization and growing local and regional autonomy, it is crucial that relevant data be provided for meaningfully defined local and regional areas. The changes proposed here will facilitate this. Such data will be needed not only for informing policy and planning decisions, including where to target interventions and where to locate services, but also for facilitating the ongoing monitoring and evaluation of the effectiveness of those decisions.

Finally, it should not be overlooked that there is much to be gained by the research community if a more meaningful settlement classification is adopted in the dissemination of census and other population data. Simple urban-rural dichotomies are very blunt instruments to attempt to capture the influence which the place where people live has on their demographic characteristics and behavior. Specified more accurately, place becomes a more sensitive variable to incorporate in any demographic analysis, and its potential to contribute toward understanding and explanations is thereby enhanced. As Mendelson (2001) observes in relation to Canada, statistical agencies have traditionally used standard administrative areas as 'containers' for a dissemination of statistical data, but this ignores the capacity of geographic structures to act as 'variables' to better analyze social and economic processes.

Recommendations

A number of recommendations arise from the considerations in the present study. These arise from a concern that the current practices used within censuses and other large scale population data collections for differentiating the places in which people live are manifestly inadequate. There has been little or no basic change in the categorization of settlement in censuses for almost half a century, while there has been nothing short of a transformation in the settlement systems themselves. Researchers are being denied access to important explanatory variables, while planners inevitably waste scarce resources because the spatial units for which they have population information are becoming increasingly irrelevant to the social and economic reality of people's lives. Hence we believe the time has come for a complete re-evaluation to be made of the settlement classification used in population data collection and analyses, especially in population censuses. This should ideally be conducted under the aegis of a relevant international body. The recommendations that follow are of three types – those for relevant international agencies, those for national statistical offices and those for the academic research community.

Firstly, with respect to *international agencies*, the following recommendations are made:

- To initiate a comprehensive review of the recommendations to be made to countries regarding the settlement classification which they use and the development of a new set of classifications to be introduced for the next round of population censuses.
- To develop a plan of action to develop the expertise and resources within countries to operationalize the new recommendations.
- To institute a number of significant changes in the way in which world urbanization data are produced. This would include doing the estimates only every four or five years instead of every two years as is presently the case. This would allow greater attention to be paid to the data for each individual country, especially in tackling problems which currently exist with long outdated boundaries of urban areas being used (as is the case, for example, with Mexico and Jakarta in the most recent data).
- Development of an international data base of the precise procedures used by countries to develop their settlement classification and provide details of those classifications.

- Stress the need for countries to tabulate urban places according to a population size hierarchy.
- Encourage a move away from administrative boundaries to a new basis of delineating urban areas and other settlement classifications as outlined above.
- To encourage data democratization which recognizes that data are an element of national infrastructure and therefore need to be made readily available to a wide cross-section of the community, not just those who can pay large sums for it.

Many of the above recommendations are also applicable to the *statistical offices of individual nations*. Additionally:

- By far the most important recommendation for allowing progress is the need to adopt small building block spatial units in their censuses, and (where possible) geocoding, in order to encourage flexibility in the spatial units for which data can be made available. This also facilitates the development of concordances to allow comparisons to be made between current and past censuses as well as with future censuses.
- There should also be the encouragement given and, where necessary, the assistance to develop an appropriate capacity in GIS and remote sensing.
- Individual nations need to develop a settlement classification system which is appropriate to their cultural, economic, social and demographic situation. This does not militate against international comparability since, even if decisions on thresholds for criteria are fitted to specific national conditions, nations will be using the same conceptual basis for delineating their settlement systems.

With respect to the *research community*, perhaps the most promising way forward is to undertake the detailed examination of the settlement systems of a selection of nations, including representatives both of large and small populations and of less developed and more developed situations across the world's major regions. Such studies should take advantage of data from the 2000 round of censuses and operationalize a new settlement classification system along the lines suggested earlier in this chapter. In conjunction with both data collectors/suppliers and data users, these would give particular attention to a number of issues in settlement classification that the present study, with its limited resources, has not been able to deal with in detail and which are in urgent need of attention. These include:

- The issue of subdividing large metropolitan areas. Are there guidelines that can assist in arriving at meaningful subdivisions of large urban centers, e.g. based on distance from city center, land use, population/housing density, accessibility to services etc.?
- The need to experiment with GIS methods of delineating hinterlands and functional urban regions, especially in less developed contexts where journey-to-work data are not available.
- The need to review the definition and meaningful subdivision of rural areas.

Conclusion

This book has argued that, despite human settlement systems having undergone dramatic changes over the last three decades, there has been a failure to capture these changes in the classification systems adopted for the collection and dissemination of population data. Such classifications have not reflected the blurring of rural and urban areas, the diversity of settlement within urban and rural contexts, the increasing scale and complexity of urban systems, and the new forms of urbanization that are emerging in both more and less developed world contexts. Moreover, they reflect only one dimension (urbanness-ruralness) of settlement systems, while data users – academic researchers, government agencies and the private sector alike – are becoming more sophisticated in technical skills and more ambitious in their aims.

This failure is unfortunate from a number of perspectives. Firstly, it has restricted the ability of researchers to incorporate in their analyses the place where people live as a causal element in shaping population characteristics and behavior. Secondly, the development and application of appropriate policies and programs is likely to be more effective if the regions to which they are applied are meaningful aggregations of people. This volume has identified the shortcomings of the current system and made a number of recommendations to improve national settlement classifications. Now that experimentation with, and the adoption of, new approaches is facilitated by contemporary developments in census taking and by advances in technological expertise, the suggestions for change should be followed up as a matter of urgency.

Bibliography

Adams, J.S. (1995), 'Classifying Settled Areas of the United States: Conceptual Issues and Proposals for New Approaches,' in D.C. Dahmann and J.D. Fitzsimmons (eds), *Metropolitan and Nonmetropolitan Areas: New Approaches to Geographical Definition*, Population Division Working Paper 12, Washington: US Census Bureau.

Adams, J.B., Smith, M.O. and Gillespie, A.R. (1993), 'Imaging Spectroscopy: Interpretaton Based on Spectral Mixture Analysis', in C.M. Pieters and A.J. Englert (eds), *Remote Geochemical Analysis: Elements and Mineralogical Composition*, Cambridge, UK: Cambridge University Press.

Adams, J.S., Van Drasek, B. and Phillips, E. (1999), 'Metropolitan Area Definition in the United States', *Urban Geography*, Vol. 20(8), pp. 695-726.

Aguilar, A. (2000) 'Megaurbanización en la Región Centro de México', *Mercado de Valores*, México, D.F., Mexico, Year LX, No. 3, pp. 77-86.

Aguilar, A.G, Graizbord, B. and Sánchez, A. (1996), *Las Ciudades Intermedias y el Desarrollo Regional en México*, Consejo Nacional para la Cultura y las Artes, México: El Colegio de México, Instituto de Geografía, UNAM.

Aguilar, A.G. and Hernández, F.R. (1995), 'Tendencias de Desconcentración Urbana en México, 1970-1990', in G. Aguilar *et al.*, *El Desarrollo Urbano de México a Fines del Siglo XX*, México: Instituto de Estudios Urbanos de Nuevo León, Sociedad Mexicana de Demografía.

Alberts, J. (1977), *Migración Hacia áreas Metropolitanas de América Latina*, Santiago de Chile: Centro Latinoamericano de Demografía (CELADE).

Aldridge, J. (1969), *Cairo*, Boston: Little, Brown and Co.

Ali, S. (1990), *Slums within Slums. A Study of Resettlement Colonies in Delhi*, New Delhi: Har-Anand and Vikas.

Ali, S. (1995), *Environment and Resettlement Colonies of Delhi*, New Delhi: Har-Anand.

Ali, S. and Singh, S.N. (1998), *Major Problems of Delhi Slums*, New Delhi: Uppal Publishing House.

Ambrose, P. (1992), 'The Rural/Urban Fringe As Battleground', in B. Short (ed.), *The English Rural Community*, Cambridge: Cambridge University Press.

Anderson, S. and Rathborne, R. (2000), *Africa's Urban Past*, Oxford and Portsmouth: James Currey and Heinemann.

Archaeomedes (ed.) (1998), *Des oppida aux métropoles*, Paris, Anthropos, coll. Villes.

Archavanitkul, K. (1989), 'Migration to Small Rural Towns in Thailand', unpublished PhD thesis, Canberra: The Australian National University.

Arnaud, M. (éd.) (1998), *Dynamique de l'Urbanization de l'Afrique au sud du Sahara*, Paris: ISTED, Ministère des Affaires étrangères, Coopération et Francophonie.

Asian Development Bank, various years, *Asian Development Outlook*, Manila: ADB.

Asuad, N. (2000), 'Transformaciones Económicas de la Ciudad de México y su Región en Los Inicios del Siglo XXI: perspectiva y políticas', *Mercado de Valores*, México, Octubre, pp. 95-104.

Australian Bureau of Statistics (2001), Australian Standard Geographical Classification (ASGC) 2001; 6. Urban Centre/Locality (UC/L) Structure: The Spatial Units. http://www.abs.gov.au/ausstats/. Accessed 2002.

Baeninger, R. (1997), 'Redistribución Espacial de la Población: Características y Tendencias del Caso Brasileño', *Notas de Población*, Santiago de Chile, Year 25, No. 65, pp. 145-202.

Bairoch P. (1996), 'Cinq Millénaires de Croissance Urbaine', in I. Sachs (ed.), *Quelles villes pour quel développement?*, Paris: PUF.

Bairoch, P., Batou, J. and Chèvre, P. (1988), *La population des villes européennes de 800 à 1850*, Genève: Droz, Centre d'histoire économique internationale.

Baker, A.R.H. (1969), 'Reversal of the Rank-Size Rule: Some Nineteenth Century Rural Settlement Sizes in France', *The Professional Geographer*, Vol. XXI(6), pp. 386-92.

Balk, D. (2002), 'The Global Urban-Rural Database: A Description with Preliminary Results', Center for International Earth Science Information Network, Columbia University, mimeo.

Bamford, E., Dunne, L. and Hugo, G. (1999), 'Accessibility/Remoteness Index of Australia (ARIA)', *Department of Health and Aged Care Occasional Papers Series No. 6*, Canberra: Commonwealth Department of Health and Aged Care.

Banerjee, B. (1986), *Rural to Urban Migration and the Urban Labour Market. A Case Study of Delhi*, Delhi: Himalaya Publishing House.

Barbary, O. and Dureau, F. (1993), 'Des citadins en mouvement. Analyse des pratiques résidentielles à Quito (Equateur)', *Cahiers des Sciences Humaines*, Vol. 29(2-3), pp. 395-418.

Basu, A., Basu, K. and Ray, R. (1987), 'Migrants and the Native Bond. An Analysis of Micro-Level Data from Delhi', *Economic and Political Weekly*, Vol. XXII (19-20-21), Annual Number, May, pp. AN-145-54.

Batty, M. (1995), 'New Ways of Looking at Cities', *Nature*, Vol. 377, p. 574.

Batty, M. (2001), 'Polynucleated Urban Landscapes', *Urban Studies*, Vol. 38, pp. 635-55.

Batty, M. and Longley, P. (1994), *Fractal Cities*, London: Academic Press.

Beale, C. (1984), 'Poughkeepsie's Complaint or Defining Metropolitan Areas.' *American Demographics*, Vol. 6(1), pp. 28-31, 46-48.

Beale, C. (1995), 'Noneconomic Value of Rural America', paper presented at the USDA Experts' Conference on the Value of Rural America, Washington, DC: US Department of Agriculture, Economic Research Service.

Beauchemin, C. (2000), 'Le Temps du Retour ? L'émigration Urbaine en Côte d'Ivoire. Une étude Géograpique', Thèse de doctorat, Université Paris VIII, Institut Français d'Urbanisme.

Beaverstock, J., Smith, R. and Taylor, P. (2000), *Globalization and World Cities*, Research Bulletin 5, GIWC Research Program, Loughborough: Loughborough University.

Bendesky, L., Godínez, V. and Mendoza, M.Á. (2001), 'La Industria Maquiladora, una Visión Regional', *Trayectorias*, Vol. 4 (7/8), pp. 133-44.

Benevolo, L. (1993), *La Ville dans l'Histoire Européenne*, Paris: Seuil.

Bentnck, J. (2000), *Unruly Urbanization of the Delhi's Fringe. Changing Patterns of Land Use and Livelihood*, Netherlands Geographical Studies, 270, Groningen: Rijksuniversiteit Groningen.

Berry, B.J.L. (1964), 'Cities as Systems Within Systems of Cities', *Proceedings of the Regional Science Association*, Vol. 13, pp. 147-63.

Berry, B.J.L. (1967), *Geography of Market Centers and Retail Distribution*, Englewood Cliffs, NJ: Prentice Hall.

Berry, B.J.L. (ed.) (1976), *Urbanization and Counterurbanization*, Beverly Hills, CA: Sage.

Berry, B.J.L. (1995), 'Capturing Evolving Realities: Statistical Areas for the American Future', in D.C. Dahmann and J.D. Fitzsimmons (eds), *Metropolitan and*

Nonmetropolitan Areas: New Approaches to Geographical Definition, Population Division Working Paper 12, Washington, DC: US Census Bureau.

Billand, C. (1990), *Delhi Case Study: Formal Serviced Land Development*, New Delhi: USAID.

Biro Pusat Statistik (1979), *Definisi Desa Urban Dalum Sensus Penduluk 1980* (Definitions of Urban Desa in the 1980 Census of Population), Jakarta: Biro Pusat Statistik (Indonesian Central Bureau of Statistics).

Black, J. and Conway, E. (1995), 'Community-Led Development Policies in the Highlands and Islands: The European Community's LEADER Programme', *Local Economy*, Vol. 10, pp. 229-45.

Boardman, J.W., Kruse, F.A. and Green, R.O. (1995). 'Mapping Target Signatures via Partial Unmixing of AVIRIS Data', *Summaries, Fifth JPL Airborne Earth Science Workshop*, JPL Publications 95-1, pp. 23-6.

Bocquier, P. and Traoré, S. (1998), 'Synthèse sur la Collecte des Données', in REMUAO (éd.), *Synthèse sur la Collecte des Données*, Rapport de recherche No. 9, Bamako: CERPOD.

Bocquier, P. and Traoré, S. (2000), *Urbanization et Dynamique Migratoire en Afrique de l'Ouest - La Croissance Urbaine en Panne*, Paris: L'Harmattan (Collection Villes et Entreprises).

Bollman, R. (ed.) (1992), *Rural and Small Town Canada*, Toronto: Thompson Educational.

Boonpratuang, C., Jones, G.W. and Taesrikul, C. (1996), 'Dispelling Some Myths About Urbanization in Thailand', *Journal of Demography* (Chulalongkorn University), pp. 21-36.

Borchert, J. (1987), *America's Heartland*, Minneapolis: University of Minnesota Press.

Bourdieu, P. (1984), *Distinction*, London: Routledge and Kegan Paul.

Bourne, L.S. (1975), *Urban Systems: Strategies for Regulation*, Oxford: Oxford University Press.

Bourne, L.S. (1992), 'Restructuring Urban Systems: Problems of Differential Growth in Peripheral Urban Systems', in *Investigaciones Geográficas. Boletín del Instituto de Geografía*. Mexico, Universidad Nacional Autónoma de México, Instituto de Geografía, special issue.

Bourne, L.S. (1999), 'The North American Urban System: The Macro-Geography of Uneven Urban Development', in F. Boal and S. Royle (eds), *North America*, London: Edward Arnold.

Bourne, L.S. (2000), 'Living on the Edge: Multiple Peripheries in the Canadian Urban System', in Y. Gradus and H. Lithwick (eds), *Developing Frontier Cities: Global Perspectives, Regional Contrasts*, Amsterdam: Kluwer Publishers.

Bourne, L.S. and Rose, D. (2001), 'The Changing Face of Canada: The Uneven Geographies of Population and Social Change', *Canadian Geographer*, Vol. 45(1), pp. 105-19.

Bourne, L.S. and Simmons, J.W. (2002), 'The Dynamics of the Canadian Urban System', in H.S. Geyer (ed.), *International Handbook of Urban Systems. The Maturing of Urban Systems in Developed and Advanced Developing Countries*, London: Edward Elgar.

Bové, J. and Dufour, F. (2001), *The World is Not For Sale*, London: Verso.

Bowler, I. (1992), 'Sustainable Agriculture as an Alternative Path of Farm Business Development', in I. Bowler, C. Bryant and M. Nellis (eds), *Rural Systems in Transition: Agriculture and Environment*, Wallingford: CAB International.

Brennan-Galvin, E.M. (2002), 'Data Needs for an Urbanizing World', paper prepared for IUSSP Working Group on Urbanization Conference on New Forms of Urbanization: Conceptualizing and Measuring Human Settlement in the Twenty-First Century, Bellagio, Italy, 11-15 March.

Bretagnolle, A., Paulus, F. and Pumain, D. (2002), 'Time and Space Scales for Measuring Urban Growth', *Cybergeo*.

Briggs, V. (1992), *Mass Immigration and the National Labor Market*, Armonk, New York: NE Sharpe, Inc.

Brockerhoff, M. (1998), 'Migration and the Fertility Transition in African Cities', in R.E. Bilsborrow (ed.), *Migration, Urbanization, and Development: New Directions and Issues*, New York and Dordrecht: United Nations Population Fund and Kluwer Academic Publishers.

Bronfenbrenner, U. (1995), 'The Bioecological Model From a Life Course Perspective: Reflections of a Participant Observer', in P. Moen, G.H. Elder, Jr. and K. Luscher (eds), *Examining Lives in Context*, Washington, DC: American Psychological Association.

Brown, D. and Lee, M. (1999), 'Persisting Inequality Between Metropolitan and Nonmetropolitan America: Implications for Theory and Policy', in P. Moen, D. Demster-McClain and H. Walker (eds.) *Diversity, Inequality, and Community in American Society*, Ithaca: Cornell University Press.

Brown, D., Fuguitt, G., Heaton, T. and Waseem, S. (1997), 'Continuities in Size of Place Preferences in the United States, 1972-1992', *Rural Sociology*, Vol. 62(4), pp. 408-28.

Brush, J. (1962), 'The Morphology of Indian Cities', in R. Turner (ed.), *India's Urban Future*, Berkeley and Los Angeles: University of California Press.

Brush, J. (1986), 'Recent Changes in Ecological Patterns of Metropolitan Bombay and Delhi', in V.K. Tewari, J.A. Weistein and V.L.S.P. Rao (eds), *Indian Cities. Ecological Perspectives*, New Delhi: Concept.

Bunle, H. (1934), 'Comparaison Internationale des Agglomérations Urbaines', Institut International de Statistiques, London, XXIIe session.

Bunting, T. and Filion, P. (eds) (2000), *Canadian Cities in Transition*, Second edition, Toronto: Oxford University Press.

Bura, S., Guérin-Pace, F., Mathian, H., Pumain, D. and Sanders L. (1996), 'Multi-Agent Systems and the Dynamics of Settlement Systems', *Geographical Analysis*, Vol. 2, pp. 161-78.

Burrough, P.A. (1996), 'Natural Objects with Indeterminate Boundaries', in P.A. Burrough and A.U. Frank, (eds), *Geographic Objects with Indeterminate Boundaries*, London: Taylor and Francis.

Bussière, R. and Stovall, H. (1978), *Systèmes Urbains et Régionaux à l'Etat d'Equilibre*, Paris: CRU.

Butler, M. and Beale, C. (1994), 'Rural-Urban Continuum Codes for Metropolitan and Nonmetropolitan Counties, 1993', *Staff Report* No. 9425, Washington, DC: US Department of Agriculture, Economic Research Service.

Byrne, D. (2001), *Understanding the Urban*. Houndmills, Basingstoke: Palgrave.

Carrasco, S. *et al.* (1997), *Chile: Reconversión Forestal y Pobreza en Comunas Seleccionadas de la Región del Biobío* (Series B, No. 115, diagnoses fascicle No.2), Santiago de Chile: Centro Latinoamericano de Demografía (CELADE).

Carrothers, G. (1956), 'An Historical Review of the Gravity and Potential Concepts of Human Interaction', *Journal of American Institute of Planners*, Vol. 22, pp. 94-102.

Carter, H. (1990), *Urban and Rural Britain*, Harlow: Longman.

Casado-Díaz, J.M. (1996), 'Mercados Laborales Locales. Análisis Preliminar del Caso Valenciano', *Revista de Estudios Regionales*, Vol. 45, pp. 129-55.

Castells, M. (1989), *The Informational City*, Oxford: Blackwell.

Castillo, M.A. and Palma, S. (1999), 'Central American International Emigration: Trends and Impacts', in R. Appleyard (ed.), *Emigration Dynamics in Developing Countries, Volume III, Mexico, Central America and the Caribbean*, Aldershot: Ashgate.

Cattan, N., Pumain, D., Rozenblat, C. and Saint-Julien, T. (1999), *Le Système des Villes Européennes*, Paris: Anthropos, coll. Villes.

CELADE (2001), 'Urbanización y Evolución de la Población Urbana de América Latina, 1950-1990', in *Boletín Demográfico* (LC/G.2140-P), Santiago de Chile, year 33, Special issue, Santiago de Chile: Centro Latinoamericano de Demografía (CELADE).

CELADE (2002), *Migración de Latinoamericanos y Caribeños en una era de Gglobalización*, Internal working paper, Santiago de Chile: Centro Latinoamericano de Demografía (CELADE).

Central Committee of CCP and the State Council (1986), 'Directive of the Central Committee of the CCP and the State Council on adjusting the organization system of cities and towns and reducing suburban area of cities' (in Chinese), in Editorial Board (ed.), *Zhongguo Renkou Nianjian (1985) (Almanac of China's Population (1985))*, pp. 96-7, Beijing: China Social Science Press.

Cervero, R. (1986), *Suburban Gridlock*, New Brunswick, NJ: Rutgers University Center for Urban Policy Research.

Cervero, R. (1989), *America's Suburban Centers: The Land Use Transportation Link*, London: Unwin Hyman.

Champaud, J. (1983), *Villes et Campagnes du Cameroun de l'Ouest*, Paris: ORSTOM.

Champion, A.G. (ed.) (1989), *Counterurbanization: The Changing Pace and Nature of Population Deconcentration*, London: Edward Arnold.

Champion, A.G. (1992), 'Population Change and Migration in Britain since 1981: Evidence for Continuing Deconcentration', *Environment and Planning A*, Vol. 26(10), pp. 1501-20.

Champion, A.G (1994), 'International Migration and Demographic Change in the Developed World', *Urban Studies*, Vol. 31(4/5), pp. 653-77.

Champion, A.G. (1997), 'The Complexity of Urban Systems: Contrasts and Similarities from Different Regions'. Invited paper presented at the IUSSP's XXIIIrd General Population Conference, Beijing, China, 11-17 October.

Champion, A.G. (1998), 'Population Distribution in Developed Countries: Has Counterurbanization Stopped?' *Population Distribution and Migration: Proceedings of the United Nations Expert Group Meeting on Population Distribution and Migration*, Santa Cruz (Bolivia), 18-22 January 1993, New York: United Nations.

Champion, T. (2001), 'Demographic Transformations', in P. Daniels, M. Bradshaw, D. Shaw and J. Sidaway (eds), *Human Geography: Issues for the 21st Century*, Harlow: Pearson Education.

Chan, K.W. (1994a), *Cities With Invisible Walls*, Hong Kong: Oxford University Press.

Chan, K.W. (1994b), 'Urbanization and Rural-Urban Migration in China Since 1982', *Modern China*, Vol. 20(3), pp. 243-81.

Chant, S. (1999), 'Population, Migration, Employment and Gender', in R. Gwynne and C. Kay (eds), *Latin America Transformed: Globalization and Modernity*, London, England, Arnold.

Chapman, M. and Prothero, R.M. (1983), 'Themes on Circulation in the Third World', *International Migration Review*, Vol. 17(4), pp. 597-632.

Chase-Dunn, C. and Jorgenson, A. (2002), 'Settlement Systems: Past and Present', *GAWC Research Bulletin 73*, http://www.lboro.ac.uk/gawc/rb/rb/73.html.

Chen, J.Y. and Huang, G.M. (1991), *Fujiansheng Jingji Dili (The Economic Geography of Fujian Province)*, Beijing: Xinhua Publishing House.

Chen, N., Valente, P. and Zlotnik, H. (1998), 'What Do We Know about Recent Trends in Urbanization?', in R.E. Bilsborrow (ed.), *Migration, Urbanization, and Development: New Directions and Issues*, New York and Dordrecht: United Nations Population Fund and Kluwer Academic Publishers.

Cheshire, P., Hay, D., Carbonaro, G. and Bevan, N. (1988), *Urban Problems and Regional Policies in the European Community*, Luxembourg: Commission of the European Communities.

China, State Council (1984a), 'Circular of the State Council Approving the Report of the Ministry of Civil Affairs Concerning the Adjustment of the Criteria for Establishing towns' (in Chinese), *Guowuyuan Gongbao (Bulletin of the State Council)*, No.30, pp. 1012-4.

China, State Council (1984b), 'Circular from the State Council Concerning the Question of Peasants Entering Towns for Settlement' (in Chinese), *Guowuyuan Gongbao (Bulletin of the State Council)*, No.26, pp. 919-20.

Christaller, W. (1966), *Central Places in Southern Germany*. Englewood Cliffs, NJ: Prentice Hall.

Chung, R. (1970), 'Space-Time Diffusion of the Demographic Transition Model: The Twentieth Century Patterns', in G.J. Demko, H.M. Rose and G.A. Schnell (eds), *Population Geography: A Reader*, New York: McGraw-Hill.

Churchill, W. (1943), 'We Shape Our Buildings'. http://www.winstonchurchill.org/bonmots.htm. Accessed 2002.

Cigler, B. (1993), 'Meeting the Growing Challenges of Rural Local Government', *Rural Development Perspectives*, Vol. 9(1), pp. 35-39.

Ciudad de México *et al.* (n.d), *Programa de Ordenación de la Zona Metropolitana del Valle de México*, México: Ciudad de México, Sedesol, Estado de México.

Claval, P. (1987), 'The Region as a Geographical, Economic and Cultural Concept', *International Social Science Journal*, Vol. 39, pp. 159-72.

Clichevsky, N. (2000), *Informalidad y Segregación Urbana en América Latina. Una Aproximación*. Serie Medio Ambiente y Desarrollo No. 28, Santiago de Chile: ECLAC.

Cloke, P. (1977), 'An Index of Rurality for England and Wales', *Regional Studies*, Vol. 11, pp. 31-46.

Cloke, P. (1980), 'New Emphases for Applied Rural Geography', *Progress in Human Geography*, Vol. 4, pp. 181-217.

Cloke, P. (1985), 'Whither Rural Studies?', *Journal of Rural Studies*, Vol. 1, pp. 1-9.

Cloke, P. (1989), 'Rural Geography and Political Economy', in R. Peet and N. Thrift (eds), *New Models in Geography. Volume One*, London: Unwin Hyman.

Cloke, P. (1993), 'The Countryside as Commodity: New Rural Spaces for Leisure', in S. Glyptis (ed.), *Leisure and the Environment*, London: Belhaven.

Cloke, P. (1994), '(En)culturing Political Economy: A Life in the Day of a "Rural Geographer"', in P. Cloke, M. Doel, D. Matless, M. Phillips and N. Thrift, *Writing the Rural*, London: Paul Chapman Publishing.

Cloke, P. and Edwards, G. (1986), 'Rurality in England and Wales, 1981: A Replication of the 1971 Index', *Regional Studies*, Vol. 20, pp. 289-306.

Cloke, P. and Goodwin, M. (1992), 'Conceptualizing Countryside Change: From Post-Fordism to Rural Structural Coherence', *Transactions of the Institute of British Geographers*, Vol. 17, pp. 321-36.

Cloke, P. and Little, J. (eds) (1997), *Contested Countryside Cultures*, London: Routledge.

Cloke, P., Goodwin, M. and Milbourne, P. (1998), 'Inside Looking Out; Outside Looking In. Different Experiences of Cultural Competence in Rural Lifestyles', in P. Boyle and K. Halfacree (eds), *Migration into Rural Areas: Theories and Issues*, Chichester: Wiley.

CONAPO (1997), *La Situación Demográfica en México, 1997*, México, D.F., México: CONAPO.

CONAPO (1998), *La Situación Demográfica en México, 1998*, México, D.F., México: CONAPO.

CONAPO (2001), *La Población de México en el Nuevo Siglo*, México, D.F., México: CONAPO.

Confédération Paysanne (1994), *L'Agriculture Paysanne: des Pratiques aux Enjeux de Société*, Paris: FPH / Confédération Paysanne.

Consejo Nacional de Población (1991), *Sistema de Ciudades y Distribución Espacial de la Población*, México: CONAPO.

Consejo Nacional de Población (1994), *Evolución de Las Ciudades en México, 1990*, México: CONAPO.

Consejo Nacional de Población (1998), *Escenarios Demográficos y Urbanos de la Zona Metropolitana de la Ciudad de México, 1990-2010*, México: CONAPO.

Conti, S. and Spriano, G. (1990), *Effetto Citta*, volumo primo: *Sistemi Urbani e Innovazione: Prospettive per l'Europa degli Anni Novanta*, Edizioni dell Fondazione Giovanni Agnelli.

Cook, P. and Mizer, K. (1994), 'The Revised ERS County Typology', *Rural Development Research Report*, No. 84, Washington, DC: US Department of Agriculture, Economic Research Service.

Coombes, M.G. (2000), 'Defining Locality Boundaries with Synthetic Data', *Environment and Planning A*, Vol. 32, pp. 1499-518.

Coombes, M.G. and Raybould, S. (2001), 'Public Policy and Population Distribution: Developing Appropriate Indicators of Settlement Patterns', *Government and Policy*, Vol. 19, pp. 223-48.

Coombes, M.G. and Wong, Y.L.C. (1994), 'Methodological Steps in the Development of Multivariate Indexes for Urban and Regional Policy Analysis', *Environment and Planning A*, Vol. 26, pp. 1297-316.

Copp, J. (1972), 'Rural Sociology and Rural Development', *Rural Sociology*, Vol. 37, pp. 515-33.

Coquery-Vidrovitch, C. (1991), 'The Process of Urbanization in Africa (from the origins to the beginning of independence)', *African Studies Review*, Vol. 34(1), pp. 1-98.

Coquery-Vidrovitch, C. (1992), *Afrique Noire: Permanences et Ruptures*, Paris: L'Harmattan.

Cordell, D.D., Gregory, J. and Piché, V. (1996), *Hoe and Wage: A Social History of a Circular Migration System in West Africa*, Boulder, CO: Westview Press (African Modernization and Development Series).

Corona, R. (2002), 'Mediciones de la Migración de Mexicanos a Estados Unidos en la Década 1990-2000', in García, B. (ed.), *Población y Sociedad al Inicio del Siglo XXI*, México: El Colegio de México.

Cosio-Zabala, M. (1998), Changements Démographique et Sociaux à la Frontiere Mexique - Etat Unis, paper presented at Sèminaire international de l' AIDELF on Régime Demographiques et Territoires: le Frontièrs en Question, la Rochelle, 22-26 Septembre.

Cowen, D.J. and Jensen, J.R. (1998), 'Extraction and Modeling of Urban Attributes Using Remote Sensing Technology', in D. Liverman, E.F. Moran, R.R. Rindfuss, and P.C. Stern (eds), *People and Pixels: Linking Remote Sensing and Social Science*, Washington, DC: National Academy Press.

Critchfield, R. (1994), *The Villagers: Changed Values, Altered Lives: The Closing of the Urban-Rural Gap*, New York: Anchor Books.

Cromartie, J. (1999), 'Rural Minorities Are Geographically Clustered', *Rural Conditions and Trends*, Vol. 9(2), pp. 14-19.

Cromartie, J. and Swanson, L. (1996), *Defining Metropolitan Areas and the Rural-Urban Continuum – A Comparison of Statistical Areas Based on County and Sub-County Geography*, ERS Staff Paper No. 9603. Washington DC: US Department of Agriculture.

Cunha, J.M.P. (2001a), 'Aspectos Demográficos da Estruturação das Regiões Metropolitanas Brasileiras', in D. Hogan, J.M.P. Cunha, R. Baeninger and R. Carmo (eds), *Migração e Ambiente nas Aglomerações Urbanas*, Campinas, SP: Nepo/Pronex-Unicamp.

Cunha, J.M.P. (2001b), 'Intraregional Mobility in the Context of Migartory Changes in Brazil between 1970 and 1991: The Case of São Paulo Metropolitan Region', in D.P. Hogan (ed.), *Population Change in Brazil: Contemporary Perspectives*, Campinas, SP: Nepo/Unicamp. (Also published in Notas de Población, no.70, Santiago de Chile, 2000, pp.149-18.)

Cunha, J.M.P. and Bitencourt, A.A. (2001), 'População e Espaço Intra-Urbano em Campinas: A Análise Sócio-Demográfica Como Subsídio Para a Política Pública Local', in D.Hogan, J.M.P. Cunha, R. Baeninger and R. Carmo (eds), *Migração e Ambiente nas Aglomerações Urbanas*, Campinas, SP: Nepo/Pronex-Unicamp.

Cunha, J.M.P. and Rodrigues, I.A. (2001), 'Transition Space: New Standpoint on São Paulo State's Population Redistribution Process', paper presented at IUSSP's XXIV General Population Conference, Salvador, BA, August.

Curran, P.J., Atkinson, P.M., Foody, G.M. and Milton, E.J. (2000), 'Linking Remote Sensing, Land Cover and Disease', in S.I. Hay, S.E. Randolph, and D.J. Rogers (eds), *Remote Sensing and Geographical Information Systems in Epidemiology*, San Diego: Academic Press.

Dahmann, D.C. (1999), 'New Approaches to Delineating Metropolitan and Nonmetropolitan Settlement: Geographers Drawing the Line', *Urban Geography*, Vol. 20(8), pp. 683-94.

Dahmann, D. and Fitzsimmons, J. (eds) (1995), *Metropolitan and Non-metropolitan Areas: New Approaches to Geographic Definition*, Working Paper No. 12, Washington DC: US Bureau of the Census.

Dai, J.L. (2000), *Zhongguo Shizhi (China's Urban Organizational System)*, Beijing: China Cartographic Publishing House.

Darley, G. (1975), *Villages of Vision*, London: Paladin.

Datta, V.N. (1986), 'Panjabi refugees and the urban development of Greater Delhi', in R.E. Frykenberg (ed.), *Delhi Through the Ages. Essays in Urban History, Culture and Society*, Delhi: Oxford University Press.

Davis, K. (1959), *The World's Metropolitan Areas*, Berkeley, California: International Urban Research, University of California.

Dax, T. (1996), 'Defining Rural Areas – International Comparisons and the OECD Indicators', *Rural Society*, Vol. 6(3), pp. 3-18.

De Mattos, C. (1998), 'The Moderate Efficiency of Population Distribution Policies in Developing Countries', in *Population Distribution and Migration*, New York: United Nations.

De Mattos, C. (2001), 'Metropolización y Suburbanización', *EURE*, Santiago de Chile, Volume XXVII, No. 80 (electronic version available: www.scielo.cl).

De Vries, J. (1984), *European Urbanization, 1500-1800*, London: Methuen.

Dear, M. and Flusty, S. (1998), 'Postmodern Urbanism,' *Annals of the Association of American Geographers*, Vol. 88(1), pp. 50-72.

Decand, G. (2000), 'Regional Policy and Statistics – A Long Story', *Sigma*, 1/2000, 6-8.

Dendrinos, D.S. and Mullaly, H. (1985), *Urban Evolution: Study in the Mathematical Ecology of Cities*, Oxford: Oxford University Press.

Denham, C. and White, I. (1998), 'Differences in Urban and Rural Britain', *Population Trends*, Vol. 91, pp. 23-34.

Department for Transport, Local Government and the Regions (2002). *Urban Policy Evaluation Strategy Consultation Document*, London: DTLR, http://www.urban.dtlr.gov.uk/consult/evalstrat/index.htm. Accessed 2002.

Dick, H.W. and Rimmer, P.J. (1998), 'Beyond the Third World City: The New Urban Geography of South-East Asia', *Urban Studies*, Vol. 35(12), pp. 2303-21.

Diwakar, A. and Qureshi, M.H. (1993), 'Physical Processes of Urbanization in Delhi', *Urban India*, Vol. XIII (2), pp. 94-106.

Doll, C.N.H., Muller, J.-P. and Elvidge, C.D. (2000), 'Night-time Imagery as a Tool for Global Mapping of Socioeconomic Parameters and Greenhouse Gas Emissions', *Royal Swedish Academy of Sciences*, Vol. 29, pp. 157-62.

Dorling, D. and Atkins, D. (1995), 'Population Density, Change and Concentration in Great Britain 1971, 1981 and 1991', *OPCS Studies on Medical and Population Subjects 58*, London: HMSO.

Douglass, M. and DiGregorio, M. (2002), *The Urban Transition in Vietnam*, Honolulu, Fukuoka and Hanoi: UNCHS/UNDP and University of Hawaii, Department of Urban and Regional Planning.

Drummond, I., Campbell, H., Lawrence, G. and Symes, D. (2000) 'Contingent or Structural Crisis in British Agriculture?', *Sociologia Ruralis*, Vol. 40, pp. 111-27.

Duncan, O.D. (1959), 'Human Ecology and Population Studies', in P.M. Hauser and O.D. Duncan (eds), *The Study of Population: An Inventory and Appraisal*, Chicago: University of Chicago Press.

Duncan, C. (1999), *Worlds Apart: Why Poverty Persists in Rural America*, New Haven: Yale University Press.

Dupont, V. (1997) 'Les "rurbains" de Delhi', *Espace, Populations, Sociétés*, Vol. 1997-2-3, pp. 225-40.

Dupont, V. (2000a), 'Residential and Economic Strategies of Houseless People in Delhi', in V. Dupont, E. Tarlo and D. Vidal (eds), *Delhi. Urban Space and Human Destinies*, Delhi: Manohar-CSH, pp. 99-124.

Dupont, V. (2000b), 'Spatial and Demographic Growth of Delhi and the Main Migration Flows', in V. Dupont, E. Tarlo and D. Vidal (eds), *Delhi. Urban Space and Human Destinies*, Delhi: Manohar-CSH.

Dupont, V. (2001a), 'Noida: nouveau pôle industriel ou ville satellite de Delhi? Le projet des planificateurs, ses Failles et son devenir', *Revue Tiers Monde*, No. 165, Janvier-Mars 2001, pp. 189-211.

Dupont, V. (2001b), 'Les nouveaux quartiers chics de Delhi. Langage publicitaire et réalités périurbaines', in H. Rivière d'Arc (ed.), *Nommer les nouveaux territoires urbains*, Paris: Editions UNESCO/Editions de la Maison des Sciences de l'Homme.

Dupont, V. and Dureau, F. (1994), 'Rôle des mobilités circulaires dans les dynamiques urbaines. Illustrations à partir de l'Equateur et de l'Inde', *Revue Tiers Monde*, Vol. XXXV(140), pp. 801-29.

Dupont, V. and Mitra, A. (1995), 'Population Distribution, Growth and Socio-economic Spatial Patterns in Delhi. Findings from the 1991 Census Data', *Demography India*, Vol. 24(1/2), pp. 101-32.

Dupont, V. and Prakash, J. (1999), 'Enquête: Mobilités spatiales dans l'aire métropolitaine de Delhi', in *Biographies d'enquête*, Paris: INED-PUF/Diffusion, Coll. Méthodes et savoirs.

Dureau, F. (1993), 'Pour une Approche non Fonctionnaliste du Milieu Urbain Africain', in *Croissance Démographique et Urbanization: Politiques de Peuplement et Aménagement du Territoire*, Paris: AIDELF.

Durkheim, E. (1951), *Suicide*, New York: Free Press.

ECLAC (1990), *Panorama Económico de América Latina 1990*, Santiago de Chile: ECLAC.

ECLAC (2000), *The Equity Gap: A Second Assessment.* Second Regional Conference in Follow-up to the World Summit for Social Development, Santiago de Chile, 15-17 May.

ECLAC (2001a), *Una Década de Luces y Sombras: América Latina y el Caribe en los Años Noventa,* Bogotá, Colombia: Editorial Alfaomega.

ECLAC (2001b), *Panorama Social de América Latina, 2000-2001* (LC/G.2138-P), Santiago de Chile: ECLAC.

ECLAC/HABITAT (2001), *El Espacio Regional: Hacia la Consolidación de los Asentamientos Humanos en América Latina y el Caribe* (LC/G.2116/Rev.1-P), Santiago de Chile: ECLAC.

ECLAC/ILPES (2000), *La Reestructuración de los Espacios Nacionales,* Series Gestión Pública, No. 7 (LC/L.1418-P), Santiago de Chile: ECLAC.

Edmondson, B. (1995), 'The Real Economy', *American Demographics,* November, p. 76.

Egler, C. (2001), *Mudanças recentes no uso e na coberta da terra no Brasil,* paper presented at seminar 'Mudanças Ambientais Globais: Perspectivas Brasileira'.

Eisenstadt, S.N. and Shachar, A. (1987), *Society, Culture and Urbanization,* Beverly Hills: Sage.

Eldridge, H.T. (1942), 'The Process of Urbanization', *Social Forces,* Vol. 20(3), pp. 311-16.

Elvidge, C.D., Baugh, K.E., Kihn, E.A., Kroehl, H.W. and Davis, E.R. (1997), 'Mapping City Lights With Nighttime Data from the DMSP Operational Linescan System', *Photogrammetric Engineering and Remote Sensing,* Vol. 63, pp. 727-34.

ESDP (1999), *European Spatial Development Perspective,* Luxembourg: European Commission.

Esparza, A.X. and Krmenec, A.J. (1996), 'The Spatial Markets of Cities Organized in a Hierarchical System', *Professional Geographer,* Vol. 48(4), pp. 367-78.

Eurostat (1992), *Study on Employment Zones,* Luxembourg: Eurostat.

Eurostat, (1995), *Pilot Project. Delimitation of European Urban Agglomerations using Remote Sensing.* Luxembourg: Eurostat.

Eurostat (1999), 'Les Zones Densément Peuplées dans l'Union Européenne, Essai de Délimitation et Caractérisation des Agglomérations Urbaines', *Statistiques en bref,* Thème 1, 2.

Fan, J. (1998), 'A Case Study on the Role of Rural Industrial Development in the Process of Urbanization in China' (in Chinese), *Dili Kexue (Scientia Geographica Sinica),* Vol. 18(2), pp. 99-105.

Faria, V. (1978), 'O Processo de Urbanização no Brasil. Algumas Notas Para Seu Estudo e Interpretação', in Encontro Nacional de Estudos Populacionais 1, 1978, Campos do Jordão, *Anais,* São Paulo: ABEP.

FIBGE (Fundação Instituto Brasileiro de Geografia e Estatística) (2000), XI Recenseamento Geral do Brasil – Manual de Delimitação dos setores de 2000. Rio de Janeiro: IBGE.

FIBGE (Fundação Instituto Brasileiro de Geografia e Estatística) (1987), Regiões de Influência das Cidades: revisão atualizada do estudo Divisão do Brasil em Regiões Funcionais Urbanas. Rio de Janeiro: IBGE.

Firman, T. (1992), 'The Spatial Pattern of Urban Population Growth in Java, 1980-1990', *Bulletin of Indonesian Economic Studies,* Vol. 28(2), pp. 95-109.

Fischer, C. (1975), 'Toward a Subcultural Theory of Urbanism', *American Journal of Sociology,* Vol. 80, pp. 1319-42.

Fisher, H. (1933), 'The Beauty of England', in Council for the Preservation of Rural England, *The Penn Country of Buckinghamshire,* London: CPRE.

Fletcher, R. (1986), 'Settlement in Archaeology: World-Wide Comparison', *World Archaeology,* Vol. 18(1), pp. 59-83.

Foot, D. and Stoffman, D. (2000), *Boom, Bust and Echo*, Toronto: Macfarland, Walter and Ross.

Fotheringham, A.S. and Wong, D.W.S. (1991), 'The Modifiable Areal Unit Problem in Multivariate Statistical Analysis', *Environment and Planning A*, Vol. 23, pp. 1025-44.

Frankhauser, P. (1993), *La Fractalité des Structures Urbaines*, Paris: Anthropos.

Frankhauser, P. and Pumain, D. (2001), 'Fractales et Géographie', in L. Sanders (ed.), *Modèles en Analyse Spatiale*, Paris: Hermès.

Frey, W.H. (1996), 'Immigration, Domestic Migration and Demographic Balkanization in America', *Population and Development Review*, Vol. 22(4), pp. 741-63.

Frey, W.H. (2001a), 'Census 2000 Shows Large Black Return to the South: Reinforcing the Region's White-Black Demographic Profile', Research Report 01-473, Ann Arbor, MI: University of Michigan, Population Studies Center.

Frey, W.H. (2001b), 'The Baby Boom Tsunami', *Milken Institute Review*, Second Quarter (August), pp. 4-7.

Frey, W.H. (2002), 'US Census Shows Different Paths for Domestic and Foreign Born Migrants', *Population Today* (August/September), Washington, DC: Population Reference Bureau.

Frey, W.H. and Farley, R. (1996), 'Latino, Asian and Black Segregation in Multi-ethnic Metro Areas: Are Multiethnic Metros Different?', *Demography*, Vol. 33(1), pp. 35-50.

Frey, W.H. and Johnson, K.M. (1996), 'Concentrated Immigration, Restructuring, and the Selective Deconcentration of the US Population', *Research Reports* No. 96-371, Ann Arbor, MI: University of Michigan Population Studies Center.

Frey, W.H. and Speare, A. Jr. (1988), *Regional and Metropolitan Growth and Decline in the United States*, New York: Russell Sage.

Frey, W.H. and Speare, A. Jr. (1991), 'The Revival of Metropolitan Population Growth in the United States: An Assessment of 1990 Census Findings', *Population and Development Review*, Vol. 18(1), pp. 109-146.

Frey, W.H. and Speare, A. Jr. (1995), 'Metropolitan Areas as Functional Communities', in D.C. Dahmann and J.D. Fitzsimmons (eds), *Metropolitan and Nonmetropolitan Areas: New Approaches to Geographical Definition*. Working Paper No. 12. Washington, DC: US Bureau of the Census, Population Division.

Friedmann, J. (1986), 'The World City Hypothesis', *Development and Change*, Vol. 17, pp. 69-83.

Fuguitt, G.V. and Heaton, T.B. (1993), 'The Impact of Migration on the Nonmetropolitan Population Age Structure, 1960-1990', unpublished manuscript. Madison, WI: University of Wisconsin, Department of Sociology.

Fuguitt, G.V., Brown, D. and Beale, C. (1989), *Rural and Small Town America*, New York: Russell Sage Foundation.

Gale, F. and McGranahan, D. (2001), 'Nonmetropolitan Areas Fall Behind in the New Economy', *Rural America*, Vol. 16(1), pp. 44-51.

García, B. (1994), *Los Determinantes de la Oferta de Mano de Obra en México*, México: Secretaría de Trabajo y Previsión Social.

Garreau, J. (1991), *Edge City: Life on the New Frontier*, New York: Doubleday.

Garza, G. (1999), 'Global Economy, Metropolitan Dynamics and Urban Policies in Mexico', *Cities*, Vol. 16(3), pp. 149-70.

Garza, G. (2000), 'La Megalópolis de la Ciudad de México Según Escenario Tendencial, 2020', om Garza, G. (ed.), *La Ciudad de México en el Fin del Segundo Milenio*, México: Gobierno del Distrito Federal, El Colegio de México.

Garza, G. (2001), 'La Urbanización de México en el Siglo XX', México: International Institute for Environment and Development (Report).

Garza, G. and Rivera, S. (1995), *Dinámica Macroecónomica de las Ciudades en México*, México: INEGI, El Colegio de México, IIS-UNAM.

Garza, G., Filion, P. and Sands, G. (2002), *Planeación Urbana en Grandes Metrópolis: Detroit, Monterrey y Toronto*, México: El Colegio de México.

Gatica, Fernando (1980). 'La Urbanización en América Latina: 1950-1970; Patrones y áreas Críticas', in *Redistribución Espacial de la Población en América Latina*. Santiago de Chile: Centro Latinoamericano de Demografía (CELADE).

Geoghegan, J., Pritchard, L. Ogneva-Himmelberger, Y., Chowdhury, R.R., Sanderson, S. and Turner, B.L. (1998), '"Socializing the Pixel" and "Pixelizing the Social"', in D. Liverman, E.F. Moran, R.R. Rindfuss and P.C. Stern (eds), *People and Pixels*, Washington, DC: National Academy Press.

George, P. (1993), *Dictionnaire de la géographie*, 5th edition, Paris: PUF.

Geyer, H.S. (ed.) 2002, *International Handbook of Urban Systems*. Cheltenham: Edward Elgar.

Ghelfi, L. and Parker, T. (1997), 'A County-Level Measure of Urban Influence', *Rural Development Perspectives*, Vol. 12(2), pp. 32-41.

Gibbs, J.P. (1963), 'The evolution of population concentration', *Economic Geography*, Vol. 39, 119-29.

Gibbs, R. (2001), 'Nonmetropolitan Labor Markets in an Era of Welfare Reform', *Rural America*, Vol. 16(3), pp. 11-21.

Gibrat, R. (1931), *Les inégalités économiques*, Paris: Sirey.

Giddens, A. (1984), *The Constitution of Society*, Cambridge: Polity Press.

Gilbert, A. (1982) 'Rural theory: the grounding of rural sociology', *Rural Sociology*, Vol. 47, pp. 609-633.

Ginsburg, N., Koppel, B. and McGee, T.G. (1991), *The Extended Metropolis: Settlement Transition in Asia*, Honolulu: University of Hawaï Press.

Glover, J., Harris, K. and Tennant, S. (1999), *A Social Health Atlas of Australia*, 2nd edition, Public Health Information Unit, University of Adelaide, South Australia. Canberra: Open Book Publishers.

Glover, T. and Openshaw, S. (1995), *The AGW Spatial Interaction Modelling Workstation*, http://www.geog.leeds.ac.uk/people/openshaw/w3agw.html.

Godard, H. (1994), 'Port-au-Prince (1982-1992): Un Système Urbain à la Dérive'. *Problèmes d'Amérique Latine*, Paris, France, No. 14, pp. 181-94.

Goldstein, S. (1978), Circulation in the Context of Total Mobility in Southeast Asia', paper of the East-West Population Institute No. 53, Honolulu, East-West Population Institute.

Goldstein, S. and Goldstein, A (1978), 'Thailand's Urban Population Reconsidered', *Demography*, Vol. 15, pp. 239-58.

Goldstein, S. and Sly, D.F. (eds) (1975a), *Basic Data Needed for the Study of Urbanization*, Working Paper 1, IUSSP Committee on Urbanization and Population Redistribution, Dolhain, Belgium: Ordina Editions.

Goldstein, S. and Sly, D.F. (eds) (1975b), *The Measurement of Urbanization and Projection of Urban Population*, Working Paper 2, IUSSP Committee on Urbanization and Population Redistribution, Dolhain, Belgium: Ordina Editions.

Goldstein, S. and Sly, D.F. (eds) (1977), *Patterns of Urbanization: Comparative Country Studies*, Working Paper 3, IUSSP Committee on Urbanization and Population Redistribution, Dolhain, Belgium: Ordina Editions.

Goodchild, M.F. and Janelle, D. (1984), 'The City Around the Clock: Space-Time Patterns of Urban Ecological Structure', *Environment and Planning A*, Vol. 10, pp. 1273-85.

Goodman, J.F.B. (1970), 'The Definition and Analysis of Local Labour Markets: Some Empirical Problems', *British Journal of Industrial Relations*, Vol. 8, pp. 179-96.

Goodman, D. and Redclift, M. (1991), *Refashioning Nature*, London: Routledge.

Gottmann, J. (1961), *Megalopolis: The Urbanized Northeastern Seaboard of the United States*, New York: Twentieth Century Fund.

Government of NCT of Delhi (Planning Department) (1996), 'Backgrounder', State Level Seminar on Approach to Ninth Five Year Plan (1997-2002), Delhi: Government of National Capital Territory of Delhi.

GRAB (Groupe de Réflexion sur l'Approche Biographique) (1999), *Biographies d'enquête, Bilan de 14 collectes biographiques*, Paris: INED-PUF/Diffusion, Coll. Méthodes et savoirs.

Graizbord, B. (1988), 'Las Necesidades de Urbanización en el Largo Plazo', in G. Bueno (ed.), *México; el Desafío de Largo Plazo*, México: Limusa.

Gray, J. (2000), 'The Common Agricultural Policy and the Re-invention of the Rural in the European Community', *Sociologia Ruralis*, Vol. 40, pp. 30-52.

Green F.H.W. (1950), 'Urban Hinterlands in England and Wales: An Analysis of Bus Services', *Geographical Journal*, Vol. 116, pp. 64-89.

Green, N.E. (1955), *Aerial Photography in the Analysis of Urban Structures, Ecological and Social*. PhD Dissertation in Sociology, University of North Carolina at Chapel Hill.

Gruen, A., Kuebler, O. and Agouris, P. (1995), *Automatic Extraction of Man-Made Objects from Aerial and Space Images*, Basel: Birkauser Verlag.

Guardian (2001), 'Extent of Farm Crisis Revealed', *The Guardian*, 11 April.

Guhry, P., Lammlen, S.B. and Ngwé, E. (1996), *Le Retour au Village: Une Solution à la Crise économique au Cameroun*, Paris: L'Harmattan, CEPED, IFORD.

Hackenberg, R.A. (1980), 'New Patterns of Urbanization in Southeast Asia: An Assessment', *Population and Development Review*, Vol. 6(3), pp. 391-419.

Hadden, J. and Barton, J. (1973), 'An Image That Will Not Die: Thoughts on the History of Anti-Urban Ideology', *Urban Affairs Annual Review*, Vol. 7, pp. 79-116.

Halfacree, K. (1992), *The Importance of Spatial Representations in Residential Migration to Rural England in the 1980s*, Unpublished PhD thesis, Lancaster University.

Halfacree, K. (1993), 'Locality and Social Representation: Space, Discourse, and Alternative Definitions of the Rural', *Journal of Rural Studies*, Vol. 9(1), pp. 23-37.

Halfacree, K. (1994), 'The Importance of "the Rural" in the Constitution of Counterurbanization: Evidence from England in the 1980s', *Sociologia Ruralis*, Vol. 34, pp. 164-89.

Halfacree, K. (1995), 'Talking About Rurality: Social Representations of the Rural as Expressed by Residents of Six English Parishes', *Journal of Rural Studies*, Vol. 11, pp. 1-20.

Halfacree, K. (1997), 'Contrasting Roles for the Post-Productivist Countryside. A Postmodern Perspective on Counterurbanization', in P. Cloke and J. Little (eds), *Contested Countryside Cultures*, London: Routledge.

Halfacree, K. (1998), 'Neo-Tribes, Migration and the Post-Productivist Countryside', in P. Boyle and K. Halfacree (eds), *Migration into Rural Areas: Theories and Issues*, Chichester: Wiley.

Halfacree, K. (1999), 'A New Space or Spatial Effacement? Alternative Futures for the Post-Productivist Countryside', in N. Walford, J. Everitt and D. Napton (eds), *Reshaping the Countryside: Perceptions and Processes of Rural Change*, Wallingford: CABI Publishing.

Halfacree, K. (2001), 'Constructing the Object: Taxonomic Practices, "Counterurbanization" and Positioning Marginal Rural Settlement', *International Journal of Population Geography*, Vol. 7, pp. 395-411.

Halfacree, K. (2003), 'Uncovering Rural Others/Other Rurals', in P. Richards and I. Robertson (eds), *Studying Landscapes*, London: Arnold.

Halfacree, K. and Boyle, P. (1998), 'Migration, Rurality and the Post-Productivist Countryside', in P. Boyle and K. Halfacree (eds), *Migration into Rural Areas: Theories and Issues*, Chichester: Wiley.

Hall, P. and Hay, D. (1980), *Growth Centres in the European Urban System*, London: Heinemann Educational Books.

Hall, S. and Thurstain-Goodwin, M. (2000), 'Geographical Information – Policy Driving, Policy Driven – Providing Statistics for the UK's Town Centres', *Statistical Journal of the United Nations Economic Commission for Europe*, Vol. 17, pp. 125-32.

Harrington, V. and O'Donoghue, D. (1998), 'Rurality in England and Wales 1991: A Replication and Extension of the 1981 Rurality Index', *Sociologia Ruralis*, Vol. 38, pp. 178-203.

Hartshorn, T.A. and Muller, P.O. (1986), *Suburban Business Centers: Employment Expectations*, Final Report for US Department of Commerce, EDA, Washington, DC: Department of Commerce.

Harvey, D. (1985), *The Urbanization of Capital*, Oxford: Blackwell.

Harvey, D. (1987), 'Flexible Accumulation Through Urbanization: Reflections on "Post-Modernism" in the American City', *Antipode*, Vol. 19, pp. 260-86.

Harvey, D. (1989a), *The Condition of Postmodernity*, Oxford: Blackwell.

Harvey, D. (1989b), *The Urban Experience*, Baltimore: The Johns Hopkins University Press.

Harvey, D. (1996), *Justice, Nature and the Geography of Difference*, Oxford: Blackwell.

Harvey, D. (2000), *Spaces of Hope*, Edinburgh: Edinburgh University Press.

Hauser, P. and Gardner, R. (1982), *Urban Future: Trends and Prospects*, Reprint No. 146, Honolulu: East-West Population Institute.

Hawley, A.H. (1971), *Urban Society: An Ecological Approach*, New York: The Ronald Press.

Heimlich, R. and Anderson, W. (2001), 'Development at the Urban Fringe and Beyond: Impacts on Agriculture and Rural Land Use', *Agricultural Economic Report*, No. 803, Washington, DC: US Department of Agriculture, Economic Research Service.

Hernández Laos, E. (2000), 'Distribución del Ingreso y la Pobreza en México', in A. Alcalde *et al.*, *Trabajo y Trabajadores en el México Contemporáneo*, México: Miguel Ángel Porrúa.

Herrin, A. and Pernia, E. (2000), 'Population, Human Resources and Employment', ANU seminar on Philippines, Canberra.

Hervieu, B. and Viard, J. (1996), *Au Bonheur des Campagnes*, Paris: Editions de L'Aube.

Hewison, R. (1987), *The Heritage Industry*, London: Methuen.

Hiernaux, D.N. (1998), 'La Economía de la Ciudad de México en la Perspectiva de la Globalización', *Economía, Sociedad y Futuro*, Volumen I, Número 4, Julio-Diciembre, El Colegio Mexiquense, México, pp. 671-94.

Higgs, G. and White, S. (2000), 'Alternatives to Census-Based Indicators of Social Deprivation in Rural Communities', *Progress in Planning*, Vol. 53, pp. 1-81.

Hines, F, Brown, D. and Zimmer, J. (1975), 'Social and Economic Characteristics of the Population in Metropolitan and Nonmetropolitan Counties, 1970', *Agricultural Economic Report*, No. 272, Washington, DC: US Department of Agriculture, Economic Research Service.

Hoggart, K. (1988), 'Not a Definition of Rural', *Area*, Vol. 20(1), pp. 35-40.

Hoggart, K. (1990), 'Let's Do Away With Rural', *Journal of Rural Studies*, Vol. 6, pp. 245-57.

Hoggart, K. (1997), 'The Middle Classes in Rural England 1971-1991', *Journal of Rural Studies*, Vol. 13, pp. 253-73.

Hope, S., Braunholtz, S., Playfair, A., Dudleston, A., Ingram, D., Martin, C. and Sawyer, B. (2001), *Scotland's People: Results From the 1998 Scottish Household Survey (Volume 1)*, http://www.scotland.gov.uk/shs/rep99-v1.pdf.

Hu, X.W., Zhou, Y.X., and Gu, C.L. (2000), *Zhongguo Yanhai Chengzhen Mijidiqu Kongjian Jiju Yu Kuosan Yanjiu (Studies on the Spatial Agglomeration and Dispersion in China's Coastal City-and-Town Concentrated Areas)*, Beijing: Science Press.

Hugo, G.J. (1980), 'New Conceptual Approaches to Migration in the Context of Urbanization: A Discussion Based on the Indonesian Experience', in P. A. Morrison (ed.), *Population Movements: Their Forms and Functions in Urbanization and Development*, Liège: Ordina Editions.

Hugo, G.J. (1982), 'Circular Migration in Indonesia', *Population and Development Review*, Vol. 8(1), pp. 59-84.

Hugo, G.J. (1987), 'Definition of Metropolitan and Urban Boundaries in Australia: Is it Time for a Change?' in A. Conacher (ed.), *Readings in Australian Geography, Proceedings of the 21ˢᵗ Institute of Australian Goegraphers' Conference, Perth 10-18 May 1986*. Perth: Institute of Australian Geographers (WA Branch) and Department of Geography, University of Western Australia.

Hugo, G.J. (1992), 'Migration and Rural-Urban Linkages in the ESCAP Region', in Economic and Social Commission for Asia and the Pacific, *Migration and Urbanization in Asia and the Pacific*, Asian Population Studies Series No. 111, New York: United Nations.

Hugo, G.J., Fenton, M., Smailes, P., Blanchfield, F. and Macgregor, C. (2001), *Defining Social Catchments in Non-Metropolitan Australia: Some Preliminary Ideas*, Second Draft of a Paper for the Servicing Regional Australia Project of the Department of Transport and Regional Services and Social Sciences Centre, Bureau of Rural Sciences, February.

Hugo, G.J., Griffith, D., Rees, P., Smailes, P., Badcock, B. and Stimson, R. (1997), *Rethinking the ASGC: Some Conceptual and Practical Issues*, Monograph Series 3, National Key Centre for Social Applications of GIS. Adelaide: The University of Adelaide.

Hummon, D. (1990), *Common Places: Community Ideology and Identity in American Culture*, Albany: SUNY Press.

ICRRC (1995), *Rural Canada: A Profile*, Ottawa: Government of Canada, Interdepartmental Committee on Rural and Remote Canada.

Ilbery, B. and Bowler, I. (1998), 'From Agricultural Productivism to Post-Productivism', in B. Ilbery (ed.), *The Geography of Rural Change*, Harlow: Longman.

INDEC (1998), *Censo Nacional de Población y Vivienda 1991. El Concepto de Localidad. Definición, Estudios de Caso y Fundamentos Técnico-Metodológicos*, Series D No. 5. Buenos Aires, Instituto Nacional de Estadística y Censos.

INEGI (2000), *Cuaderno Estadístico de la Zona Metropolitana de la Ciudad de México*, México: INEGI, Gobierno del Distrito Federal, Gobierno del Estado de México.

International Institute of Statistics (1962 and 1980), *Annuaire de Statistiques Internationales des Grandes Villes*, La Haye: Institut International de Statistiques.

IPEA/IBGE/NESUR (1999), *Caracterização e Tendências da Rede Urbana do Brasil*, Campinas, SP: Instituto de Economia/Unicamp.

Jain, A.K. (1990) *The Making of a Metropolis. Planning and Growth of Delhi*, New Delhi: National Book Organisation.

Janelle, D.G. (1969), 'Spatial reorganization: a model and a concept', *Annals of the Association of American Geographers*, Vol. 59, pp. 348-64.

Jencks, C. (2001), 'Who Should Get In?' *New York Review of Books*, Part One, Nov. 29, 2001; Part Two, Dec. 20, 2001.

Jensen, J.R. and Cowen, D.C. (1999), 'Remote Sensing of Urban/Suburban Infrastructure and Socio-Economic Attributes', *Photogrammetric Engineering and Remote Sensing,* Vol. 65, pp. 611-24.

Johnson, K. and Beale, C.L. (1995), 'The Rural Rebound Revisited', *American Demographics* (July), pp. 46-54.

Johnston, R.J. (1971), *Urban Residential Patterns,* New York: Praeger Publishers.

Johnston, R.J. (1986), *On Human Geography,* Oxford: Basil Blackwell.

Jones, G.W. (1965), 'The Employment Characteristics of Small Towns in Malaya', *Malayan Economic Review,* Vol. X(1), pp. 44-72.

Jones, G.W. (1984), 'Links Between Urbanization and Sectoral Shifts in Employment in Java', *Bulletin of Indonesian Economic Studies,* Vol. XX(3), pp. 120-57.

Jones, N. (1993), *Living in Rural Wales,* Llandysul: Gomer.

Jones, O. (1995), 'Lay Discourses of the Rural: Developments and Implications for Rural Studies', *Journal of Rural Studies,* Vol. 11, pp. 35-49.

Jones, G.W. (1997), 'The Thoroughgoing Urbanization of East and Southeast Asia', *Asia Pacific Viewpoint,* Vol. 38(3), pp. 237-50.

Jones, G.W. and Rao, N.R. (2001), 'Proposal for a New Classification of Urban and Rural Areas in Cambodia', Mission Report, Phnom Penh: National Institute of Statistics.

Jones, G. and Visaria, P. (1997), 'Urbanization of the Third World Giants', in G.W. Jones and P. Visaria (eds), *Urbanization in Large Developing Countries,* Oxford: Clarendon Press.

Jones, G.W., Tsai, C.-L. and Bajracharya, B. (2000), 'Demographic and Employment Change in the Mega-cities of South-East and East Asia', *Third World Planning Review,* Vol. 22(2), pp. 119-46.

Jordán, R. and Simioni, D. (1998), *Ciudades intermedias en América Latina y el Caribe: Propuesta Para la Gestión Urbana* (LC/l.1117), Santiago de Chile: ECLAC.

Katzman, R. (2001), 'Seducidos y Abandonados: el Aislamiento Social de Los Pobres Urbanos', *Revista de la CEPAL,* Santiago de Chile, No. 75, pp. 171-189.

Kellogg Foundation (2002), *Perceptions of Rural America,* Battle Creek, MI: Kellogg Foundation.

Kirananda, Thienchay and Surasiengsunk, Suwanee (1985), *Population Policy Background Paper Study on Economic Consequences of Urbanization in Thailand, 1987-2001,* Bangkok: Thailand Development Research Institute.

Klove, R.C. (1952), 'The Definition of Standard Metropolitan Areas,' *Economic Geography,* Vol. 28(2), pp. 95-104.

Knox, P. (ed.) (1993), *The Restless Urban Landscape,* New York: Prentice Hall.

Knox, P. and Taylor, P. (eds) (1995), *World Cities in a World System,* New York: Wiley.

Korcelli, P. (1984), 'The Turnaround in Urbanization in Developed Countries', *Population Distribution, Migration and Development*: Proceedings of the Expert Group on Population Distribution, Migration and Development, Hammamet (Tunisia), 21-25 March 1983, New York: United Nations.

Kraybill, D. and Lobao, L. (2001), *County Government Survey: Changes and Challenges in the New Millennium,* Washington, DC: National Association of Counties.

Kundera, M. (1985), *The Unbearable Lightness of Being,* London: Faber and Faber.

Lago, L.C. (1998), 'Estruturação urbana e mobilidade espacial: uma análise das desigualdades socioespaciais na metrópole do Rio de Janeiro', Tese de doutoramento, São Paulo: Faculdade de Arquitetura e Urbanismo (FAU), USP.

Lago, L.C. (2000), 'Divisão Sócio-Espacial e Mobilidade Residencial: Reprodução ou Alteração das Fronteiras Espaciais?', in XII Encontro Nacional de Estudos Populacionais, 2000), *Anais,* Caxambú, MG: ABEP.

Landstreet, B. and Mundigo, A. (1981), 'Internal Migration and Changing Urbanization Patterns in Cuba', paper presented at the Annual Meeting of the Population Association of America, Washington, DC.

Lang, M. (1986), 'Redefining Urban and Rural for the US Census of Population: Assessing the Need for Alternative Approaches', *Urban Geography*, Vol. 7(2), pp. 118-34.

Laschewski, L., Teherani-Kroenner, P. and Bahner, T. (2002), 'Recent Rural Restructuring in East and West Germany: Experiences and Backgrounds', in K. Halfacree, I. Kovach and R. Woodward (eds), *Leadership and Local Power in European Rural Development*, Aldershot: Ashgate.

Lattes, A.E. (1984), 'Territorial Mobility and Redistribution of the Population: Recent Developments', in *International Conference on Population, 1984. Population Distribution, Migration and Development*, New York: United Nations.

Lattes, A.E. (1998), 'Population Distribution in Latin America: Is There a Trend towards Population Deconcentration?' in United Nations, *Population Distribution and Migration*, New York: United Nations.

Lee, Y.S.F. (1989), 'Small Towns and China's Urbanization Level', *China Quarterly*, Vol. 120, pp. 771-86.

Lefebvre, H. (1991), *The Production of Space*, Oxford: Blackwell.

Léger, D. and Hervieu, B. (1979), *Le Retour à la Nature. Au Fond de la Forêt, l'Etat*, Paris: Le Seuil.

Le-Gleau, J.P., Pumain, D. and Saint-Julien, T. (1996), 'Villes d'Europe: A Chaque Pays sa Definition; Regard Socioeconomique sur la Structuration de la Ville', *Economie-et-Statistique*, 294-95, pp. 9-23.

Lewis, C. and Lewis, K. (1997), *Delhi's Historic Villages. A Photographic Evocation*, Delhi: Ravit Dayal Publisher.

Lewis, M. (1991), 'Elusive Societies: A Regional-Cartographical Approach to the Study of Human Relatedness', *Annals of the Association of American Geographers*, Vol. 18(4), pp. 605-26.

Lillesand, T.M. and Kiefer, R.W. (2000), *Remote Sensing and Image Interpretation*, Fourth Edition, New York: John Wiley.

Little, J. (1999), 'Otherness, Representation and the Cultural Construction of Rurality', *Progress in Human Geography*, Vol. 23, pp. 437-42.

Liu, W. (2001), 'The Reform in *Hukou* System Should Not be Delayed' (in Chinese), *Dushu (Reading)*, No. 12, pp. 99-104.

Lo, C.P. (1995), 'Automated Population and Dwelling Unit Estimation from High-Resolution Satellite Images: A GIS Approach', *International Journal of Remote Sensing*, Vol. 16, pp. 17-34.

Lo, C.P., Quatrochi, D. and Luvall, J. (1997), 'Application of High-Resolution Thermal Infrared Remote Sensing and GIS to Assess the Urban Heat Island Effect', *International Journal of Remote Sensing*, Vol. 18, pp. 287-304.

Lo, Fu-chen and Salih, K. (1987), 'Structural Change and Spatial Transformation: Review of Urbanization in Asia, 1960-1980', in R.J. Fuchs, G.W. Jones and E.M. Pernia (eds), *Urbanization and Urban Policies in Pacific Asia*, Boulder: Westview Press.

Logan, J. (1996), 'Rural America As A Symbol of American Values', *Rural Development Perspectives*, Vol. 12(1), pp. 24-28.

Long, L. and DeAre, D. (1988), 'US Population Redistribution: A Perspective on the Nonmetropolitan Turnaround', *Population and Development Review*, Vol. 14, pp. 433-50.

Lungo, M. (1993), 'Las Ciudades y la Globalización: una Mirada Desde Centroamérica y el Caribe', *Revista Interamericana de Planificación* (San Antonio, TX), Vol. 26(104), pp. 7-22.

Ma, L.J.C. and Cui, G.H. (1987), 'Administrative Changes and Urban Population in China', *Annals of the Association of American Geographers*, Vol. 77(3), pp. 373-95.

Madhavan, B.B., Kubo, S., Kurisaki, N. and Sivakumar, T.V.L.N. (2001), 'Appraising the Anatomy and Spatial Growth of the Bangkok Metropolitan Area Using a Vegetation-Impervious-Soil Model through Remote Sensing', *International Journal of Remote Sensing*, Vol. 22, pp. 789-806.

Majumdar, T.K. (1983), *Urbanising the Poor. A Sociological Study of Low-Income Migrant Communities in the Metropolitan City of Delhi*. New Delhi: Lancers Publishers.

Mamas, S.G.M., Jones, G.W. and Sastrasuanda, T. (2001), 'A Zonal Analysis of Demographic Change in Indonesia's megacities', *Third World Planning Review*, Vol. 23(2), pp. 155-74.

Marsden, T., Murdoch, J., Lowe, P., Munton, R. and Flynn, A. (1993), *Constructing the Countryside*, London: UCL Press.

Martin, P. and Midgley, E. (1994), 'Immigration to the United States: Journey to an Uncertain Destination', *Population Bulletin*, Vol. 49(2), Washington, DC: Population Reference Bureau.

Martinez, J. (1999), *La Migración Interna y Sus Efectos en Dieciséis Ciudades de Chile*, (LC/DEM/R.302), Santiago de Chile: ECLAC / CELADE.

Marx, L. (1964), *The Machine in the Garden*, Oxford: Oxford University Press.

Marx, K. (1976), *Capital*, Vol. I, London: Penguin NLR.

Massey, D. and Allen, J. (1984), *Geography Matters!* Cambridge: Cambridge University Press.

Mather, P.M. (1999), 'Land Cover Classification Revisited', in P.M. Atkinson and N.J. Tate (eds), *Advances in Remote Sensing and GIS Analysis*, Chichester: Wiley.

Mathieu, N. and Gajewski, P. (2002), 'Rural Restructuring, Power Distribution and Leadership at National, Regional and Local Levels: The Case of France', in K. Halfacree, I. Kovach and R. Woodward (eds), *Leadership and Local Power in European Rural Development*, Aldershot: Ashgate.

Matless, D. (1998), *Landscape and Englishness*, London: Reaktion.

Matos, R. (1995), 'Dinâmica Migratória e Desconcentração Populacional na Macrorregião de Belo Horizonte', Tese de Doutorado, Belo Horizonte: CEDEPLAR/UFMG.

McGee, T.G. (1991), 'The Emergence of Desakota Regions in Asia: Expanding a Hypothesis', in N. Ginsburg, B. Koppel and T.G. McGee (eds), *The Extended Metropolis: Settlement Transition in Asia*, Honolulu: University of Hawaii Press.

McGranahan, D. (1999), 'Natural Amenities Drive Rural Population Change', *Agricultural Economic Report* No. 781. Washington, DC: US Department of Agriculture, Economic Research Service.

McNiven, C., Puderer, H. and Janes, D. (2000), 'Census Metropolitan Area and Census Agglomeration Influenced Zones (MIZ): A Description of the Methodology', *Geography Working Paper Series No. 2000-2*, Ottawa: Statistics Canada.

Mejia-Raymundo, Corazon (1983), 'Population Growth and Urbanization', in M.B. Concepcion (ed.), *Population of the Philippines: Current Perspectives and Future Prospects*, Manila: National Economic and Development Authority.

Mendelson, R. (2001), 'Geographic Structures as Census Variables: Using Geography to Analyse Social and Economic Processes', *Geography Working Paper Series, No. 2001-1*, Ottawa: Statistics Canada.

Merrifield, A. (1993) 'Place and Space: A Lefebvrian Reconciliation', *Transactions of the Institute of British Geographers*, Vol. 18, pp. 516-31.

Mesev, V. (1998), 'The Use of Census Data in Urban Image Classification', *Photogrammetric Engineering and Remote Sensing*, Vol. 64, pp. 431-38.

Mesev, V., Gorte, B. and Longley, P. (2001), 'Modified Maximum-Likelihood Classification Algorithms and Their Application to Urban Remote Sensing', in J.-P. Donney, M.J. Barnsley and P. Longley (eds), *Remote Sensing and Urban Analysis*, London: Taylor and Francis.

Meuriot, P. (1897), *Des Agglomérations Urbaines dans l'Europe Contemporaine*, Paris: Belin.

Milbourne, P. (ed.) (1997), *Revealing Rural 'Others'*, London: Pinter.

Milton, E.J. (1999), 'Image Endmembers and the Scene Model', *Canadian Journal of Remote Sensing*, Vol. 25, pp. 112-20.

Morales, P. (1989), *Indocumentados Mexicanos*, México: Grijalbo.

Moriconi-Ébrard, F. (1993), *L'urbanisation du Monde*, Paris: Anthropos, coll. Villes.

Moriconi-Ébrard, F. (1994), *Geopolis, Pour Comparer les Villes du Monde*, Paris: Anthropos, coll. Villes.

Moriconi-Ébrard, F. (2001), 'Identifier les Territoires de la Métropolisation: Un état de la Question', in *Cahiers de la Métropolisation*, No. 2, pp. 7-27.

Mormont, M. (1987), 'Rural nature and urban natures', *Sociologia Ruralis*, Vol. 27, pp. 3-20.

Mormont, M. (1990), 'Who is Rural? or, How to Be Rural: Towards a Sociology of the Rural', in T. Marsden, P. Lowe and S. Whatmore (eds), *Rural Restructuring*, London: David Fulton.

Morrill, R., Cromartie, J. and Hart, G. (1999), 'Metropolitan, Urban, and Rural Commuting Areas: Toward a Better Depiction of the United States Settlement System', *Urban Geography*, Vol. 20(8), pp. 727-48.

Moscovici, S. (1981), 'On Social Representation', in J. Forgas (ed.), *Social Cognition: Perspectives on Everyday Understanding*, London: Academic Press.

Moscovici, S. (1984), 'The Phenomenon of Social Representations', in R. Farr and S. Moscovici (eds), *Social Representations*, Cambridge: Cambridge University Press.

Moss, M.L. and Townsend, A.M. (1998), 'Spatial Analysis of the Internet in US Cities and States,' paper prepared for the Technological Futures – Urban Futures Conference, Durham, England, 23-28 April. http://urban.nyu.edu/ archives/spatial-analysis/spatial-analysis.pdf. Accessed January 2002.

Muller, T.O. (1981), *Contemporary Suburban America*, Englewood Cliffs, NJ: Prentice-Hall.

Murdoch, J. and Day, G. (1998), 'Middle Class Mobility, Rural Communities and the Politics of Exclusion', in P. Boyle and K. Halfacree (eds), *Migration into Rural Areas: Theories and Issues*, Chichester: Wiley.

Murdoch, J. and Marsden, T. (1994), *Reconstituting Rurality*, London: UCL Press.

Murdoch, J. and Pratt, A. (1993), 'Rural Studies: Modernism, Postmodernism and the "Post-Rural"', *Journal of Rural Studies*, Vol. 9, pp. 411-27.

National Bureau of Statistics of China (1998), *Zhongguo Tongji Nianjian (1998) (China Statistical Yearbook (1998)*, Beijing: China Statistics Press.

National Bureau of Statistics of China (2001), *Zhongguo Tongji Nianjian (2001) (China Statistical Yearbook (2001)*, Beijing: China Statistics Press.

National Research Council (2003), *Cities Transformed: Demographic Change and Its Implications in the Developing World*. Panel on Urban Population Dynamics. M.R. Montgomery, R. Stren, B. Cohen and H.E. Reed (eds), Committee on Population, Division on Behavioral and Social Sciences and Education. Washington DC: The National Academies Press.

National Statistics Office (2001), *The 2000 Population and Housing Census. Advance Report*, Bangkok: National Statistics Office.

Naudet, J.-D. (1996), 'Crise de l'économie Réelle et Dynamique de la Demande en Afrique de l'Ouest', in J. Coussy and J. Vallin (éds.), *Les Études du CEPED*, Paris: CEPED (Les Études du CEPED).

NCR Planning Board (1988), *Regional Plan 2001, National Capital Region*, Delhi: National Capital Region Planning Board, Ministry of Urban Development, Government of India.

NCR Planning Board (1996), *National Capital Region. Growth and Development*, New Delhi: Har-Anand.

NIUA (1988), *National Capital Region. A Perspective on Patterns and Process of Urbanization*, Research Studies Series No 29, National Institute of Urban Affairs, New Delhi.

Noin, D. (1970), *La Population Rurale du Maroc*, Paris: Presses Universitaires de France.

Nordregio (1999), *European Spatial Development Perspective: Towards Balanced and Sustainable Development of the Territory of the European Union*, Luxembourg: European Commission.

NUREC (1994), *Atlas of Agglomerations in the European Union*. 3 vols, Duisburg: NUREC.

OCDE (1988), *Statistiques Urbaines dans les Pays de l'OCDE*, Programme des Affaires Urbaines.

OECD (1994), *Creating Rural Indicators for Shaping Territorial Policy*, Paris: OECD.

OECD (2001), *Functional Regions: a Summary of Definitions and Usage in OECD Countries*, Paris: OECD.

Office for National Statistics and Coombes, M.G. (1998), *1991-Based Travel-to-Work Areas*, London: ONS, http://curdsweb1.ncl.ac.uk/files/4753maindoc.pdf

Office for Population Censuses and Surveys (1984), *Key Statistics for Urban Areas*, London: HMSO.

OMB (Office of Management and Budget) (1998), 'Alternative Approaches to Defining Metropolitan and Nonmetropolitan Areas,' *Federal Register*, Vol. 63(244), pp. 70526-61.

OMB (Office of Management and Budget) (1999), 'Recommendations From the Metropolitan Area Standards Review Committee to the Office of Management and Budget Concerning Changes to the Standards for Defining Metropolitan Areas,' *Federal Register*, Vol. 64(202), pp. 56628-44.

OMB (Office of Management and Budget) (2000a), 'Final Report and Recommendations From the Metropolitan Area Standards Review Committee to the Office of Management and Budget Concerning Changes to the Standards for Defining Metropolitan Areas,' *Federal Register*, Vol. 65(163), pp. 51060-77.

OMB (Office of Management and Budget) (2000b), 'Standards for Defining Metropolitan and Micropolitan Statistical Areas,' *Federal Register*, Vol. 65(249), pp. 82228-38.

OPCS (1991), *Key Population and Vital Statistics: Local and Health Authority Areas 1990*, Office of Population Censuses and Surveys. London: HMSO.

Ortega, E. (1992), 'La Trayectoria Rural de América Latina y el Caribe', *Revista de la CEPAL*, Santiago de Chile, No. 46, pp. 131-65.

Ortega, L. (1998), *Los Vínculos Rurales con Ciudades Intermedias* (LC/R.1835), Santiago de Chile: ECLAC.

Palmer-Moloney, L.J. (2001), 'Cairo, Egypt: Centuries of Landscape Change', *Geocarto International*, Vol. 16, pp. 89-94.

Patarra, N. (2000), *Do Urbano às Novas Territorialidades: Conceitos e Questões*, Rio de Janeiro: IPEA, Relatório de Pesquisa.

Peirce, N.R. (1993), *Citistates. How Urban America Can Prosper in a Competitive World*, Washington, DC: Seven Locks Press.

Pfeffer, M.J. (1993), 'Black Migration and the Legacy of Plantation Agriculture', in J. Singelmann and F.A. Deseran (eds), *Inequalities in Local Labor Markets*, Boulder: Westview Press.

Phillips, M., Fish, R. and Agg, J. (2001), 'Putting Together Ruralities: Towards a Symbolic Analysis of Rurality in the British Mass Media', *Journal of Rural Studies*, Vol. 17, pp. 1-27.

Pisarski, A. (1996), *Commuting in America II*, Lansdowne, Virginia: Eno Transportation Foundation, Inc.

Policy Research Office of the Provincial Committee of CCP and Department of Construction of Fujian Province (2001), *Fujian chengshihua fazhan yantaohui chailiao (Materials for the symposium on Fujian's urbanization development)*, unpublished manuscript.

Potter, J. and Wetherell, M. (1987), *Discourse and Social Psychology*, London: Sage.

Potts, D. (1997), 'Urban Lives: Adopting New Strategies and Adapting Rural Links', in C. Rakodi (ed), *The Urban Challenge in Africa: Growth and Management of its Large Cities*, Tokyo: United Nations University Press.

Poulain, M. (1985), 'La Migration, concept et méthodes de mesure', in *Migrations internes. Méthodes d'observation et d'analyse*, Louvain: UCL.

Pratt, A. (1996), 'Discourses of Rurality: Loose Talk or Social Struggle?', *Journal of Rural Studies*, Vol. 12, pp. 69-78.

Pred, A. (1977), *City Systems in Advanced Economies*, Chichester: Wiley.

Pred, A. (1984), 'Place as Historically Contingent Process: Structuration and the Time-Geography of Becoming Places', *Annals of the Association of American Geographers*, Vol. 74, pp. 279-97.

Pumain, D. (1982), *La Dynamique des Villes*, Paris: Economica.

Pumain, D. (2000), 'Settlement Systems in the Evolution', *Geografiska Annaler*, 82B(2), pp. 73-87.

Pumain, D. and Saint-Julien, T. (1991), *Le Concept Statistique de Ville en Europe*, Luxembourg: Eurostat.

Pumain, D. and Saint-Julien, T. (eds) (1996), *Urban Networks in Europe*, Paris: John Libbey/INED, Congresses and colloquia n°15.

Pumain, R., Saint-Julien, T., Cattan, N. and Rozenblat, C. (1991), 'The Statistical Concept of the Town in Europe', *Report 0673002 by the Network for Urban Research in the European Community*, Luxembourg: Eurostat.

Ramírez, M.A. (1998), *Desarrollo Sustentable en áreas Rurales Marginadas: Entre la Sobrevivencia y la Conservación*, Papeles de Población, Nueva Época, Year 4, No. 18, México, D.F.

Rashed, T., Weeks, J.R., Gadalla, M.S. and Hill, A.G. (2001), 'Revealing the Anatomy of Cities Through Spectral Mixture Analysis of Multispectral Imagery: A Case Study of the Greater Cairo Region, Egypt', *Geocarto International*, Vol. 16, pp. 5-16.

Rees, P.H. (1970), 'The Axioms of Intra-Urban Structure and Growth', in B.J.L. Berry and F.E. Horton (eds), *Geographical Perspectives on Urban Systems*, Englewood Cliffs NJ: Prentice-Hall.

Rendón, T. and Salas, C. (2000), 'La Evolución del Empleo', in A. Alcalde *et al.* (eds), *Trabajo y Trabajadores en el México Contemporáneo*, México: Miguel Ángel Porrúa.

Ridd, M. (1995), 'Exploring a V-I-S (Vegetation-Impervious Surface-Soil) Model or Urban Ecosystem Analysis Through Remote Sensing: Comparative Anatomy of Cities', *International Journal of Remote Sensing*, Vol. 16, pp. 2165-85.

Rigg, J. (1997), *Southeast Asia: The Human Landscape of Modernization and Development*, London: Routledge.

Rigotti, J.I. (1994), 'Distribuição Espacial da População na Região Metropolitana de Belo Horizonte', Dissertação de Mestrado, Belo Horizonte, MG: CEDEPLAR/UFMG.

Rindfuss, R.R. and Stern, P.C. (1998), 'Linking Remote Sensing and Social Science: The Need and the Challenges', in D. Liverman, E.F. Moran, R.R. Rindfuss, and P.C. Stern (eds), *People and Pixels: Linking Remote Sensing and Social Science*, Washington, DC: National Academy Press.

Roberts, D.A., Batista, G.T., Pereira, J.L.G., Waller, E.K. and Nelson, B.W. (1998), 'Change Identification Using Multitemporal Spectral Mixture Analysis: Applications in Eastern Amazonia', in C.M. Elvidge and R.S. Lunetta (eds), *Remote Sensing Change Detection: Environmental Monitoring Applications and Methods*, Ann Arbor, MI: Ann Arbor Press.

Robinson, I.M. (1995), 'Emerging Spatial Patterns in ASEAN Mega-urban Regions: Alternative Strategies', in T.G. McGee and I.M. Robinson (eds), *The Mega-Urban Regions of Southeast Asia*, Vancouver: UBC Press.

Robson, B. (1973), *Urban Growth, An Approach*, London: Methuen.

Rodenbeck, M. (1999), *Cairo: The City Victorious*, New York: Alfred A. Knopf.

Rodríguez, J. and Villa, M. (1997), 'Dinámica Sociodemográfica de las Metrópolis Latinoamericanas Durante la Segunda Mitad del Siglo XX', *Notas de Población*, Santiago de Chile, No. 65, pp. 17-110.

Ru, X., Lu, X.Y. and Zhu, R.P. (2001), *Chengshihua: Sunan Xiandaihua De Xinshijian (Urbanization: the New Effort in the Modernization Drive of Southern Jiansu)*, Beijing: China Social Sciences Press.

Rubin, J. (1969), 'Function and Structure of Community: A Conceptual and Theoretical Analysis', *International Review of Community Development*, Vol. 21-22, pp. 111-19.

Ruiz Chiapetto, C. (1994), 'Hacia un País Urbano', in F. Alba and G. Cabrera (eds), *La Población en el Desarrollo Contemporáneo de México*, México: El Colegio de México, pp. 159-81.

RuPRI (1995), *1995 National RuPRI Poll: Differential Attitudes of Rural and Urban America*, Columbia, Missouri: Rural Policy Research Institute.

Ryznar, R. (1998), *Urban Vegetation and Social Change: An Analysis Using Remote Sensing and Census Data*, PhD Dissertation, Department of Geography, University of Michigan, Ann Arbor.

Sabatini, F. (1999), *Tendencias de la Segregación Residencial Urbana en Latinoamérica: Reflexiones a Partir del Caso de Santiago de Chile*, paper presented at the seminar 'Latin America: Democracy, markets and equity at the Threshold of New Millenium', University of Uppsala, Sweden.

Sample Registration System, *Sample Registration Bulletin*, Office of the Registrar General, Vol. 6 (1972) to Vol. 35 (2001).

Sampson, R., Morenoff, J. and Earls, F. (1999), 'Beyond Social Capital: Spatial Dynamics of Collective Efficacy for Children', *American Journal of Sociology*, Vol. 92(1), pp. 27-63.

Samuel, R. (1989), 'Introduction: Exciting to be English', in R. Samuel (ed.), *Patriotism, Volume 1*, London: Routledge.

Sassen, S. (1991), *The Global City: New York, London, Tokyo*, Princeton: Princeton University Press.

Sassen, S. (2000), *Cities in the World Economy*, Thousand Oaks: Pine Forge Press.

Sawyer, D. (1986), 'Urbanização da Fronteira Agrícola no Brasil', in L. Lavinas (ed.) *A Urbanização da Fronteira*, Rio de Janeiro: IPPUR/UFRJ.

Sayer, A. (1989), 'The "New" Regional Geography and the Problems of Narrative', *Environment and Planning D: Society and Space*, Vol. 7, pp. 253-76.

Sayer, A. (1991), 'Beyond the Locality Debate: Deconstructing Geography's Dualisms', *Environment and Planning A*, Vol. 23, pp. 283-308.

Schnore, L.F. (1965), *The Urban Scene*, New York: Free Press.

SCOT (1997) Remote sensing and urban boundaries. http://www.scot-sa.com/urbain.

Sforzi, F., Openshaw, S. and Wymer, C. (1997), 'Le Procedura di Identificazione dei Sistemi Locali del Lavoro', in F. Sforzi (ed.), *I Sistemi Locali del Lavoro 1991*, Rome: ISTAT.

Shand, R.T. (ed.) (1986), *Off-Farm Employment in the Development of Rural Asia: Papers Presented to a Conference in Chaingmai, Thailand, 23 to 26 August 1983*, Canberra: National Centre for Development Studies, Australian National University.

Shields, R. (1990), *Places on the Margin*, London: Routledge.

Shields, R. (1999), *Lefebvre, Love and Struggle*, London: Routledge.

Short, J. (1991), *Imagined Country*, London: Routledge.

Shotter, J. (1993), *Cultural Politics of Everyday Life*, Buckingham: Open University Press.

Silva, J.G. (1997), 'O Novo Rural Brasileiro', *Nova Economia* (Belo Horizonte, MG), Vol. 7(1), pp. 43-81.

Simmons, J. W. (1986), 'The Urban System: Concepts and Hypotheses', in J.G. Borchert *et al.* (eds), *Netherlands Geographical Studies*, No. 16, Utrecht.

Smailes, A.E. (1966), *The Geography of Towns*, Chicago: Aldine Publishing Company.

Smailes, P.J. (2000), 'The Diverging Geographies of Social and Business Interaction Patterns: A Case Study of Rural South Australia', *Australian Geographical Studies*, Vol. 38(2), pp. 158-79.

Small, K.A. and Song, S. (1992), '"Wasteful" Commuting: A Resolution', *Journal of Political Economy*, Vol. 100, pp. 888-98.

Smith, N. (1984), *Uneven Development*, Oxford: Blackwell.

Smith, N. (1987), 'Dangers of the Empirical Turn: Some Comments on the CURS Initiative', *Antipode*, Vol. 19, pp. 59-68.

Snrech, S. (1994), *Pour Préparer l'Avenir de l'Afrique de l'Ouest: Une Vision à l'Horizon 2020 - Document de Synthèse*, Paris: WALTPS, (Synthèse n°SAH/D(94)439).

Sobrino, L.J. (2000), *Productividad y Ventajas Competitivas en el Sistema Urbano Nacional*, PhD thesis, México: Universidad Nacional Autónoma de México, Facultad de Arquitectura.

Socolow, S.M. and Johnson, L.L. (1981), 'Urbanization in Colonial Latin America', in *Journal of Urban History*, Vol. 8(1), pp. 27-60.

Soni, A. (2000), 'The Urban Conquest of Outer Delhi: Beneficiaries, Intermediaries and Victims. The Case of the Mehrauli Countryside', in V. Dupont, E. Tarlo and D. Vidal (eds), *Delhi. Urban Space and Human Destinies*, Delhi: Manohar-CSH.

Soper, K. (1981), *On Human Needs*, Brighton: Harvester Press.

Speare, A. Jr. (1993), *Changes in Urban Growth Patterns, 1980-90*, Cambridge, MA: Lincoln Institute of Land Policy.

Speare, A., Liu, P.K.C. and Tsay, C.L. (1988), *Urbanization and Development: The Rural-Urban Transition in Taiwan*, Boulder: Westview Press.

Sposati, A. (ed.) (2000), *Mapa da Exclusão/Inclusão Social da Cidade de São Paulo/2000: Dinâmica Social dos Anos 90*, São Paulo, SP: PUC-SP/POLIS/INPE, CD-Rom.

Sposati, A. (2001), *Cidade em Pedaços*, São Paulo, SP: Brasiliense.

Spotila, J.T. (1999), 'Recommendations from the Metropolitan Area Standards Review Committee to the Office of Management and Budget concerning changes to the standards for defining metropolitan areas', *Federal Register*, Vol. 64, pp. 56628-44 (October 20).

Stanback, T.M. Jr. (1991), *The New Suburbanization: Challenge to the Central City*, Boulder, CO: Westview Press.

Statistical Bureau of Fujian Province (2001), *Fujian Tongji Nianjian (2001) (Fujian Statistical Yearbook (2001))*, Beijing: China Statistics Press.

Statistics Canada (1997), *1996 Census Dictionary*, Catalogue No. 92-351-XPE, Ottawa: Industry Canada.

Statistics Canada (1998), *A National Overview: Population and Dwelling Counts, 1996*, Catalogue No. 93-357-XPB, Ottawa: Industry Canada.

Statistics New Zealand (1997), *1996 New Zealand Census of Population and Dwellings: Introduction to the Census*, Wellington: Statistics New Zealand.

Steinberg, J. (1993), 'Le périurbain: définition, délimitation et spécificité', *Métropolisation et périurbanisation*, Cahiers du CREPIF, No. 42, Paris: Université de Paris Sorbonne.

Sternlieb, G. and Hughes, J. (eds) (1988), *America's New Market Geography*, New Brunswick, NJ: Rutgers.

Sundaram, K.V. (1978), 'Delhi: the National Capital', in P.P. Misra (ed.), *Million Cities of India*, New Delhi: Vikas Publishing House.

Suri, P. (1994), *Urban Poor. Their Housing Needs and Government Response*, Delhi: Har-Anand Publications.

Sutton, P., Roberts, D., Elvidge, C. and Baugh, K. (2001), 'Census from Heaven: An Estimate of the Global Human Population Using Night-Time Satellite Imagery', *International Journal of Remote Sensing*, Vol. 22(16), pp. 3061-76.

Symes, D. (1981), 'Rural Community Studies in Great Britain', in J.-L. Durand-Drouhin, L.-M. Szwengrub and I. Mihailescu (eds), *Rural Community Studies in Europe*, Oxford: Pergamon.

Tabb, W. (2001), *The Amoral Elephant*, New York: Monthly Review Press.

Tanaka, S., Takasaki, K., Yamanokuchi, T. and Kameda, K. (1999), 'RADARSAT and TM Data Fusion for Urban Structure Analysis', *Canadian Journal of Remote Sensing*, Vol. 25, pp. 80-4.

Tang, Z., and Kong, X. Z. (2000), *Zhongguo Xiangzhen Qiye Jingjixue Jiaocheng (A Course on the Economics of China's Township and Village Enterprises)*, Beijing: People's University of China Press.

Tarlo, E. (1996), *Clothing Matters. Dress and Identity in India*, London: C. Hurst and Co.

Tarlo, E. (2000), 'Welcome to History: A Resettlement Colony in the Making', in V. Dupont, E. Tarlo and D. Vidal (eds), *Delhi. Urban Space and Human Destinies*, Delhi: Manohar-CSH.

Taylor, P. (1993), *Political Geography*, Third edition, Harlow: Longman.

Taylor, J.R., and Banister, J. (1991), 'Surplus Rural Labor in the People's Republic of China', in G. Veeck (ed.), *The Uneven Landscape: Geographic Studies in Post-Reform China*, Baton Rouge LA: Geoscience Publications, Department of Geography and Anthropology, Louisiana State University.

Taylor, D., Bozeat, N., Parkinson, M. and Belil, M. (2000), *The Urban Audit: Towards the Benchmarking of Quality of Life in 58 European Cities, (Volume 3: The Urban Audit Manual)*, Luxembourg: European Communities.

Thomas-Hope, E. (1999), 'Emigration Dynamics in the Anglophone Caribbean', in R. Appleyard (ed.), *Emigration Dynamics in Developing Countries, Volume III: Mexico, Central America and the Caribbean*, Aldershot: Ashgate.

Thrift, N. (1986), 'Little Games and Big Stories', in K. Hoggart and E. Kofman (eds), *Politics, Geography and Social Stratification*, London: Croom Helm.

Thrift, N. (1987), 'Introduction: The Geography of Late Twentieth-Century Class Formation', in N. Thrift and P. Williams (eds), *Class and Space*, London: Routledge and Kegan Paul.

Thrift, N. (1989), 'Images of Social Change', in C. Hamnett, L. McDowell and P. Sarre (eds), *The Changing Social Structure*, London: Sage.

Thrift, N. (1996), *Spatial Formations*, London: Sage.

Tisdale, H. (1942), 'The Process of Urbanization.' *Social Forces*, Vol. 20, pp. 311-16.

Tolbert, C.M. and Sizer, M. (1996), *US Commuting Zones and Labour Market Areas – 1990 Update*, ERS Staff Paper No. 9614, Washington, DC: US Department of Agriculture, Economic Research Service Rural Economy Division.

Tompkins, S., Mustard, J.F., Pieters, C.M. and Forsyth, D.W. (1997), 'Optimization of Endmembers for Spectral Mixture Analysis', *Remote Sensing of Environment*, Vol. 59, pp. 472-89.

Torres, H. da G. (1997), *Desigualdade Ambiental na Cidade de São Paulo*, tese de doutorado, Campinas, SP: IFCH/Unicamp.

Torres, H. (2001), 'Cambios Socioterritoriales en Buenos Aires Durante la Década de 1990', *EURE*, Vol. 27(80), pp. 33-56.

Treadway, R.C. (1990), 'Central Cities, Suburbs and the Metropolitan Core', paper presented at the Annual Meeting of the Population Association of America, Toronto, Canada.

Treadway, R.C. (1991), 'Alternative Definitions of the Metropolitan Core', paper presented at the Annual Meeting of the American Statistical Association, Atlanta, GA.

Treble, R. (1989), 'The Victorian Picture of the Country', in G. Mingay (ed.), *The Rural Idyll*, London: Routledge.

Tsay, C.L. (1982), Urban Population Growth and Distribution in the Taiwan Region (in Chinese), *Zhongyang Yanjiuyuan Sanminzhuyi Yanjiusuo Congkan (Journal of the Institute of the Three Principles of the People of Academia Sinica)*, No. 9, pp. 207-42.

Tuirán, R. (2000), 'Tendencias Recientes de la Movilidad Territorial en Algunas Zonas Metropolitanas de México', *Mercado de Valores*, México, D.F., México, year LX, No. 3, pp. 47-61.

UN (1950), *Data on Urban and Rural Population in Recent Censuses*. Lake Success, New York: United Nations (Population Studies No. 8).

UN (1952), *Demographic Yearbook 1952*, New York: United Nations.

UN (1958), *Principles and Recommendations for National Population Censuses*, Statistical Papers Series M No. 27, New York: United Nations.

UN (1967), *Principles and Recommendations for the 1970 Population Censuses*, Statistical Papers Series M No. 44, New York: United Nations.

UN (1969), *Growth of the World's Urban and Rural Population, 1920-2000*, Population Studies No. 44, New York: United Nations.

UN (1973), *Demographic Yearbook 1972*, New York: United Nations.

UN (1980), *Patterns of Urban and Rural Population Growth*, New York: United Nations.

UN (1981), *Modalidades del Crecimiento de la Población Urbana y Rural*, New York: United Nations.

UN (1996), *An Urbanizing World: Global Report on Human Settlements*, Oxford and New York: Oxford University Press.

UN (1998), *Principles and Recommendations for Population and Housing Censuses, Revision 1*, Statistical Papers, Series M. No. 67/ Rev. 1, New York: United Nations.

UN (2001a), *Demographic Yearbook 1999*, New York: United Nations.

UN (2001b), *World Urbanization Prospects: The 1999 Revision*, New York: United Nations.

UN (2003) *World Urbanization Prospects: The 2001 Revision*, New York: United Nations.

UN Secretariat (1975), 'Statistical Definitions of Urban Population and Their Uses in Applied Demography', in S. Goldstein and D.F. Sly (eds.), *Basic Data Needed for the Study of Urbanization*, Working Paper 1, IUSSP Committee on Urbanization and Population Redistribution, Dolhain, Belgium: Ordina Editions.

Unikel, L., Ruiz, C. and Garza, G. (1976), *El Desarrollo Urbano de México. Diagnóstico e Implicaciones Futuras*, México: El Colegio de México.

Urry, J. (1995a), *Consuming Places*, London: Routledge.

Urry, J. (1995b), 'A Middle-Class Countryside?', in T. Butler and M. Savage (eds), *Social Change and the Middle Classes*, London: UCL Press.

Urzúa, R. *et al.* (1982), *Desarrollo Regional, Migraciones y Concentración Urbana en América Latina: Una investigación comparativa,* Mimeo, Santiago de Chile: Centro Latinoamericano de Demografía (CELADE).

Van den Berg, L. (1987), *Urban Systems in a Dynamic Society*, Aldershot: Gower.

Van den Berg, L., Drewett, R., Klaassen, L.H., Rossi, A. and Vijverberg, C.H.T. (1982), *Urban Europe, A Study of Growth and Decline*, Oxford: Pergamon Press.

Vapñarsky, C.A. (1968), *La Población Urbana Argentina. Revisión de Los Resultados Censales de 1960*, Buenos Aires: Centro de Estudios Urbanos y Regionales del Instituto Torcuato Di Tella, Editorial del Instituto.

Vapñarsky, C.A. (1978), 'Toward Scientific Foundations for the Determination of Localities in Population Censuses', *GENUS*, Vol. XXXIV(1-2), pp. 79-130.

Vapñarsky, C.A. (1999), 'La Distribución de la Población Aglomerada y Dispersa Sobre el Territorio Argentino. Reflexiones en Vísperas del Censo del año 2000', paper presented at *V Jornadas Argentinas de Estudios de Población*, AEPA, Luján, Argentina.

Veiga, José Eli da (2002), *Desenvolvimento Territorial do Brasil: Do Entulho Varguista ao Z.E.E.* http://www.fea.usp/professores/zeeli/ultimos.

Walcott, S.M. and Wheeler, J.O. (2001), 'Atlanta in the Telecommunications Age: The Fiber-Optic Information Network', *Urban Geography*, Vol. 22(4), pp. 316-39.

Waldinger, R. (1996), *Still the Promised City? African Americans and New Immigrants in Post-Industrial New York*, Cambridge, MA: Harvard University Press.

Wang, S.J. and Zhou, Z.G. (1996), 'Potential Urbanization and Its Forming Mechanism, Regional Differences, and Prospects', in S.J. Wang (ed.), *Zhongguo Chengshihua Quyu Fazhan Wenti Yanjiu (Studies on Regional Development of Urbanization in China)*, Beijing: Higher Education Press.

Ward, D., Phinn, S.R. and Murray, A.T. (2000), 'Monitoring Growth in Rapidly Urbanizing Areas Using Remotely Sensed Data', *The Professional Geographer,* Vol. 52, pp. 371-85.

Weber, A.F. (1968), *The Growth of Cities in the Nineteenth Century*, third printing, Ithaca, NY: Cornell University Press.

Weber, M. (1968), *Economy and Society*, New York: Bedminister.

Weeks, J.R. (2002), *Population: An Introduction to Concepts and Issues,* eighth edition, Belmont, CA: Wadsworth Publishing Co.

Weeks, J.R. (2003), 'Does Night-Time Lighting Deter Crime? An Analysis of Remotely-Sensed Imagery and Crime Data', in V. Mesev (ed.), *Remotely Sensed Cities*, London: Taylor and Francis.

Weeks, J.R., Gadalla, M.S., Rashed, T., Stanforth, J. and Hill, A.G. (2000), 'Spatial Variability in Fertility in Menoufia, Egypt, Assessed Through the Application of Remote Sensing and GIS Technologies', *Environment and Planning A*, Vol. 32, pp. 695-714.

Whatmore, S. (1993), 'On Doing Rural Research (or Breaking the Boundaries)', *Environment and Planning A*, Vol. 25, pp. 605-7.

Wheeler, J.O., Aoyama, Y., and Warf, B. (eds) (2000), *Cities in the Telecommunications Age: The Fracturing of Geographies*, New York: Routledge.

Williams, R. (1973), *The Country and the City*, London: Hogarth Press (1985 edn.).

Willits, F. and Bealer, R. (1967), 'An Evaluation of a Composite Index of Rurality', *Rural Sociology*, Vol. 32(2), pp. 165-77.

Willits, F., Bealer, R. and Timbers, V. (1990), 'Popular Images of Rurality: Data From a Pennsylvania Survey', *Rural Sociology*, Vol. 55(4), pp. 559-78.

Wilson, G. (2001), 'From Productivism to Post-productivism… and Back Again? Exploring the (Un)changed Natural and Mental Landscapes of European Agriculture', *Transactions of the Institute of British Geographers*, Vol. 26, pp. 77-102.

Winter, M. (1996), *Rural Politics*, London: Routledge.

Wirth, L. (1938), 'Urbanization As a Way of Life', *American Journal of Sociology*, Vol. 44(1), pp. 1-29.

Wright, P. (1985), *On Living in an Old Country*, London: Verso.

Wright, S. (1992), 'Image and Analysis: New Directions in Community Studies', in B. Short (ed.), *The English Rural Community*, Cambridge: Cambridge University Press.

Zelinsky, W. (1971), 'The Mobility Transition', *Geographical Review*, Vol. 59, pp. 143-57.

Zhang, B. (2000), 'An Outlet for the Gradual Reform of the Planning System: A Discussion on Enacting Urban and Rural Planning Act' (in Chinese), *Chengshi Guihua (City Planning Review)*, Vol. 24(10), pp. 8-13.

Zhou, Y.X., and Shi, L. (1995), 'Towards Establishing the Concept of Physical Urban Area in China' (in Chinese), *Dili Xuebao (Acta Geographica Sinica)*, Vol. 50(4), pp. 289-301.

Zhou, Y.X., and Sun, Y. (1992), 'An Analysis of the Proportion of the Population of Cities and Towns in the Fourth National Population Census of China' (in Chinese), *Renkou Yu Jingji (Population and Economics)*, No. 1, pp. 21-7.

Zhu, Y. (1998), '"Formal" and "Informal Urbanization" in China: Trends in Fujian Province', *Third World Planning Review*, Vol. 20(3), pp. 267-84.

Zhu, Y. (1999), *New Paths to Urbanization in China: Seeking More Balanced Patterns*, New York: Nova Science.

Zhu, Y. (2000), '*In situ* Urbanization in Rural China: Case Studies from Fujian Province', *Development and Change*, Vol. 31(2), pp. 413-34.

Zhu, Y. (2001a), 'Globalization, "Floating Population", and Household Strategies in Migration: Regional Integration Through Migration in China', paper presented to the international conference 'Migration and the "Asian Family" in a Globalising World', Singapore, 16-18 April.

Zhu, Y. (2001b), 'The Transformation of Townships into Towns and Their Roles in China's Urbanization: Evidence from Fujian Province', paper presented to IUSSP's XXIV General Population Conference, Bahia, Brazil, August 2001.

Index

Printed and bound by CPI Group (UK) Ltd, Croydon, CR0 4YY

21/10/2024

01777082-0009